Power Systems

Antonio Moreno-Muñoz (Ed.)

Power Quality

Mitigation Technologies in a Distributed Environment

Springer

Antonio Moreno-Muñoz, PhD
Área de Electrónica
Universidad de Córdoba
Campus de Rabanales
14071 Córdoba
Spain

British Library Cataloguing in Publication Data
Power quality : mitigation technologies in a distributed
  environment. - (Power systems)
  1. Electric power transmission - Reliability 2. Distributed
  resources (Electric utilities) - Reliability 3. Electric
  power failures
  I. Moreno-Munoz, Antonio
  621.3'19
ISBN-13: 9781846287718

Library of Congress Control Number: 2007925389

Power Systems Series ISSN 1612-1287
ISBN 978-1-84628-771-8   e-ISBN 978-1-84628-772-5   Printed on acid-free paper

© Springer-Verlag London Limited 2007

Chapter 5: Transient Mitigation Methods on ASDs © 2002 IEEE. Reprinted, with permission, from IEEE Transactions on Power Electronics, vol. 17, no. 5, pp. 799–806, Sept. 2002

LabVIEW™ is a trademark of National Instruments Corporation, 11500 N Mopac Expressway, Austin, TX 78759-3504, USA. http://www.ni.com/

Lineator™ is a trademark of Mirus International Inc., 6805 Invader Crescent, Unit #12, Mississauga, Ontario L5T 2K6, Canada. http://www.mirusinternational.com/

Matlab® and Simulink® are registered trademarks of The MathWorks, Inc., 3 Apple Hill Drive, Natick, MA 01760-2098, USA. http://www.mathworks.com

Apart from any fair dealing for the purposes of research or private study, or criticism or review, as permitted under the Copyright, Designs and Patents Act 1988, this publication may only be reproduced, stored or transmitted, in any form or by any means, with the prior permission in writing of the publishers, or in the case of reprographic reproduction in accordance with the terms of licences issued by the Copyright Licensing Agency. Enquiries concerning reproduction outside those terms should be sent to the publishers.

The use of registered names, trademarks, etc. in this publication does not imply, even in the absence of a specific statement, that such names are exempt from the relevant laws and regulations and therefore free for general use.

The publisher makes no representation, express or implied, with regard to the accuracy of the information contained in this book and cannot accept any legal responsibility or liability for any errors or omissions that may be made.

9 8 7 6 5 4 3 2 1

Springer Science+Business Media
springer.com

Dedicated to our ever patient, supportive and loving families

Preface

Electrical energy is one of the main elements for the economical development of society. The just aspirations of modern societies to economical growth have forced us to secure more continual energy resources. On the other hand, climate change and international treaties aiming to reduce greenhouse gas emissions have prompted all of us to be increasingly concerned about energy efficiency and conservation.

With the rapid growth of the information-based economy, widespread expansion of electronic devices has become a prevalent phenomenon in both the public and private sectors. Their suitability to perform various functions such as storage, management, processing and exchange of digital data and information, are essential support for information and communication technology (ICT). Although the consumption of energy may increase, influenced by the increase in the necessary communications infrastructure, the use of renewable energy may contribute to a more rational consumption of energy, reducing the impact on the environment.

The current trend toward miniaturization in microelectronics, increased processing speed and greater functionality results in a particular sensitiveness to certain kinds of electromagnetic perturbations. Thus, this situation is not only bringing about a greater demand for electricity, but in addition higher levels of power quality and reliability (PQR) needs, in quantities and time frames that have not been experienced before.

The reliability of the power supply delivered by utilities varies considerably and depends on a number of external factors. This infrastructure was designed primarily to serve "analog electrical devices", which are generally tolerant to voltage fluctuations in the power supply. However, the present electric power grid is unable to consistently provide the PQR level required by the "digital devices" of our information-based economy.

The problems associated with the presence of disturbances on power-distribution systems are not just the power-quality problems but also affect the energy efficiency of the plant. As far as energy efficiency is concerned in a building power-distribution system, the two dominant factors in power quality are its unbalanced distortion and harmonic distortion. Unbalanced distortion in three-phase supply voltages will create negative sequence components causing additional power losses in conductors and motors. On the other hand, harmonics created mainly by the nonlinear electronic loads, can damage the system components.

In the future, the use of PQ mitigation technologies and their incorporation into electrical systems and electronic equipment will also prevent an increase in the number of problems detected. Then, it will also contribute to an improvement in the efficiency of our daily activities.

This book discusses the modern mitigation technologies employed in power quality and recent challenges. The book starts with an introductory chapter and moves to cover topics including:

- how to deal with the characteristics of power-quality analysis;
- how common power disturbances can affect circuit and process design;
- the implications that electromagnetic compatibility has on the equipment and installation design;
- how power-electronic devices can improve the power quality of the distribution supply; and,
- advances in IT that support a better power quality.

This book introduces advanced concepts to engineers and students and is a reference for facility electrical power engineers and maintenance technicians. One very attractive feature of this book is the breadth of literature the authors refer to in each chapter.

The book is organized in 13 chapters. As follows:

Chapter 1 is an introduction to the power-quality issue. It presents the different definitions of the term and the implications of the related standards, from the perspective of electromagnetic compatibility. Finally, the chapter provides an overview of the causes and effects of poor power quality in power-distribution systems and presents some of the currently available mitigation technologies.

Chapter 2 presents instrumentation for the measurement of conducted disturbances in power systems. From the first instrumentation designed for general-purpose measurements, up to the current highly improved transient recorders, this kind of device has continuously evolved, becoming increasingly more specialised. This chapter not only describes the evolution of the hardware, but also the significant activity in all topics related to the development of software for the analysis of measurements.

Chapter 3 presents joint time frequency analysis techniques. For nonstationary signals, a representation in a similar way to Fourier analysis was the goal of many previous studies. However, the impossibility of the classical power spectrum to reveal important information about the evolution over the time of a real signal's frequency content has motived the development of models capable to combine effectively both time and frequency domains. In this chapter, the mathematical background and basic methods of the joint time frequency analysis techniques is introduced.

Chapter 4 describes the main issues related to the measurement and analysis of voltage events in power systems. The chapter presents the definitions and the different magnitude and duration thresholds used in the main power-quality standards to distinguish between the different types of voltage events. The main sections are devoted to the study of the method for detection and analysis of voltage events, analysing the effects and performance of equipment to these voltage variations.

In Chapter 5, new approaches to mitigate nuisance tripping of adjustable speed drives due to utility capacitor-switching transient events is proposed. The chapter analyses these transients in detail, reviews the traditional mitigation techniques and proposes new approaches to reduce the overvoltage in the adjustable speed drives dc-link voltage.

Chapter 6 discusses general issues concerning supply quality in AC systems, and especially methods and selected countermeasures for power-quality improvement. The chapter allows the comparison and evaluation of various modern mitigation technologies. It also characterizes the potential of modern high-power electronics converters for electrical power engineering.

Chapter 7 focuses on power-electronics-based solutions used in distribution network for reduction of supply-voltage-quality deterioration. Various custom power-compensating controllers can be used in distribution systems. In general, there are three types of static shunt PE voltage-quality controllers: The distribution static var compensator (D-SVC), distribution static synchronous compensator (D-STATCOM) and various hybrid arrangements. Their main purpose is voltage-regulation, load compensation, but also voltage-profile improvement.

Chapter 8 discusses the most important responses and characteristics of static series controllers, designed on the basis of a distribution static synchronous series compensator (D-SSSC), specifically the dynamic voltage restorer (DVR) and AC/AC voltage regulators (VR). In addition, it also outlines selected static shunt–series controllers, preserving a sensitive load from deterioration of supply voltage and at the same time protecting the electrical system from load impact.

Chapter 9 is devoted to active power line conditioners. The chapter presents the more common configurations: from shunt and series to hybrid passive-active filter.

In particular, the control strategy and practical design considerations from a shunt arrangement is discussed.

Chapter 10 elucidates the general impact of distributed generation on power-quality. The widespread use of distributed resources has created a series of new problems. The necessary coordination between overcurrent protection and sensitive equipment voltage-sag immunity is addressed in different scenarios. In addition, the chapter studies the impact of distributed generation on recloser–fuse coordination. Finally, the harmonics and flicker problem of distributed generation is introduced.

Chapter 11 presents results from two exemplary power-quality audits conducted; the first in a biotechnological university campus and the second in a highly automated factory. It was found that the main problems for the equipment installed were voltage sags and harmonics. The chapter analyses the capabilities of modern power supplies and the convenience of an "embedded solution" is also discussed. Finally, the role of the Standards on the protection of electronic equipment and the implications for the final costumer is addressed.

Chapter 12 is a presentation of a general integral assessment of the power-transfer quality of a three-phase network by means of a new indicator designated as the power-quality factor (PQF). In the chapter, the PQF considers various quality aspects notably the current and voltage harmonic levels, the phase displacements between corresponding phase voltages and currents at the fundamental frequency, and the degree of unbalance in the different phase voltages and currents. The advantages of the measurement techniques of the PQF are studied and, finally, illustrative use of the power-quality factor is described.

Chapter 13 introduces IEC 61850, an approved international standard for communications in substations. This chapter describes the standard, its model and functional hierarchy, the communication protocol that it prescribes and the implications to power-quality analysis in power distribution.

Antonio Moreno Muñoz
Córdoba, Spain

Acknowledgements

Many people have helped to make this book a reality. We have benefited from the advice and support of our professional colleagues, our students, friends and families and we thank you all.

We are grateful to Anthony Doyle, Senior Editor of this book and to our Editorial Assistant, Simon Rees and the Springer editorial staff for their unprecedented help.

Contents

List of Contributors xix

1 **Introduction** 1
   1.1 Introduction 1
   1.2 Electromagnetic Compatibility 2
   1.3 Power-quality 6
   References 14

2 **Power-quality Monitoring** 15
   2.1 Introduction 15
   2.2 State-of-the-art 16
      2.2.1 Historical Background 16
      2.2.2 General-purpose Instrumentation 16
      2.2.3 Specialised-purpose Instrumentation 17
   2.3 Instrumentation Architecture 20
      2.3.1 Safety Use of PQ Instrumentation 21
      2.3.2 Number of Channels 23
      2.3.3 Common Unified Channels 24
      2.3.4 Independent Input Channels 25
      2.3.5 Transducers 25
      2.3.6 Signal-conditioning Module 29
      2.3.7 Analog-to-digital Converter 30
      2.3.8 Signal-processing Module 30
   2.4 PQ Instrumentation Regulations 31
   2.5 Harmonic Monitoring 33
   2.6 Flicker Monitoring 35
   2.7 Data Postprocessing 35
   2.8 Management of PQ Files 37
      2.8.1 COMTRADE 37
      2.8.2 PQDIF 37

Contents

- 2.9 Summary ... 38
- References ... 38

3 Joint Time Frequency Analysis of the Electrical Signal ... 41
- 3.1 Introduction ... 41
- 3.2 Application of JTFA to Electrical Signals ... 44
- 3.3 Review of Fundamental Mathematical Tools ... 46
  - 3.3.1 Fourier Theory ... 47
- 3.4 Time Frequency Analysis Limits: Uncertainty Principle ... 49
- 3.5 JTFA Linear Methods ... 52
  - 3.5.1 Windowed Fourier Transform (STFT) and Gabor Expansion ... 52
  - 3.5.2 Adaptive Representation and Adaptive Transform ... 56
  - 3.5.3 Wavelet Theory ... 57
- 3.6 Time-dependent Spectrum: Quadratic Transforms ... 62
  - 3.6.1 STFT Spectrogram ... 62
  - 3.6.2 Wigner–Ville Distribution and Pseudo–Wigner Ville Distribution ... 64
  - 3.6.3 Cohen Class ... 65
  - 3.6.4 Choi Williams Distribution ... 66
  - 3.6.5 Conic Distribution ... 66
  - 3.6.6 Gabor Spectrogram ... 66
  - 3.6.7 Adaptive Spectrogram ... 67
- 3.7 Algorithms Summary ... 68
- References ... 71

4 Measurement and Analysis of Voltage Events ... 73
- 4.1 Introduction ... 73
- 4.2 Monitoring of Voltage Events ... 75
  - 4.2.1 Performance of the IEC Standard Method in the Detection and Evaluation of Voltage Events ... 76
  - 4.2.2 Other Methods for the Detection and Evaluation of Voltage Events ... 80
- 4.3 Effects of Voltage Events on Equipment ... 90
  - 4.3.1 Voltage Tolerance of Equipment ... 91
  - 4.3.2 Measurement System ... 92
  - 4.3.3 Effect of Harmonic Distortion ... 94
- 4.4 Voltage-event Surveys ... 96
- References ... 100

5 Transient Mitigation Methods on ASDs ... 103
- 5.1. Introduction ... 103
- 5.2. Transient analysis ... 105
  - 5.2.1 Analysis of CST Events ... 105
- 5.3. Traditional Transient Mitigation Methods on ASDs ... 111
  - 5.3.1 Review of Power System Mitigation Techniques ... 111
- 5.4. New Mitigation Methods ... 114

	5.4.1	Proposed Approaches to Mitigate Nuisance Tripping of CST Events	114
5.5	Example of Experimental Results		119
	5.5.1	Design Example	119
	5.5.2	Simulation Results	121
	5.5.3	Experimental Results	122
5.6.	Conclusions		127
References			127

6 Modern Arrangement for Reduction of Voltage Perturbations ... 129
6.1 Introduction ... 129
6.2 Influence of Load on the Power-quality ... 130
    6.2.1 Investigation Results ... 133
    6.2.2 Other Important Influences of Loads ... 136
6.3 Reduction on Load Influence on the Voltage Profile ... 140
    6.3.1 Principle Compensation of the Load Influence ... 141
    6.3.2 Review of Selection Problems ... 144
6.4 Mitigation of the Voltage Disturbance ... 155
    6.4.1 Basic Concept and Methodological Questions ... 155
    6.4.2 Modern Protection and Reconfiguration Devices ... 156
    6.4.3 Compensation Devices ... 160
    6.4.4 Immunisation – Standby Power-supply Devices ... 164
6.5 Usability of the Modern Power Electronics ... 167
    6.5.1 General Model of Power-electronics Converters ... 168
    6.5.2 Basic 3-phase Power-electronics Converters with AC Output ... 171
    6.5.3 Multilevel Voltage Inverters as Arrangement to MV Grid ... 176
6.6 Summary ... 179
References ... 179

7 Static Shunt PE Voltage-quality Controllers ... 183
7.1 Fundamentals of Shunt Compensation ... 183
7.2 Distribution Static Var Compensator (D-SVC) ... 186
    7.2.1 Simple D-SVC Controllers ... 186
    7.2.2 Combined D-SVC Controllers ... 187
7.3 Distribution Static Synchronous Compensator (D-STATCOM) ... 188
    7.3.1 Topology ... 188
    7.3.2 Principle of Operation ... 190
    7.3.3 Load Compensation ... 192
    7.3.4 Voltage Regulation ... 195
7.4 Other Shunt Controllers Based on D-STATCOM ... 197
    7.4.1 Hybrid Arrangements ... 198
    7.4.2 Controllers with Energy-storage Systems ... 200
7.5 Summary ... 202
References ... 203

8 Static Series and Shunt-series PE Voltage-quality Controllers ... 205
8.1 Distribution Static Synchronous Series Compensators ... 206

         8.1.1   Identification of Separate Components of the
                 Supply-terminal Voltage 207
         8.1.2   Harmonic Filtration and Balancing of the Voltage
                 in 3-wire Systems 210
   8.2   Dynamic Voltage Restorer (DVR) 213
         8.2.1   What It Is a DVR 213
         8.2.2   Control Strategies of the DVR Arrangements 213
         8.2.3   Comparison of the DVR Types 216
   8.3   AC/AC Voltage Regulators 221
         8.3.1   Electromechanical Voltage Regulators 222
         8.3.2   Step-voltage Regulators 223
         8.3.2   Continuous-voltage Regulators 224
   8.4   Summary 227
   References 228

9 **Active Power Line Conditioners** 231
   9.1. Introduction 231
   9.2. Power-quality and Active Power Filters 233
         9.2.1. Distribution Static Compensator, DSTATCOM 234
         9.2.2. Series Active Filters 235
         9.2.3. Hybrid Filters 236
         9.2.4. Unified Power-quality Conditioner 237
   9.3. Power-electronic Inverters in APLCs 238
         9.3.1. Voltage-source Inverter Topologies 240
         9.3.2. Control of Voltage-source Inverters 250
   9.4. Strategies for Load Static Compensation 262
         9.4.1. Instantaneous Reactive Power Theory 262
         9.4.2. Instantaneous d-q Theory 267
   9.5. Practical Design 268
         9.5.1. Component-design Considerations 269
         9.5.2. Simulation Analysis 272
   9.6. APLC Prototyped Trough PC Acquisition Board 276
         9.6.1. Experimental System 277
         9.6.2. Results of a Practical Case 283
   References 287

10 **Distributed Generation** 293
   10.1 Introduction 293
   10.2 General Impact of Distributed Generation
          on Power-quality to Strong or Weak Electrical Systems 296
   10.3 Dissimilar Effect of
          the Different Distributed Resources Technologies 296
   10.4 Coordination between Overcurrent Protection
          and Sensitive Equipment Voltage Sag Immunity 296
          10.4.1 Application of Specific Energy Concept 298
          10.4.2 Transformation of Protective Device Time–Current Curves into
                   Time–Voltage Curves, for Voltage-sag Coordination Studies ... 300

   10.4.3 Mitigation of Distant Voltage Sag Penetration
      into Industrial Premises by Using Semirigid Connection 314
   10.4.4 New Overcurrent-protection Schemes using Intelligence 317
 10.5 Impact of DG on Recloser-Fuse Coordination 319
 10.6 Harmonics Generated by Distributed Generators 320
 10.7 Flicker Due to Wind Gusts and Tower Shadow 321
 10.8 Ferroresonant Overvoltages 322
 10.9 Conclusions 323
 References 323

11 Electronic Loads and Power-quality 325
 11.1. Introduction 325
 11.2. Electromagnetic Disturbances 328
 11.3. The Rabanales Campus Case Study 328
 11.4. The Infrico Case Study 331
 11.5. Mitigations Technologies 335
 11.6. The Improvement of Electronic Power Supplies 339
 11.7. Conclusion 349
 References 350

12 Power-quality Factor for Electrical Networks 353
 12.1 Quality of the Electrical Signal 355
 12.2 Quantitative Formulations of Power-quality Aspects 356
   12.2.1 System-frequency Variations 356
   12.2.2 Total Current and Voltage Harmonic Distortion 357
   12.2.3 Degree of Unbalance 358
   12.2.4 Phase Displacements Between Corresponding
      Fundamental Voltage and Currents 359
 12.3 Voltage-quality Factor and Power-quality Factor 361
   12.3.1 Definition of the Power-quality Factor 361
   12.3.2 Definition of the Voltage-quality Factor 362
 12.4 Measurement of PQF and VQF 363
 12.5 Illustrative Use of the Power-quality Factor 365
 12.6 Quantitative Formulations of Power-quality Aspects
   Under Transient State Conditions 369
   12.6.1 Fourier Analysis Versus Time Frequency
      analysis for Power-quality 369
   12.6.2 Time Frequency-based Transient Quality Aspects (TQA) 371
   12.6.3 Procedure to Obtain the Transient Quality Aspect 373
   12.6.4 Application Example of Transient Disturbance 374
 References 375

13 IEC 61850 and Power-quality Monitoring and Recording 379
 13.1 Introduction 379
 13.2 What is IEC 61850? 380
 13.3 Logical Interfaces and Distributed Applications 384
 13.4 Functional Hierarchy 386

13.5 The IEC 61850 Model 387
13.6 Distribution and Modelling of Functions
 in Power-quality Monitoring Devices.. 392
   13.6.1 Logical Nodes for Measurements 396
   13.6.2 Logical Nodes for Power-quality Events 398
   13.6.3 Logical Nodes for Recording.. 399
13.7 Power-quality Event Analysis in IEC 61850-based Systems................. 399
13.8 Recording of Power-quality-events 403
   13.8.1 Waveform Recording.. 404
   13.8.2 High and Low-speed Disturbance Recording 404
   13.8.3 Periodic Measurement Logging 405
13.9 Recording Systems for Power-quality Event Analysis 405
13.10 Performance Requirements.. 408
13.11 High-speed Peer-to-peer Communications Applications..................... 409
References... 414

Index... 417

List of Contributors

Alexander Apostolov
OMICRON Electronics,
2950 Bentley Ave., Unit 4,
Los Angeles,
California, USA
Email:
alex.apostolov@omicronusa.com

Julio Barros
Dept. Electronics and Computers,
University of Cantabria,
Av. de los Castros,
39005 Santander, Spain
Email: barrosj@unican.es

Grzegorz Benysek
Department of Electrical Engineering,
Informatics and Telecommunications
University of Zielona Góra,
50 Podgórna,
Zielona Góra, Poland
Email: G.Benysek@iee.uz.zgora.pl

Juan-Carlos Bravo,
Dpto. Ingeniería eléctrica,
Escuela Universitaria Politécnica,
Universidad de Sevilla,
Virgen de África 7,
41011-Sevilla, Spain

Dolores Borrás
Dpto. Ingeniería eléctrica,
Escuela Universitaria Politécnica,
Universidad de Sevilla,
Virgen de África 7,
41011-Sevilla, Spain

Ramón I. Diego
Dept. Electronics and Computers,
University of Cantabria,
Av. de los Castros,
39005 Santander, Spain
Email: diegori@unican.es

José Luis Durán-Gómez
División de Estudios de Posgrado
e Investigación,
Instituto Tecnológico de Chihuahua,
Ave. Tecnológico No. 2909,
Chihuahua, Chih. México 31310
Email: jlduran@itchihuahua.edu.mx

Juan Carlos Gomez Targarona
Instituto de Protecciones de Sistemas
Eléctricos de Potencia,
Departamento de Electricidad y
Electrónica,
Universidad Nacional de Río Cuarto,
Ruta 8 y 36, Km. 603,
(5800) Río Cuarto, Córdoba, Argentina
Email: jcgomez@ing.unrc.edu.ar

List of Contributors

Juan José González-de-la-Rosa
Área de Electrónica,
Dpto. ISA, TE y Electrónica,
Universidad de Cádiz,
Avda. Ramón Puyol, S/N,
E-11202-Algeciras,
Cádiz, Spain

Mario Mañana Canteli
Department of Electrical Engineering,
University of Cantabria,
Santander, Spain
Email: mananam@unican.es

Juan-Carlos Montaño
Laboratorio de Electrónica,
Institute for Natural Resources and
Agricultural Research (IRNAS),
Spanish Research Council (CSIC),
Reina Mercedes Campus, POB 1052,
41080-Sevilla, Spain
Email: montano@irnas.csic.es

Medhat M. Morcos
Department of Electrical and
Computing Engineering
Kansas State University
289 Rathbone Hall
Manhattan, KS 66506-5204, USA
Email: morcos@ksu.edu

Antonio Moreno-Muñoz
Área de Electrónica,
Departamento de Arquitectura de
Computadoras, Electrónica y
Tecnología Electrónica,
Universidad de Córdoba,
Campus de Rabanales,
E-14071 Córdoba, Spain
Email: amoreno@uco.es.

Prasad N. Enjeti
Department of Electrical and Computer
Engineering,
Texas A&M University,
214 Zachry Engineering Center
College Station,
TX, USA 77843-3128
Email: enjeti@ee.tamu.edu

Enrique Pérez
Dept. Electronics and Computers,
University of Cantabria,
Av. de los Castros,
39005 Santander, Spain.
Email: eperez25@boj.cnice.mecd.es

Patricio Salmerón
Department of Electrical Engineering,
EPS, Huelva University,
Ctra de Palos de la Frontera, 21819
Huelva, Spain
Email: patricio@uhu.es

Ryszard Strzelecki
Department of Electrical Engineering,
Gdynia Maritime University,
81-87 Morska Str,
Gdynia, Poland
Email: rstrzele@am.gdynia.pl

Jesús R. Vázquez
Department of Electrical Engineering,
EPS, Huelva University,
Ctra de Palos de la Frontera ,
21819 Huelva. Spain
Email: vazquez@uhu.es

Daniel Wojciechowski
Department of Electrical Engineering,
Gdynia Maritime University,
81 87 Morska Str,
Gdynia, Poland
Email: dwojc@am.gdynia.pl

1
Introduction

Antonio Moreno-Muñoz

Área de Electrónica,
Departamento de A. C., Electrónica y T.E,
Universidad de Córdoba,
Campus de Rabanales,
E-14071 Córdoba, Spain.
Email: amoreno@uco.es.

1.1 Introduction

Nowadays, the growth of the digital economy implies a widespread use of electronic equipment not only in the industrial and commercial sectors, but in the domestic environment too. Studies undertaken in different countries on the contribution of information and communication technology (ICT) to the consumption of electricity conclude that office and telecommunication equipment used in the nonresidential sector represents about 3 or 4% of the annual consumption of electricity. This will not only bring about a greater demand for power, but in addition a higher level of power quality and reliability (PQR), in quantities and time frames that have not been experienced before. It has been estimated that more than 30% of the power currently being drawn from the utility companies is now heading for sensitive equipment, and this is increasing [1].

The reliability of the power supply delivered by utilities varies considerably and depends on a number of external factors. Things like lightning, large switching loads, nonlineal load stresses or accidents can disrupt the electric power grid. This infrastructure was designed primarily to serve "analog electric devices" like lights or motors. These devices are generally tolerant to voltage fluctuations in the power supply. However, the electric power grid is unable to consistently provide the level of digital quality power required by our digital manufacturing assembly lines, information systems, and soon even our home appliances.

The electronic devices demand for higher PQR stems from the fact that semiconductor components require low-voltage direct current and are highly sensitive to short power interruptions, voltage surges and sags, harmonics, and other waveform distortions.

Today, for every microprocessor inside a computer, there are 30 more in stand-alone applications, resulting in the digitization of society. In applications ranging from industrial sensors to home appliances, microprocessors now number more than 12 billion in the US alone [2].

1.2 Electromagnetic Compatibility

Any electromagnetic phenomenon that may degrade the performance of a device, equipment or system, or adversely affect living or inert matter is considered an electromagnetic disturbance. Equipment and systems performance can be adversely impacted by electromagnetic disturbances and, conversely, any electrotechnical equipment is, itself, more or less an electromagnetic-disturbance generator. Disturbances cause undesirable problems, thus, avoiding electromagnetic interference (EMI) became an important issue in electrical engineering. In this sense, it is possible to consider "power quality" as a dedicated subset of electromagnetic compatibility (EMC), limited to the area of low-frequency conducted phenomena.

According to the International Electrotechnical Vocabulary (IEV161-01-07), EMC is the ability of an equipment or system to function satisfactorily in its electromagnetic environment without introducing intolerable electromagnetic disturbances to anything in that environment.

Achieving EMC requires us to deal with two different aspects:

- The disturbance emission level: the level of an electromagnetic disturbance of a given form, measured under particular conditions. The maximum permissible electromagnetic disturbance level is denominated the disturbance limit.
- The immunity level: the maximum level of a given electromagnetic disturbance on a particular device, equipment or system for which it remains capable of operating at a required degree of performance. The electromagnetic susceptibility is the inability of a device, equipment or system to perform without degradation in the presence of an electromagnetic disturbance.

Reliable, trouble-free performance is usually equated with a supply of high quality. However, equipment performance has as much to do with its susceptibility to supply disturbances as it has with the actual characteristics of the supply that it operates on. There is a growing need to ensure electric service compatibility between end-user equipment and the utility power system. It is also the case that supply characteristics will be influenced by the "quality" of the load connected to it.

In the whole power system, interference inevitably occurs on some occasions and therefore there is some overlapping between the distributions of disturbance and immunity levels. Voltage characteristics may be equal to or higher than the compatibility level; they are specified by the European Standard EN 50160 [3], which will be analysed below. Planning levels may be equal to or lower than the compatibility level; they are specified by the owner of the network. Immunity test levels are specified by relevant standards or agreed upon between manufacturers and users.

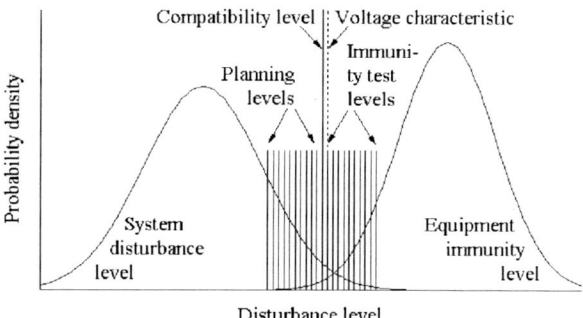

Figure 1.1. Illustration of basic voltage-quality concepts with time/location statistics covering the whole system

In 1989, the European Council issued its so-called EMC Directive 89/336, which came into force on 1 January 1996. The main features of this directive are:

- Application of the relevant standards is compulsory for all EU countries and all other countries that intend to put their products on the EU market.
- The EMC Directive concerns emission from as well as immunity of electrical and electronic devices, and it applies basically to product standards.
- All the products must be tested and certified according to specific rules.

Table 1.1. Part 6 generic EMC standards

EMC Standard	Europe	International
Emission. Residential, Commercial, Light industry environment	EN 61000-6-3	IEC 61000-6-3
Emission. Industrial environment	EN 61000-6-4	IEC 61000-6-4
Immunity. Residential, Commercial, Light industry environment	EN 61000-6-1	IEC 61000-6-1
Immunity. Industrial environment	EN 61000-6-2	IEC 61000-6-2

In order to support the EMC Directive with relevant technical standards, the European Committee for Electrotechnical Standardization (CENELEC), received a mandate from the European Commission to ensure that all products and systems coming within the scope of the directive were adequately covered. The CENELEC standards assist in achieving adequate EMC. They achieve this by providing a framework within which the supply environment (i.e. power quality), the susceptibility of equipment and the emissions from equipment (which can influence power quality) are all defined. Most of the international standards are set by the International Electrotechnical Committee (IEC). And the CENELEC introduced the practice of relying only on internationally published standards. Thus, a coherent family of standards such as the EN 61000 series establishes a means of promoting EMC as well as power quality. Essentially, they concern:

- general subjects like terminology and safety;
- descriptions of the electromagnetic environment: phenomena and levels;
- recommendations for the limitation of emission of electromagnetic disturbances;
- guidance values for immunity tests;
- measurement techniques;
- testing techniques;
- installation guidelines;
- mitigation methods.

The EN 61000 consistes of nine parts and the present structure is as follows:

- Part 1. General: general considerations (introduction, fundamental principles, safety), definitions and terminology.
- Part 2. Environment: description of the environment, classification of the environment and compatibility levels.
- Part 3. Limits: emission limits, immunity limits (insofar as they do not fall under the responsibility of product committees)
- Part 4. Testing and measurement techniques.
- Part 5. Installation and mitigation guidelines: Installation guidelines and mitigation methods and devices.
- Part 6. Generic standards.
- Part 9. Miscellaneous.

Table 1.2. EMC–power-quality standards

EMC Standard	Europe	International
Characterisation of the supplied power	EN 50160	IEC 50160
LF conducted disturbances. Compatibility levels in public LV power systems	EN 61000-2-2	IEC 61000-2-2
LF conducted disturbances. Limits for harmonic current emissions ($n \leq 40$), $I \leq 16$ A, LV	EN 61000-3-2	IEC 61000-3-2

Table 1.2. (continued)

LF conducted disturbances. Limitation of voltage fluctuations & flicker, $I \leq 16$ A	EN 61000-3-3	IEC 61000-3-3
LF conducted disturbances. Limits for harmonic current emissions ($n = 40$), $I \leq 6$ A	EN 61000-3-4	IEC 61000-3-4
LF conducted disturbances. Limitation of voltage fluctuations & flicker, $I > 16$ A	EN 61000-3-5	IEC 61000-3-5
LF conducted disturbances. Limits for harmonic emissions in MV & HV power systems	EN 61000-3-6	IEC 61000-3-6
LF conducted disturbances. Limitation of voltage fluctuations & flicker in MV & HV power systems	EN 61000-3-7	IEC 61000-3-7
Overview of immunity tests	EN 61000-4-1	IEC 61000-4-1
LF conducted disturbances. Voltage dips, short interruptions AC	EN 61000-4-11	IEC 61000-4-11
Electrostatic discharge immunity test	EN 61000-4-2	IEC 61000-4-2
HF radiated disturbances. EM fields, 80 1000 MHz	EN 61000-4-3	IEC 61000-4-3
HF conducted disturbances. Fast transients (bursts), 5/50 ns	EN 61000-4-4	IEC 61000-4-4
HF conducted disturbances. Surges 1.2/50 µs 8/20 µs	EN 61000-4-5	IEC 61000-4-5
HF conducted disturbances. Induced currents, 0.15 80 (230) MHz	EN 61000-4-6	IEC 61000-4-6
LF conducted disturbances. Guide on harmonics and interharmonics measurements and instrumentation	EN 61000-4-7	IEC 61000-4-7
Flickermeter – functional and design specification	EN 61000-4-15	IEC 61000-4-15
Power-quality measurement	EN 61000-4-30	IEC 61000-4-30

Product standards have reached a state of development where equipment survival in the field is adequately addressed, but the more subtle immunity to unavoidable disturbances is not addressed. Conversely, efforts to limit emissions of disturbances into the power system caused by normal operation of the equipment have faced a difficult challenge in achieving consensus, nationally as well as internationally [4].

1.3 Power Quality

The European Council directive on Product Liability (85/374/EEC) explicitly qualifies electricity as a product and, like any other product, should satisfy the proper quality requirements. However, electrical energy is a very specific product. The possibility for storing electricity in any significant quantity is very limited so it is consumed at the instant it is generated. Measurement and evaluation of the quality of the supplied power has to be made at the instant of its consumption.

There are many approximations to the term "power quality". A well-known definition based on the principle of EMC, is as follows [5]. The term "power quality" refers to a wide variety of electromagnetic phenomena that characterize voltage and current at a given time and at a given location on the power system. It is possible to link it with power-supply reliability, service quality and supply quality. Usually it has been sufficient to distinguish between: "voltage quality" and "continuity of supply" [6].

Voltage quality (internationally used term: "power quality" (PQ)) is concerned with the technical characteristics of the electricity at a given point on an electrical system, evaluated against a set of reference technical parameters [7]. In this term the deviations of the voltage (or the current) from the ideal are analyzed, considering electricity as a single frequency sine wave of constant amplitude and frequency.

Continuity of supply is concerned with the probability of satisfactory operation of a power system over the long term. It denotes the ability to supply adequate electrical service on a nearly continuous basis, with few interruptions over an extended period of time. It is covered by reliability indices; the three most common are referred to as SAIFI, SAIDI, and CAIDI, defined in [8].

These definitions can be complemented by the following [9]:
- Quality of supply, which is a combination of voltage quality and the non-technical aspects of the interaction from the power network to its customers.
- Quality of consumption, as the complementary term to quality of supply.

In contrast to the term "reliability" applied to intervals of minutes, typical power quality issues include short-term events such as voltage sags, swells or even transients with a duration of less than a few seconds. Power system harmonic and flicker issues also fall into the category of power quality, even though these issues tend to occur over much longer intervals than sags and transients.

Thus, the PQ requirements can be divided into two categories: steady-state variations and disturbances. The first category refers to the deviations of the normal voltage or current supplied to a facility. It is basically a measure of the magnitude by which the voltage or current may vary from the nominal value, plus distortion and the degree of unbalance between the 3 phases. The second category refers to the events. Events are larger deviations that only occur occasionally, *e.g.* voltage interruptions or load-switching currents. Events are disturbances that start and end with a threshold crossing. Table 1.3 defines the PQ deviations.

Table 1.3. Classification of power quality variation

Characteristic of the voltage wave	Denomination	Description
Frequency	Frequency deviations	Deviation of the power system fundamental frequency from its specified nominal value (*e.g.* 50 Hz or 60 Hz).
Amplitude	Long-duration voltage variations	Measured voltage having a value greater or less than the nominal voltage for a period of time greater than 1 min. The first one is called overvoltage, typical values are 1.1 to 1.2 pu., and the second one undervoltage, typical values are 0.8 to 0.9 pu.
	Voltage fluctuations	Systematic variations of the voltage envelope or a series of random voltage changes, the magnitude of which does not normally exceed the voltage ranges of 0.95 to 1.05 pu.
		The impact of the voltage fluctuation on lighting intensity and the subsequent impression of unsteadiness of visual sensation induced is denominated flicker.
	Sag (dip)	A decrease to between 0.1 and 0.9 pu in rms voltage at the power frequency for durations of 0.5 cycle to 1 min. When the decrease is to essentially zero volts (less than 0.1 pu) it is considered a short interruption.
	Swell	An increase in rms voltage or current at the power frequency for durations from 0.5 cycles to 1 min.
	Transient	A sudden, nonpower frequency change in the steady-state condition of voltage. When unidirectional in polarity it is an impulsive transient, when it includes both positive and negative polarity values it is an oscillatory transient.
Waveform	Harmonic distortion	The presence of frequencies at integer multiples of the fundamental system frequency (usually 50 Hz or 60 Hz). Harmonics combine with the fundamental voltage or current, and produce waveform distortion.
Symmetry of the three-phase system	Voltage unbalance	The maximum deviation from the average of the three phase voltages, divided by the average of the three phase voltages, expressed as a percentage.

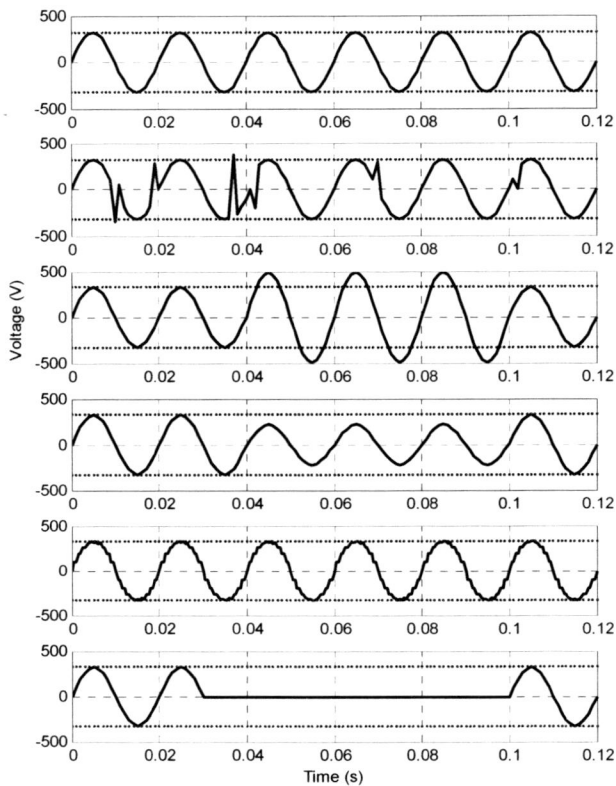

Figure 1.2. Example of the most usual power disturbances

Probably the most exhaustive classification can be found in the North American Standard IEEE 1159 [5], as shown in Table 1.4. However, the limits have all been quantified only in the EN 50160 [3]. This is not another EMC standard, but rather a product standard that defines the quality of electricity. Being mandatory in all European countries since 2003, it is used as a base for utility client contracts in the European Union and for small power-generation contracts. The Standard EN 50160 [3] defines maximum acceptable levels at the customer's supply terminals for medium ($\leq$ 35 kV) and low ($\leq$ 1 kV) voltage electricity distribution systems under normal operating conditions. These are quasi-guaranteed limits (at least for some parameters) covering any location of the power system. These limits are equal to or slightly higher than the compatibility levels.

Table 1.4. Categories of power-quality variation according to IEEE 1159

Categories	Spectral content	Duration	Magnitudes
1.0 Transients			
1.1 Impulsive			
1.1.1 Nanoseconds	5 ns rise	<50 ns	
1.1.2 Microseconds	1 μs rise	50 ns to 1 ms	
1.1.3 Miliseconds	0.1 ms rise	> 1 ms	
1.2 Oscillatory			
1.2.1 Low frequency	< 5 kHz	0.3 to 50 ms	0 to 4 pu
1.2.2 Medium frequency	5 to 500 kHz	20 μs	0 to 8 pu
1.2.3 High frequency	0.5 to 5MHz	5 μs	0 to 4 pu
2.0 Short duration variations			
2.1 Instantaneous			
2.1.1 Sag		0.5 to 30 cycles	0.1 to 0.9 pu
2.1.2 Swell		0.5 to 30 cycles	1.1 to 1.8 pu
2.2 Momentary			
2.2.1 Interruption		0.5 cycles to 3seg	< 0.1 pu
2.2.2 Sag		30 cycles to 3seg	0.1 to 0.9 pu
2.2.3 Swell		30 cycles to 3seg	1.1 to 1.4 pu
2.3 Temporary			
2.3.1 Interruption		3 seg to 1 min	< 0.1 pu
2.3.2 Sag		3 seg to 1 min	0.1 to 0.9 pu
2.3.3 Swell		3 seg to 1 min	1.1 to 1.2 pu
3.0 Long duration variations			
3.1 Interruption, sustained		> 1 min	0.0 pu
3.2 Undervoltage		> 1 min	0.8 to 0.9 pu
3.3 Overvoltage		> 1 min	1.1 to 1.2 pu

Table 1.4. (continued)

4.0 Voltage imbalance		Steady state	0.5 to 2%
5.0 Waveform distortion			
5.1 DC offset		Steady state	0 to 0.1%
5.2 Harmonics	0 to 100th H	Steady state	0 to 20%
5.3 Interharmonics	0 to 6 kHz	Steady state	0 to 2%
5.4 Notching		Steady state	
5.5 Noise	Broadband	Steady state	0 to 1%
6.0 Votage fluctuations	< 25 Hz	Intermittent	0.1 to 7%
7.0 Frequency variations		<10 seg	

In order to confirm that the disturbances in a power supply remain within allowable boundaries, the results of the power-quality measurements must be evaluated against the limits defined in the EN 50160 [3]. Voltage variation is characterized in two different ways: descriptive indices for events and statistically obtained values for steady-state variations.

Descriptive indices are used for power-quality events that have a highly random nature, can vary in time and are highly dependent on the topology of the system. Those events happen occasionally. They cannot be ascribed limits and only approximate figures are given. This approach is applied for describing: a rapid voltage change, voltage dips (sags), voltage swells, short interruption, long interruption and transient overvoltage.

The assessment method of the actual characteristics at a given point of the network (to be compared to the specified characteristics) is based on time statistics exclusively. Statistically obtained values are used for the assessment of a phenomenon that can be measured for a specific time period: power frequency, magnitude of supply voltage, supply-voltage variations, flicker severity, supply-voltage unbalance, harmonic voltage and interharmonic voltage. In these cases, cumulative frequency is the method used for statistical evaluation of the measured values. For example, considering harmonic voltage magnitudes under normal operating conditions, the measurement period is one week and 95% of the RMS values of individual harmonics on successive 10-min periods shall not exceed the specified limits.

Table 1.5. Supply-voltage characteristics according to EN 50160

Parameter	Low voltage (LV)	Medium voltage (MV)
Frequency	50 Hz ± 1% (10 s/95% of week)	
	50 Hz +4%/6% (10 s/ 100% of week)	
Magnitude of the supply voltage	Nominal voltage of the system U_n (rms)	Declared supply voltage U_c ($1 \leq U_c \leq 35$ kV) (rms)
Supply voltage variations	$U_n \pm 10\%$ 10 m/95%/week)	$U_c \pm 10\%$ (10 m/95%/week)
Rapid voltage changes	5% of U_n normal; 10% of U_n infrequent.	4% of U_c normal; 6% of U_c infrequent.
	Flicker: $Plt \leq 1$ (10 m/95%/week)	Flicker: $Plt \leq 1$ (10 m/95%/week)
Voltage dips	Majority: depth <60%, duration < 1s.	Majority: depth < 60%, duration < 1 s.
	Some locations: 1000/year, depth $10 \leq 50\%$	Some locations: 1000/year, depth $10 \leq 15\%$
Harmonics voltages	$U3 \leq 5\%$; $U9 \leq 1.5\%$; $U15 \leq 0.5\%$; $U21 \leq 0.5\%$;	
	$U5 \leq 6\%$, $U7 \leq 5\%$, $U11 \leq 3.5\%$, $U13 \leq 3\%$, $U17 \leq 2\%$, $U19,...U23 \leq 1.5\%$;	
	$U2 \leq 2\%$ $U4 \leq 1\%$ $U6...U24 \leq 0.5\%$; $THD \leq 8\%$; (10 m/95%/week)	
Voltage unbalance	2% for 95% of week, 10min rms. 3% in some locations	
Short interruptions of supply voltage	20 to 500/year. Duration 1 s; 100%	20 to 500/year. Duration 1 s; 100%
Long interruptions of supply voltage	(longer than 3 min) <10 to 50/year	
Temporary overvoltage	1.5 kV rms	170% (solidly or impedance earth) 200% (unearthed or resonant earth)
Transient overvoltage	6 kV rms	

According to standard EN 50160 [3], the supplier is the party who provides electricity via a public distribution system, and the user or customer is the purchaser of electricity from a supplier. The user is entitled to receive a suitable quality of power from the supplier. In practice, the level of PQ is a compromise between user and supplier.

Table 1.6 list the causes, effects and solutions for the majority of perturbations.

Table 1.6. Causes, effects and solutions for the PQ perturbations

Perturbation	Causes	Typical effects	Solutions
Voltage variations	Load variations and other switching events that cause long-term changes in the system voltage	Premature ageing, preheating or malfunctioning of connected equipment	Line-voltage regulators UPS Motor–generator set
Voltage fluctuations (Flicker)	Arcing condition on the power system (*e.g.* resistance welder or an electric arc furnace)	Disturbing effect in lighting systems, TV and monitoring equipment.	Installation of filters, static VAR systems, or distribution static compensators
Transients	Direct lightning strike to the building Induced in the distribution circuits by a nearby lightning strike. Switching events (*e.g.* capacitor, load switching) Switching from fault clearing	Upsets barely noticeable, with self-recovery like a click in a sound system or a flash on a video screen; upset permanent and noticeable, requiring, manual reset: blinking, clocks and VCRs; upset permanent but not readily noticeable: data corruption Damaged components, repairable or too costly to repair and irreparable damage requiring complete replacement of the equipment, such as internal equipment fire (that could set other objects afire)	Transient suppressors
Sag (dip) Short interruptions of supply voltage	Fault in the network or by excessively large inrush currents	Malfunctions of electronic drives, converters and equipment with an electronic input stage. Relays and contractors can drop out. Asynchronous motor can draw a current higher than its starting current at dip recover	UPS Constant-voltage transformer Energy storage in electronic equipment New energy-storage technologies (SMES, flywheels...)

Table 1.6. (continued)

Swell	Single-line ground failures (SLG), upstream failures, switching off a large load or switching on a large capacitor	Trip-out of protective circuitry in some power-electronic systems	UPS Power conditioner
Long interruptions of supply voltage	Distribution faults Installation failures	Current data can be lost and the system can be corrupted After interruption is over, the reboot process, especially on a large and complex system, can last for several hours.	UPS Distributed energy sources
Harmonic distortion	Nonlineal industrial loads: variable-speed drives, welders, large UPS systems, lighting systems Non lineal residential and commercial loads: Computers, electronic office equipment, electronic devices and lighting	Overheating and fuse blowing of power-factor-correction capacitors Overheating of supply transformers Tripping of overcurrent protection Overheating of neutral conductors and transformers	Passive and active filter
Voltage unbalance	Less than 2% is unbalanced single-phase loads on a three-phase circuit, capacitor bank anomalies such as a blown fuse on one phase of a three-phase bank. Severe (greater than 5%) can result from single-phasing conditions.	Overheating of motors Skipping some of the six half-cycles that are expected in variable-speed drives	To reassess the allocation of single-phase loads from the three-phase system

The increasing trends in the consumption of power-disturbance-mitigation equipment reflects the growing consumer implication in this issue, which must be encouraged with education through the distribution of guidelines and related information. Before considering any further levels of power protection, consumers should know that the power supply in his system could already afford him a substantial amount of protection. What is more, consumers must have a basic knowledge about the power-protection devices available and under what circumstances they should be used.

References

[1] Ward, DJ. Power quality and the security of electricity supply. Proceedings of the IEEE 2001; 89(12): 1830–1836.
[2] Gellings C W, Lordan R J. The Power Delivery System of the Future. The Electricity Journal 2004; 17(1): 70–80
[3] EN 50160. Voltage Characteristics of Electricity Supplied by Public Distribution System (1994).
[4] Martzloff FD. Performance criteria for power-system compatibility. Proceedings of the Applied Power Electronics 1992; 1: 287–292
[5] IEEE Std.1159. IEEE recommended practice for monitoring electric power quality (1995).
[6] Bartak G. Evaluation of Responses to the 2nd Questionnaire. Power Quality-Service Level Report n° 2004-030-0703, Nov. 2004.
[7] EN 61000-4-30. EMC, Part 4. Testing and measurement techniques: Power Quality measurement methods.
[8] IEEE Std.1366. IEEE Standard Guide for Power Distribution Reliability (2003).
[9] Bollen MHJ. What is power quality? Electric Power Systems Research 2003; 66: 5–14

2

Power-quality Monitoring

Mario Mañana Canteli

Department of Electrical Engineering
University of Cantabria
Santander, Spain
Email: mananam@unican.es

2.1 Introduction

Instrumentation for the measurement of conducted disturbances in power systems has undergone great development during the last decade. From the first instrumentation designed for general-purpose measurements, up to the current highly improved transients recorders, this kind of device has continuously evolved, becoming increasingly more specialised. In addition to the evolution of the hardware, there has also been significant activity in all topics related to the development of software for the analysis of measurements. In fact, managing records of power-quality (PQ) events is a problem that is growing day by day. Power-quality (PQ) monitoring should consider some basic questions:

- When to monitor. It is easy to program a power-quality survey after a problem has appeared. The difficulty lies in being able to do that before the problem arises, using a predictive approach.
- Where to connect. A correct choice of the instrumentation location in the power system is essential in order to draw valid conclusions.
- What instrument should be used. The choice between hand-held, portable or fixed equipment has to be made as a function of the time the instrument has to measure, the number of channels and the kind of disturbances we are looking for.
- What magnitudes should be measured. Sometimes a general survey has to be done, so all the power-quality indices have to be measured. In other cases, we are only interested in specific parameters.
- How to postprocess the registered data. After the measurement has been done, raw data and events have to be analysed in order to obtain conclusions.

2.2 State-of-the-art

In the spring of 1989 the European Union published the Directive 89/338/CEE [1,2], which started the countdown for the application of harmonized regulations relative to electromagnetic compatibility in the European Union. Currently, the regulatory process has still not finished, but there is an extensive set of standards that covers almost all aspects of electromagnetic compatibility (EMC) regulations.

It is obvious that EMC is an up-to-date topic, and not only for political reasons. The spectacular growth that has occurred during the last two decades in the utilization of power electronics in almost all kinds of electronic devices has given rise to an increase in the distortion of the distribution network. This growth in the number of electronic devices has not been accompanied, in many cases, by an improvement in the quality of the electronic designs. What is worse, most of them are extremely sensitive to the existing disturbances in the distribution network, so they exhibit malfunction during their operation under disturbed networks. These and other reasons are the motivation behind active research work on PQ instrumentation.

2.2.1 Historical Background

Two decades ago, the available instrumentation for power-quality assessment did not exist, and at best, they had a general purpose like oscilloscopes or spectrum analysers. The use of general-purpose instruments provided raw data that had to be postprocessed in order to obtain any conclusion. In other cases, the engineers and technicians were equipped with real root mean square (rms) voltmeters and ammeters, so the analysis used to be a difficult task only available to experts.

The existing instrumentation that can be utilized for power-quality evaluation could be classified into two sets according to their degree of specialization:

- General-purpose instrumentation.
- Specific-purpose instrumentation.

2.2.2 General-purpose Instrumentation

Basically, general-purpose instrumentation includes oscilloscopes and spectrum analysers. Spectrum analysers can be also divided according to the procedure of analysis:

- Digital signal analysers. These utilize the fast Fourier transform (FFT) or similar techniques to compute the spectrum of the signal.
- Analog signal analysers. These are based on parallel banks of analog filters than can be tuned in order to obtain the value of the spectrum components.

The analysers that use the FFT have standardized features, and in commercial format they are provided with bandwidths from 0 Hz up to 20, 100 and even 200 kHz. These equipments usually have 1 or 2 input channels, though some of them can have 4, 8 or 16: it being possible to carry out graphical representations of the signals in both time and frequency domains. Another characteristic to evaluate, in this type of meter, is the possibility of doing zooms of specific ranges of the spectrum of the signals. In many cases, it is also possible to use different windowing (rectangular, triangular, *etc.*) during the sampling process.

The analysers based on parallel banks of filters were, historically, the first ones to appear. The basic principle behind them is quite simple. A set of analogical bandpass filters divide the spectrum into bands whose union makes it possible to reconstruct the spectrum of the original signal. The main drawbacks are the cost and the complexity, reasons for which their use is reduced to applications of very high accuracy in which cost is a secondary factor. The bandwidth of these analysers is in the 100 kHz range, and they are capable of achieving resolutions of 1 Hz.

Whatever the structure of the meter, it is important to be able to have offline access to the measured information, in order to postprocess the measured data. The majority of the systems that use the FFT provide an output of information in RS232 or IEEE-488 format, with the possibility of programming the instrumentation externally, and to do some type of remote control. Nowadays, many of them also have an ethernet interface, so that they can be connected to a local area network or The internet.

Another important feature is the number of different windows that can be used with them in the process of sampling, usually between 2 and 12. The resolution of the analog-to-digital converter (ADC) is another parameter that changes from one device to another. The majority have between 12 and 16 bits, though many of the existing oscilloscopes have only 8 bits.

With regard to the possibility of analysing interharmonics, some instrumentation devices provide up to 3200 spectral lines free of aliasing and distortions in a bandwidth of 128 kHz. Working in real time, the bandwidths allow the majority of the equipments to go beyond 10 kHz.

2.2.3 Specialised-purpose Instrumentation

Two decades ago the instrumentation to monitor power quality was only a prototype that one could find in some research laboratories. Today, there is a competitive market.

Table 2.1 summarizes the most important features of the instrumentation designed specifically for power-quality measuring. Also included here are the characteristics of an open platform named MEPERT that has been developed at the

Electrical Engineering Department of the University of Cantabria [3–5]. MEPERT is shown in Figure 2.1.

Table 2.1. Typical parameters of a power-quality meter

Type of application	Hand-held
	Portable
	Fixed installation
User interface	Alphanumeric
	Graphic
	Oscilloscope
	Text
	Blackbox
Measured parameters	DC voltage and current
	Harmonics and interharmonics
	Ground resistivity
	Power factor
	Flicker
	Power / energy
	Transients (> 200 μs)
	Impulses (< 200 μs)
	Dips
	Overvoltages
	Imbalance
	Frequency
	Other disturbances
Type of meter	Trends
	Energy
	Spectrum analyser
	Transients recorder
Type of communication	RS-232
	Ethernet
	TCP/IP
	Modem
	Power-line communication
	Other

In general, this type of system makes it possible to carry out the evaluation of any type of conducted disturbance: variations of the nominal frequency of the supply, variations in the magnitude of the voltage supply, transients, flicker, imbalance, harmonics, interharmonics, dips and interruptions as defined by the standard EN 50160 [6]. In many cases they have, in addition, specialized software able to analyse the stored measurements [7,8].

In 1990, the Electric Power Research Institute (EPRI), which is the organization that coordinates the research on electrical systems in the USA, had already established projects with Electrotek Concepts for the development of an integral software for disturbances analysis [9].

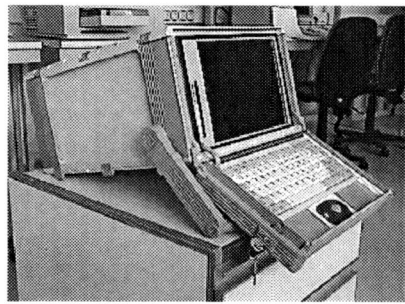 

Figure 2.1. PQ meter open platform designed at the Department of Electrical Engineering, University of Cantabria

In the above-mentioned project, an instrumentation developed by BMI was in use. Power-quality surveys are not new. In fact, there are references [10,11] from 1993 that summarize the state of the distribution network in the eastern part of the USA.

Since the beginning of the 1990s, it has been possible to find commercial power-quality meters that do not limit their analysis to harmonics. This study includes more than 5400 points/month of information over two years. The implemented software was designed at Electrotek Concepts by a research team led by E. Gunther.

Nevertheless, and in spite of the fact that there exist documents that specify the requirements for the measurement of harmonics and interharmonics (IEC 61000-4-7) [12], flicker (IEC 61000-4-15) [13], and some other disturbances, there is still no document that draws together all the aspects and specifications necessary for the development of a global power-quality meter. Figure 2.2 includes the results obtained from the comparison of a set of commercial power-quality meters from different manufacturers [5].

From the point of view of the available communications, it is possible to say that in the 50% of the cases the remote management of the instrumentation is implemented by means of modems. The remaining 50% is distributed in almost equal parts between one native ethernet interface and TCP/IP.

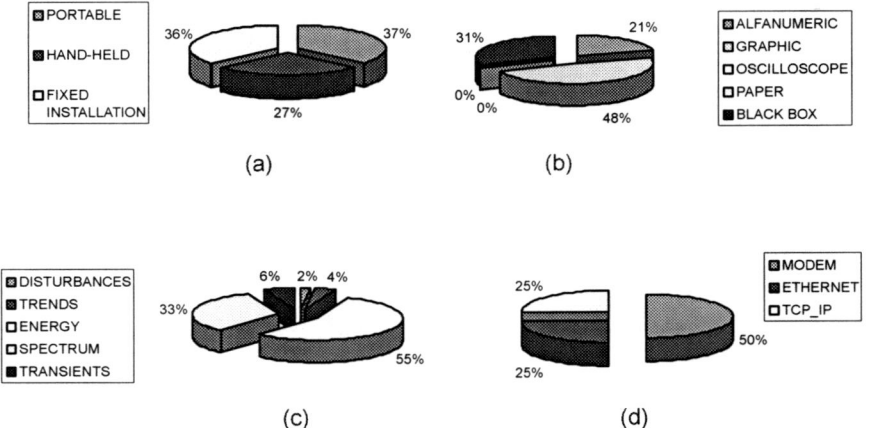

Figure 2.2. Classification of commercial PQ meters by different criteria. (*a*) Type of meter; (*b*) Type of user's interface; (*c*) Type of measurements; (*d*) Type of communication interface

2.3 Instrumentation Architecture

The voltages and currents to be measured can be accessible directly or indirectly in the case of LV and MV/HV systems, respectively. EURELECTRIC [14] established a basic architecture of the measuring system that is shown in Figure 2.3.

The term instrumentation spans, theoretically, V_s to G_e, though in general it is assumed that a power-quality meter includes all the elements from V_m to G_e. This means that the measuring transformers can be considered as an independent system.

This dichotomy of the measuring system can be also observed in the standards, which define specific documents for the instrument transformers and the rest of the electronic system. In addition, Figure 2.4 shows the basic architecture of a power-quality meter.

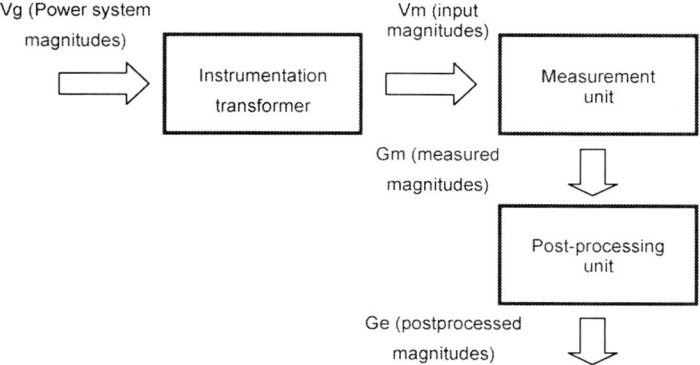

Figure 2.3. Main elements of a PQ instrumentation system

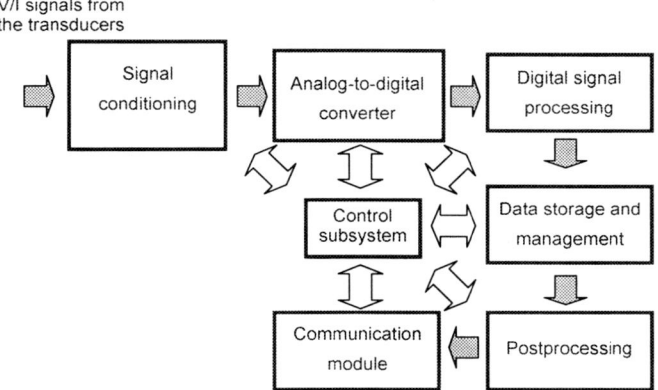

Figure 2.4. Basic architecture of a power quality-meter

2.3.1 Safety Use of PQ Instrumentation

The safety of both the operators and the instrumentation used in the measurement of electrical magnitudes is an aspect that is considered to be an annex to the intrinsic problem of measurement, since in general it is transparent to the user. It seems to be evident that if a manufacturer wants to sell an instrument to monitor the evolution of the rms value of an intensity up to a maximum value of 100 A, the equipment will be capable of supporting at the very least the above-mentioned intensity without suffering any type of damage or malfunction either transitorily or permanently. This type of guarantee is, in many cases, obvious for the users, but perhaps not so much for the designers and manufacturers of the instrument, who must submit the designs to a series of tests that guarantee their safety.

The measure of voltages and currents with portable equipment constitutes generally risky work, so that it is necessary to reduce or to eliminate any situation of potential risk for the user, the instrumentation or the installation. In many cases the problem can be worse, because there could appear overvoltages or overcurrents that go over and above the nominal values established by the standards, so that it is necessary to introduce additional measures of protection. In order to guarantee the safety of persons and equipment, there exist some regulations that must be followed when the instrument is acquired.

The equipment must indicate clearly the maximum voltage for every type of measurement (DC or AC). In addition, the test points must also indicate the maximum voltage that they support, since, in general, they are not permanently connected to the instrumentation, and therefore, they can be exchanged by others of different characteristics. The equipments have, in general, protection fuses, which will have to clearly indicate the maximum current and the performance speed.

The probes must be suitably isolated, and at the same time the connections with the instrumentation must assure the separation of parts under voltage. If it is necessary to replace some internal element such as fuses or batteries, it will be necessary to give clear warning of the obligation to disconnect the equipment from parts with voltage.

All the previous points must be taken into account for designers and users, who must not forget the existence of regulations relative to safety in equipment. The basic sources of information are: European Union (CENELEC), specific regional Procedure of Canada (CSA), United Kingdom (UL), Germany (VDE), Spain (UNESA) or Switzerland (GS) just to mention some, and that must be certified in order to be sold in the above-mentioned countries; in the case of the European Union, the equipments have to bear the mark CE.

As an example, the standard IEC 61010-1 [15] relative to the safety requirements for electrical equipment for measuring, controlling or use in a laboratory establishes a classification according to different criteria, based on the type of isolation:

- *Basic*, which refers to equipment whose parts under voltage, in normal working conditions, have levels below 30 rms volts, 42.4 peak volts or 60 DC volts.
- *Double*, for all other equipment, or those that are not connected to ground.

2.3.2 Number of Channels

From the point of view of instrumentation design, one of the most important aspects to consider is related to the input channels. First, it is necessary to establish the number of channels. Due to economic criteria, the choice of the number of channels is based on a specification of minimal needs. Figure 2.5 establishes a criterion for the classification of the number of measuring channels according to the application.

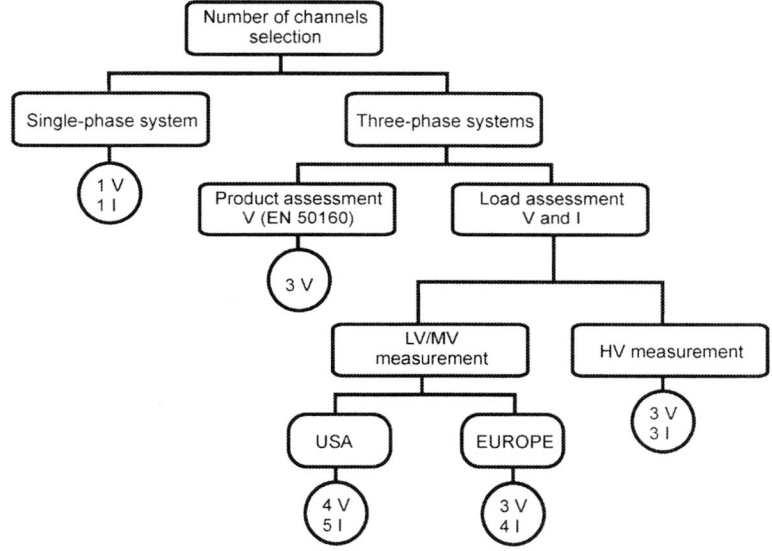

Figure 2.5. Diagram for number and type of input channels

One of the most interesting aspects of the structure shown in Figure 2.5 focuses on the difference in measurement needs on low voltage networks between USA and Europe. The distribution networks in USA are based on the connection of single-phase loads or unbalanced three-phase loads, principally in an open triangle form, which originates the need to control the current of the neutral. In addition, it is quite common for the distribution in medium voltage also to be based on four wires. On the other hand, in Europe, the distribution is based on three-wire systems. This means that if there is any current flowing to ground it can be considered a fault.

In the USA, it is quite usual to install the protection wire, so it is necessary to monitor the voltage between neutral and ground in order to detect certain faults. The measurement of this voltage requires a channel with a very small range of measurement. In Europe, it is normal to meet systems of isolated neutral or neutral connected to ground by means of a high impedance, which is why the measurement of these voltages and the associated current to ground is not important in many cases.

Voltage measurement can be classified according to two different approaches: common unified and independent channels.

2.3.3 Common Unified Channels

In this type of measurement, all voltage channels have the same reference. From a practical point of view they are characterized by having $N+1$ measurement wires for the monitored voltage, with N being the number of channels. It is a measurement system quite commonly used at present, especially in the USA, where the distribution systems have the neutral conductor connected to ground.

This structure is valid only for distribution systems with a neutral conductor. In some cases, if the neutral conductor is not accessible, an artificial one is created. Nevertheless, the utilization of an artificial neutral for the measurement of phase to neutral voltages is subjected to the hypothesis of having a symmetrical voltage.

Figure 2.6 shows a three-phase system where only points A, B and C are accessible. To be able to measure the phase to neutral voltages, a high-impedance wye-balanced three-phase load is used, which does not modify the network.

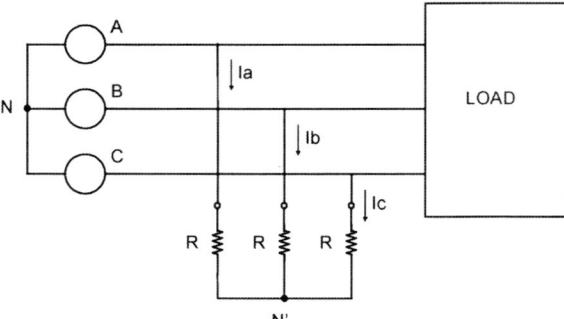

Figure 2.6. Equivalent circuit of virtual wye load for phase to neutral measurements

If the three-phase system, the voltages for which we are trying to measure, is balanced, the point N is equal to N' and then the neutral point of the fictitious load does not suffer any displacement with regard to the neutral of the generating system or, in other words, the phase voltages are the distribution network voltages.

If the voltages to be measured constitute an unbalanced three-phase system, the measured phase voltages are different from the distribution-network phase voltages. In this case, infinite solutions can be found for the phase voltages, thus it is not possible to measure the real ones.

Since, in general, it is not possible to establish whether a distribution network is balanced from the measurement of the phase to phase voltages, the conclusions obtained from these kinds of measurements must be avoided.

2.3.4 Independent Input Channels

This type of input-channel arrangement constitutes the most widely used architecture. From a practical point of view it also enables the unification of all the input-channel references. In a general case, completely independent measurements can be done without additional elements. The latter possibility can be really interesting in cases in which different voltages without common references have to be monitored, since it can be the case of different measurement points in substations (*e.g.* different potential transformers).

2.3.5 Transducers

The aim of transducers is to provide a small voltage signal (with typical amplitudes between −10 and 10 V), which are proportional to the levels of voltage and current that have to be measured. In general, a correct utilization of the transducers means that their frequency response has to be completely characterised, both in magnitude and phase.

Although according to EURELECTRIC [14], voltage and current transducers are physically independent from the instrumentation, it is very important to have a good understanding of the basic criteria for their correct selection and utilization. An appropriate selection of current transducers has to take into consideration, among others, the following features: nominal values, geometry, maximum value, accuracy, range of frequencies to measure, *etc*.

There are some different technologies available for current measurement. Table 2.2 summarizes a comparison between current transducers for low-voltage usage.

Potential transformers (PT) usually have an accuracy greater than 1% between 40 Hz and 1500 kHz. At MV and HV the capacitive effects become more dominant. From a practical point of view, the accuracy itself is not as important as their linearity. This is because voltage harmonics are computed as a percentage of fundamental, and not as absolute, values. The phase displacement in that range is typically over 40 minutes or more.

On the other hand, current transformers (CT) have an accuracy of over ±0.5% for frequencies from 20 to 1000 Hz. The performance at 1 kHz starts to decrease and the maximum bandwidth is about 10 kHz.

Independently of the kind of current transducer being used, all of them have a set of common parameters that have to be understood. Figure 2.7 shows the equivalent electric circuit for a generic current transducer.

Table 2.2. Current transducer comparison

	Input Range	Over-range Capacity	Accuracy	Sensitivity	Bandwidth	Security
Resistive Shunt	+	+	++++	++++	++++ DC-100 MHz	+ No isolation
Inductive	++++	+++	++ ±0.5% 0.5°	++	++ 20 Hz–1 kHz	+++ Single isolation
Hall effect	++++	+++	+ ±1% 1.5°	++ 1.5-3.0 V/FSA	+++ DC–50 kHz	++++ Double isolation
Rogowski Coil	++++	+++	+ ±1% 1.5°	+ 0.2 V/FSA	+++ 1 Hz–10 MHz	++++ Double isolation
Optical	+++++	+++++	++	+	++++	+++++ Double isolation

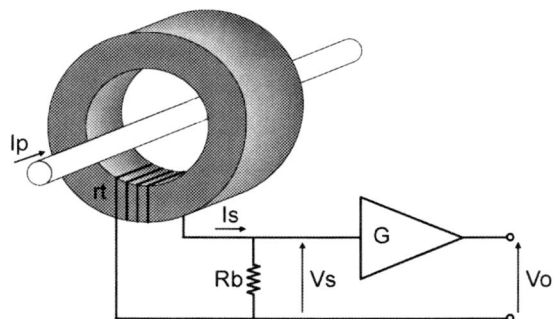

Figure 2.7. Equivalent circuit of a generic current transducer

The circuit shown in Figure 2.7 allows us to obtain the relation between the current I_p that has to be measured and the proportional output voltage V_o,

$$I_p = \frac{V_o r_t}{GR_p} \quad (2.1)$$

where,

I_p is the current to be measured,
V_s is the voltage before the amplifier,
V_o is the output voltage,
G is the amplifier gain.

From the point of view of the user of the instrumentation, the most important aspects are: transformation ratio, measurement range, bandwidth and geometric dimensions. The manufacturer is required to provide the curves of magnitude and phase versus frequency. In [16-18] some procedures are commented on in order to accomplish the contrast of PQ meters and transducers. Figure 2.8 shows the Fluke 6100A, which is a commercial test system that can be used for instrumentation and current transducer tests.

One of the problems that is still not completely solved in the design of instrumentation for power-quality measurements in high-voltage systems is the direct measurement of voltage and current, especially when transitory phenomena have to be registered.

Classical instrument transformers include both current transformer (CTs) and potential or voltage transformers (VTs). These kinds of devices are designed for 50 or 60 Hz, so their behaviour at high frequency could exhibit nonlinearities and other unwanted effects.

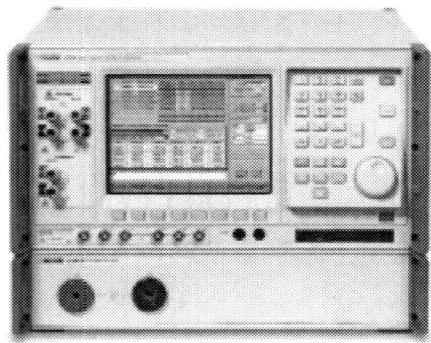

Figure 2.8. PQ meters and transducers test and calibration unit Fluke 6100A

One of the solutions to this problem can be the utilization of optical transducers. In the early stages, optical elements were introduced only in the signal transmission between the classical transducer and the instrumentation. In recent years, much effort has been made towards the design of totally optical transducers [19].

In formal terms, the current that is passing through a conductor can be expressed as the integral of the magnetic field in a closed path that includes the conductor. This is,

$$I_p = \oint H dl \qquad (2.2)$$

where,

I_p is the current to be measured,
H is the magnetic field,
dl is the closed path of integration.

The previous equation serves as a background for the design and manufacturing of the classic transducers that are used nowadays, and it also, certainly, constitutes the theoretical basis for optical transducers.

In a basic form, optical transducers are physical devices that modify some of the properties of the light that travels through them, because they have been immersed in a magnetic field, the intensity of which changes with time. One of the most used phenomena is named the "Faraday's effect" or the "magneto-optic effect" in the technical literature.

To understand the mechanism that gives rise to this effect, it is necessary to know that a beam of light with an arbitrary polarization can be expressed as a combination of two orthogonal components polarized linearly, or as a combination of two orthogonal components circularly polarized. An analogy can be established between these circular basic elements and the components of positive and negative sequence in a three-phase system.

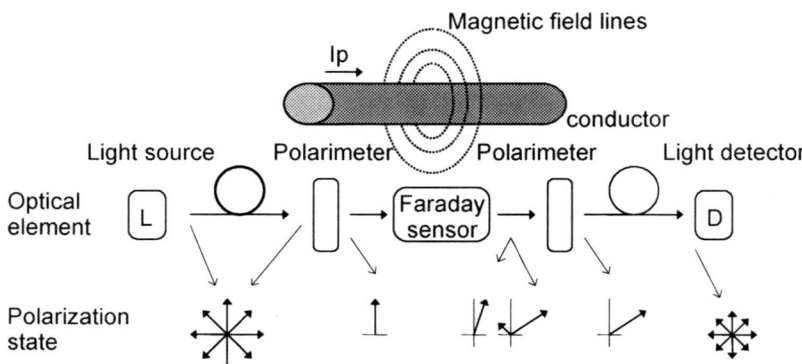

Figure 2.9. Structure of a current transducer based on the Faraday effect

The Faraday effect is, basically, a process of optical modulation, that is, a process of rotation of the plane of polarization of a beam of light polarized linearly, and proportional to the magnetic field that is passing through the material. The variation between the angle of rotation and the intensity that flows through the conductor is a constant named Verdet's constant. The law that defines the angle of rotation is,

$$\theta = \mu_r v \int H \cdot dl \qquad (2.3)$$

where,

θ is the rotation that the light beam component undergoes when it passes through the material,
μ_r is the relative permeability of the medium,
ν is Verdet's constant.

The following problem to solve is rooted in the fact that it is difficult to measure the degree of polarization of the light directly, hence it is necessary to perform a previous transformation of the degree of polarization of the light in optical power (square of the field E). This transformation is done by means of a light polarimeter. Figure 2.9 shows the block diagram of a Faraday-effect sensor. The measure of the intensity is computed as a ratio between the input and output optical power.

There are also optical voltage transformers. These are based on an electro-optical effect that produces a rotation of the polarization plane.

As a conclusion, it can be said that optical sensors for current measurement are small, accurate and do not suffer from the isolation problems, lack of accuracy and poor frequency response of the classic transformers. Nevertheless, they have not led, at least at present, to a reduction in volume compared with the conventional instrument transformer, since the electrical companies demand elements similar to the classical ones, in order to be able to replace them without mechanical changes.

2.3.6 Signal-conditioning Module

The signal-conditioning module acts as an interface between the transducers and the analog-to-digital converter. It can be seen as a glue element between the analog and digital worlds. Their main objectives can be summarized as follows:

- To provide galvanic isolation between the power system and the user of the instrumentation.
- To amplify and/or to attenuate adequately the voltage or current signals in order to obtain the maximum dynamic signal range, guaranteeing in this way a high signal-to-noise ratio.
- To avoid aliasing problem by means of low-frequency filters.

Figure 2.10 shows a block diagram of this module, where the described elements can be seen.

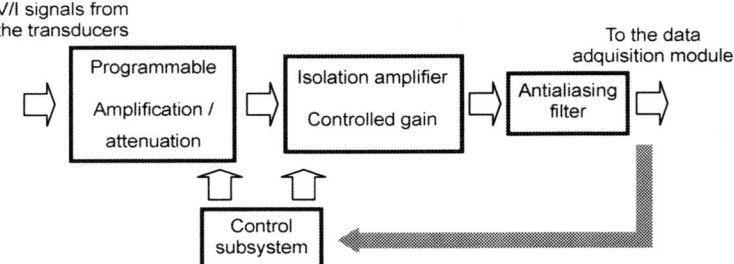

Figure 2.10. Diagram block of the signal-conditioning module

2.3.7 Analog-to-digital Converter

The analog-to-digital converter (ADC) is the module that has the responsibility of translating the signal from the analog to the digital domain. It carries out two basic tasks [20]:

- i) Signal discretization. This is the process of periodic sampling of the signal using the Nyquist–Shannon theorem [21]. It allows us to obtain a sequence of samples $x[n]$ obtained from a continuous-time signal $x(t)$.
- ii) Signal quantization. This is a mathematical transformation that assigns a fixed-point binary number to a sampled signal value that belongs to the real numbers.

The resolution of analog-to-digital converters in power-quality instrumentation has values between 8 and 24 bits. Other values can also be found, but are not typical.

From a numerical point of view, quantification is a nonlinear transformation that produces an error. The error range depends on the number of bits of the ADC and the input range.

2.3.8 Signal-processing Module

The signal-processing module is the core of the instrumentation. It includes the task manager, which controls the device and the set of algorithms that compute the power-quality indexes, the file manager and the communication facilities. The main features of this module are:

- Reprogrammed firmware. This characteristic allows us to programme and to reschedule the functionality of the system without hardware modification, which reduces the development costs enormously. For instance, a system can be updated to measure harmonics with different algorithms or standards without changing the hardware.

- System stability. This involves repeatability of the implementation and of the response. It is easy to understand that a digital system provides the same answer to the same question. This is not so with analog systems, where the response is a function of the temperature, age, humidity, *etc.*
- Suitable for implementing adaptive algorithms and special functions such as linear phase filters. The utilization of numerical algorithms allows external errors due to changes in the operating conditions to be corrected dynamically.
- Able to compress and store measurement data. It is not necessary to highlight here the importance of computers in the storage, treatment and recovery of information.
- User-friendly interfaces. The utilization of graphical user interfaces (GUIs) facilitates the interaction of the user with the instrument.
- Low power consumption. These kinds of devices have a power consumption less than 3 VA.

2.4 PQ Instrumentation Regulations

Standards provide a common reference framework that allows us to compare the qualities of the products that we, as consumers, use in our daily lives. The set of standards that regulate power-quality measurements belongs to different groups of documents. Firstly, the whole regulation that defines the technical characteristics of the instrumentation. Secondly, the measuring procedures and finally, the legal limits of the power -quality indices.

In Spain, the legal framework began with Law 54/1997, which was later given greater depth in Law 1955/2000.

Almost all the technical documents are generated by IEC and CENELEC, and have to be applied in all of the European Union. In 1996 the European Commission ratified the directive 89/336/EEC [1]. This has been applied since 1996 and can be considered as the first step in a series that facilitates the globalization of power-quality procedures. The compliance with these standards is shown by the CE marking of the product. This global framework enables the European power-quality-meter manufacturers to gain access to a wider market. Among others, the most active groups that define and publish standards and reference documents are:

- IEC. International Electrotechnical Commission.
- CENELEC. European Committee for Electrotechnical Standardisation.

There are other standards within the European union, at the local level, like:

- UNESA. Spanish Standards Institute.
- BSI. British Standards Institute.

Figure 2.11 shows the standards structure both in Europe and Spain. The countries that are included in the European Union have to harmonize the CENELEC documents and requirements to their national standards. On the other hand, CENELEC also adopts, in some cases, international regulations, like those generated by the International Electrotechnical Commission (IEC). This fact can be readily observed if the standard reference is compared in the IEC and CENELEC environments. The same happens with the Spanish standard editor UNESA. A good example could be the standard IEC 61000-4-7, which defines the characteristics of the instrumentation for harmonic and interharmonic measurement. At the European level, this standard has been adopted by CENELEC in Europe as EN 61000-4-7 and by UNESA in Spain as UNE EN 61000-4-7.

In the USA, the Institute of Electrical and Electronic Engineers (IEEE) is working on the new standard IEEE 1159 [22] entitled "Recommended Practice on Monitoring Electric Power Quality". This work is sponsored by the Power Quality Subcommittee of the IEEE Power Engineering Society.

Figure 2.12 shows the Dranetz PX5, a power-quality meter that fulfils the standard IEC 61000-4-30 [23]. Among other available synchronization techniques, the utilization of the GPS could be a good tool for synchronizing power-quality meters in different locations.

Generic standards are general-purpose documents that all electrical equipment should comply with. One example is the standard IEC 61000-6, which has been adapted by CENELEC as EN 50081 (generic emission standard) and EN 50082 (generic immunity standard).

From the electrical point of view, a device can be considered a product or part of the distribution network.

From one point of view, product standards cover the aspects of specific product lines where the application of the generic standard would not be the right choice. From another, network standards include the documents that cover different aspects related with the power system: general considerations, description and classification of the environmental levels, emissions and immunity limits, testing and measuring techniques, installation and mitigation guidelines, generic standards and miscellaneous topics.

As an example, IEC 61000 (EN 61000) covers almost all aspects related with low and high-frequency disturbances.

The IEC 61000-4-30 [23] defines the methods for measurement and interpretation of results for power-quality parameters in 50/60 Hz AC power-supply systems.

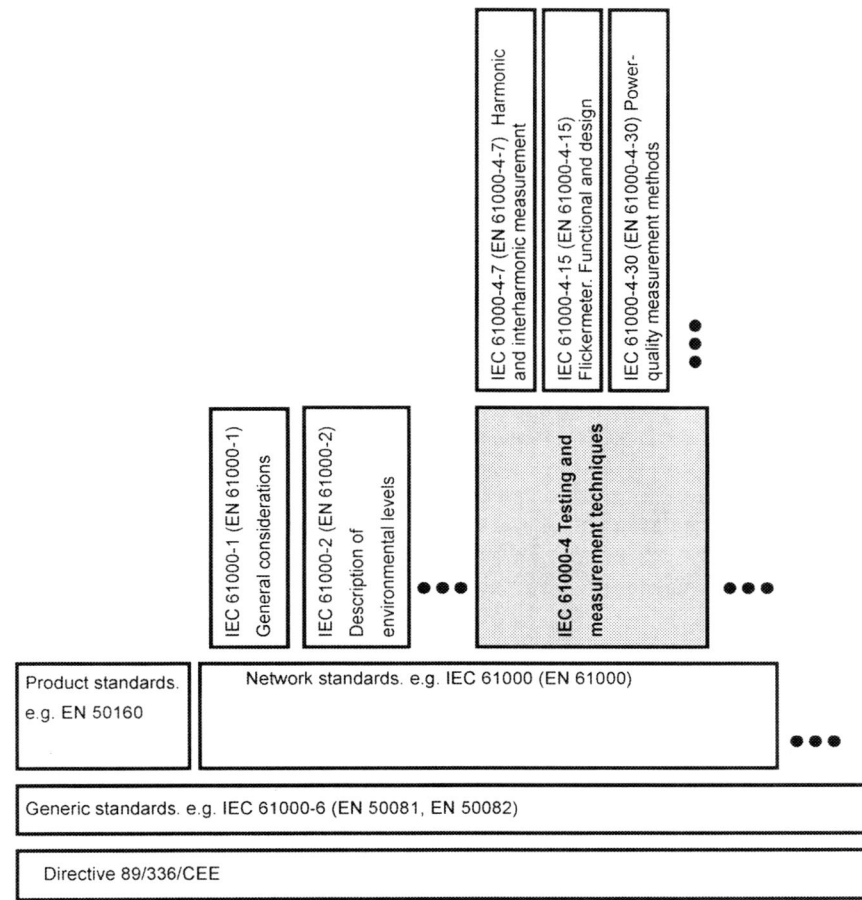

Figure 2.11. Structure of the standards related to PQ measurement

The basic aim of this document is to define measurement methods that will make it possible to obtain reliable, repeatable and comparable results independently of the power-quality-meter manufacturer. The standard does not provide information about instrumentation design. This kind of information is devoted to other standards like IEC 61000-4-7 and IEC 61000-4-15.

2.5 Harmonic Monitoring

Both harmonics and interharmonics can be measured using various techniques. In order to obtain results that can be compared between different equipments and manufacturers, a standard document has been published by the IEC, CENELEC and other regional agencies like the Spanish AENOR. The standard IEC 61000-4-7 [12] defines a general guide on harmonics and interharmonics measurements and instrumentation, for power-supply systems and equipment connected thereto. The basic architecture of digital equipment based on the FFT is shown in Figure 2.13.

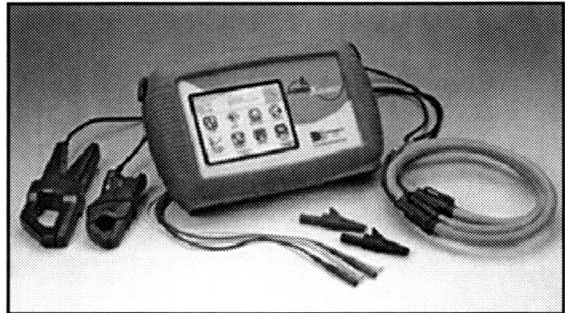

Figure 2.12. Dranetz-BMI PX-5 power-quality meter

In a first step the input signal is low filtered in order to fulfill the Nyquist–Shannon sampling theorem. This element is also named an antialiasing filter. The cutoff frequency is fixed to the 50th harmonic. The filtered signal is sampled and stored using a frequency that is obtained with a phase-locked loop (PLL).

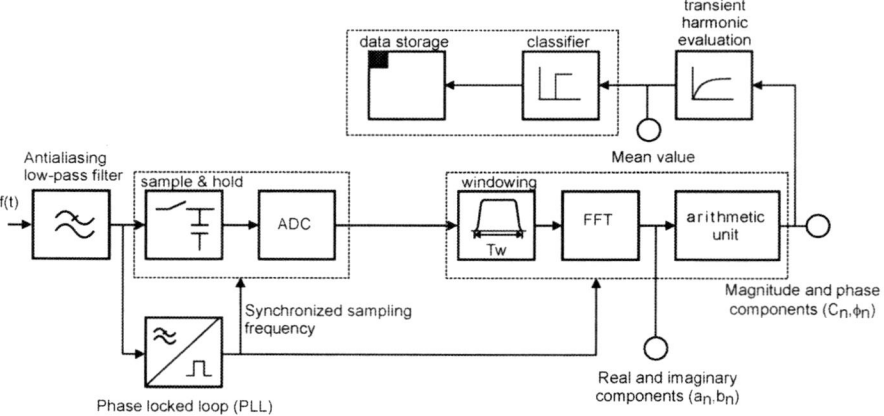

Figure 2.13. Block diagram of the FFT harmonic meter proposed by the IEC 61000-4-7

The FFT is then applied to a block of 2^k samples included in a period $T_w=NT_0$, where N is an integral number and T_0 is the fundamental period of the signal. The sampling frequency is then $f_s=2^k/(NT_0)$. The block of 2^k samples is sometimes multiplied by a window function that reduces the spectral leakage. The FFT subsystem computes the Fourier coefficients a_n and b_n of frequencies $f_h=h/T_w$ for $h=0, 1, 2, ..., 2^{k-1}$. The terms $n=h/N$ with $n \in Z$ are the nth harmonic of the fundamental frequency.

The standard defines almost all the basic aspects of the harmonics and interharmonics measurement: accuracy, type of application, hardware architecture, FFT application, evaluation methods, data-aggregation method, immunity test, *etc*.

2.6 Flicker Monitoring

Flicker measurement can be done according to the IEC 61000-4-15 [13]. This document defines the architecture and some test procedures that have to be followed in order to obtain the harmonized flicker perception P_{st} and P_{lt}. The block diagram of the IEC flickermeter is shown in Figure 2.14.

The flickermeter has two main parts. The first one attempts to simulate the behaviour of the set lamp-eye-brain, considering the lamp as incandescent with a power of 60 W. The second part is focused on the statistical analysis of the instantaneous flicker perception.

It produces the short-term flicker severity index P_{st} with a 10-min period between samples and the long-term flicker severity index P_{lt} every 2 h.

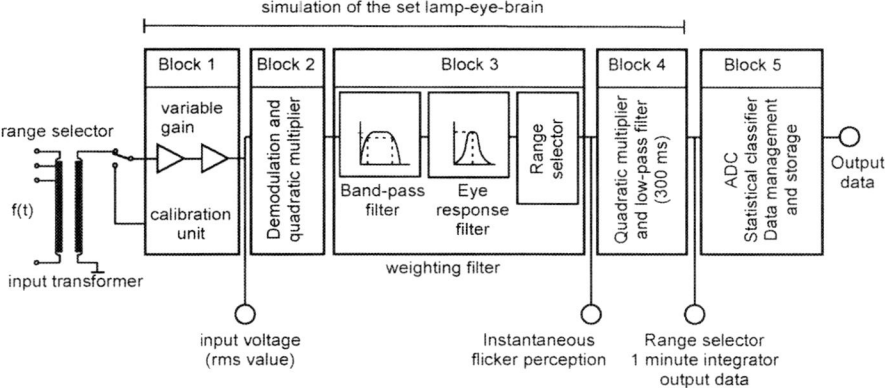

Figure 2.14. Block diagram of a flickermeter based on the IEC 61000-4-15

2.7 Data Postprocessing

Power-quality instrumentation should have not only good hardware architecture, but also powerful software that permits the measured data to be managed and analysed. Figure 2.15 shows the architecture of a postprocessing application.

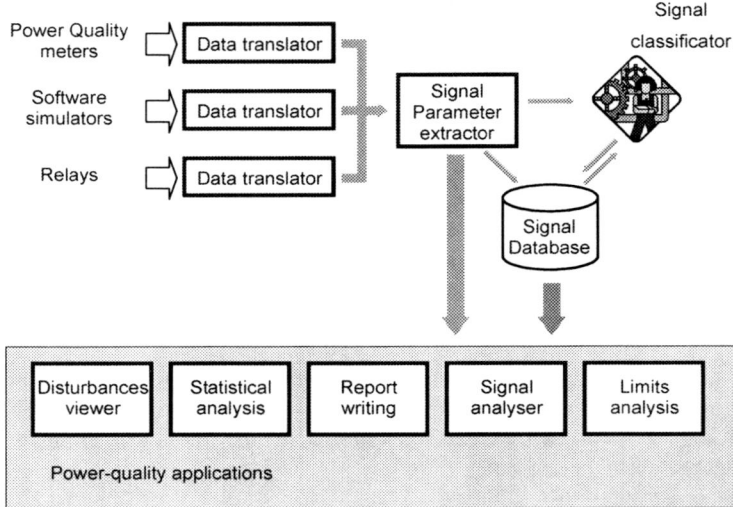

Figure 2.15. Architecture of a power-quality postprocessing system

In a complementary way, a power-quality survey involves the postprocessing analysis after the measurements. The analysis can be supported by specialized tools with some degree of automation [24–34]. The most powerful tools include some kind of automatic disturbance classification.

Disturbance classification includes all aspects related with the problem of determining the degree of similarity between a signal and a set of classes that act as a dictionary of reference. This problem involves three main actors: i) the unknown signal to be classified. ii) the database with the classes of reference and iii) the set of classification algorithms.

In spite of the fact that multiple classification methods or data-mining techniques can be used, it is possible to compile them into three large families:

- Statistical classification or theoretical decision methods. The foundation of this type of method consists in obtaining an index that allows the degree of similarity between the unknown signal and the information stored in the database to be quantified. Euclidean distance classifiers belong to this set of groups, where they use the Euclidean distance between two points in a multidimensional space. This space can be obtained as the Fourier transform or the wavelet transform of the signal.
- Syntactic classification, which uses a structural representation of the disturbance and an expert system in order to model human know-how.
- Classification based on nonlinear methods. The utilization of neural nets is included here. Supervised learning can be considered a classic method of classification. Correlation methods can be also included here. They exhibit

a good behaviour when the signals are distorted with white noise. Another method to be included here is the use of the geometrical properties of the phase-space representation of the signal.

2.8 Management of PQ Files

Some years ago, the interchange of information stored by different equipments was difficult and almost impossible. This came about because the manufacturers had their own format files. In order to solve this problem, some standards have been defined, providing a way to store and interchange power-quality data with a well-known structure. The most important formats are the IEEE Standard C37.111 "COMTRADE" [35] and the power-quality data-interchange format "PQDIF" [36].

2.8.1 COMTRADE

The IEEE defined in 1991 the standard C37.111 entitled "IEEE Standard Common Format for Transient Data Exchange (COMTRADE) for Power Systems". This work was sponsored by the Power System Relaying Committee of the IEEE Power Engineering Society.

The main purpose of this standard was to define a common format for the data files and exchange medium needed for the interchange of various types of fault, test, or simulation data. Among others, the standard defines as sources of data the following: digital fault recorders, analog tape recorders, digital protective relays, transient simulation programs and analog simulators.

2.8.2 PQDIF

The power-quality data-interchange format, also known as PQDIF was developed by the Electric Power Research Institute (EPRI) and one of its contractors, Electrotek Concepts, Inc. The main aim of this project was to develop a vendor-independent interchange format for power-quality-related information. This power-quality data-interchange format (PQDIF) has been in use since 1995.

In 1996, EPRI and Electrotek placed PQDIF in the public domain to facilitate the interchange of power-quality data between interested parties. EPRI and Electrotek have also offered the format, sample source code, and documentation to the IEEE 1159.3 task force as a possible initial format to meet that group's requirements.

2.9 Summary

The evaluation of the future evolution of this type of instrumentation is always a complex topic, since it involves, besides the classical problems of electronic instrumentation, new aspects that are independent from the process of measurement.

From a technical point of view, current meters are capable of registering all the phenomena of interest: harmonics, flicker, dips, transients, imbalance, *etc.*, with the possibility of registering the measurement in a constant way or by means of trigger mechanisms such as thresholds, slopes, logical events, *etc.*

An aspect to improve is the synchronization between different equipments for the accomplishment of measurements synchronized in geographically distributed environments. In this respect there are already developing projects that are starting to use advantages such as the GPS (global positioning system), which is a set of satellites provided with synchronized clocks, and that act in a way that might be seen as a universal time base, which is accessible from almost any part of the world.

As regards control and remote monitoring, it is important to be able to include the equipment in an integral system of measurement. In order to do that, it is necessary to establish common protocols for all the manufacturers, and to facilitate the interchange of data between them.

The current trend is focusing on the use of standardized protocols such as MMS (manufacturing message specification), which is simply a protocol meeting the OSI stack. It has been designed for the control and remote monitoring of industrial devices such as PLCs (programmable logic controller), or instrumentation devices, allowing remote access to variables, programs, tasks and events.

However, it is necessary not only to standardize the communication protocols, but also to establish mechanisms that define the types of allowed information and their structure. In this sense, PQDIF constitutes a good example that can be used as a starting point.

In short, power-quality meters have evolved enormously over the last few decades, so the next steps will have to be directed towards the standardization of information and protocols and towards the development of software for the automatic analysis of the measurements according to international standards.

References

[1] Directive EMC 89/336/EEC of 23 May 1989 on the approximation of the laws of the Member States relating to electromagnetic compatibility. Official Journal of the European Union. N° L 139/19.

[2] Directive 92/31/EEC of 28 April 1992 amending Directive 89/336/EEC on the approximation of the laws of the Member States relating to electromagnetic compatibility. N° L 126/11.
[3] Eguíluz L.I., Mañana M., Lara; P., Lavandero J.C., Benito P. MEPERT I: Electric disturbance and energy meter. International Conference on Industrial Metrology. Zaragoza. October 1995.
[4] Eguiluz L.I., Manana M., Lavandero J.C., Voltage distortion influence on current signatures in non-linear loads. Proceedings of the IEEE Power Engineering Society Transmission and Distribution Conference. Seattle. 2000; 2: 1165–1170.
[5] Mañana M. Contributions to the representation, detection and classifications of power disturbances. PhD Thesis (in Spanish). University of Cantabria. 2000.
[6] CENELEC EN 50160. Voltage characteristics of electricity supplied by public distribution systems (1999).
[7] Gunther E., Grebe T. Visualization of power system data using a PC-based GUI interfaz". First European Conference on Power System Transients. Lisbon. 1993.
[8] Ribeiro P.F., Celio R. Advanced techniques for voltage quality analysis: unnecessary sophistication or indispensable tools. Ref. A-2.06. PQA'94 Amsterdam 1994.
[9] Gunther E.W., Thompson J.L., Dwyer R., Mehta, H. Monitoring Power Quality Levels on Distribution Systems. PQA'92. Atlanta, Georgia 1992.
[10] Gunther E.W., Sabin D.D., Mehta H. Update on the EPRI Distribution Power Quality Monitoring Project. PQA'93. San Diego. November 1993.
[11] Sabin D., Grebe T., Sundaram A. Preliminary Results for Eighteen Months of Monitoring from the EPRI Distribution Power Quality Project. PQA'95. New York. May 1995.
[12] IEC 61000-4-7. Electromagnetic Compatibility (EMC). Part 4: Testing and Measurement Techniques. Section 7: General Guide on Harmonics and Interharmonics measurements and Instrumentation for Power Supply Systems and Equipment Connected thereto (2002).
[13] IEC 61000-4-15. Electromagnetic compatibility (EMC) - Part 4: Testing and measurement techniques - Section 15: Flickermeter - Functional and design specifications (2003).
[14] EURELECTRIC. Measurement Guide for Voltage Characteristics (1995).
[15] IEC 61010-1. Safety requirements for electrical equipment for measurement, control and laboratory use Part 1: General requirements (2001).
[16] Ramboz J. Machinable Rogowski Coil, Design, and Calibration. IEEE Transactions on Instrumentation and Measurement 1996; 45(2):
[17] Arseneau R. Calibration system for power quality instrumentation. Proceedings of the IEEE Power Engineering Society Transmission and Distribution Conference 2002. pp. 1686-1689.
[18] Svensson S. Verification of a calibration system for power quality instruments. IEEE Transactions on Instrumentation and Measurement 1998; 47(5): 1391–1394.
[19] Emerging Technologies Working Group; Fiber Optic Sensors Working Group. "Optical Current Transducer for Power Systems: A review". IEEE Transactions on Power Delivery 1994; 9(4).
[20] Gordon B. Linear Electronic A/D Conversion Architectures, Their Origins, Parameters, Limitations and Applications. IEEE Transactions on CAS 1978; 25(7).
[21] Oppenheim A.; Schafer R.; Discrete-Time Signal Processing. Prentice-Hall. (1989).
[22] IEEE 1159. Recommended Practice for Monitoring Electric Power Quality (1995).
[23] IEC 61000-4-30. Electromagnetic compatibility (EMC) - Part 4-30: Testing and measurement techniques - Power quality measurement methods (2003).

[24] Daniels R. Power Quality Monitoring Using Neural Networks. Proceedings of the First International Forum on Applications of Neuronal Networks to Power Systems 1991; pp 195-197.
[25] Collins J.J., Hurley W.; Application of Expert Systems and Neural Networks to the Diagnosis of Power Quality Problems. PQA'94. Amsterdam. A-2.03. 1994.
[26] Gaouda A., Salama M., Sultan M.; Automated Recognition System for Classifying and Quantifying The Electric Power Quality. 8th International Conference on Harmonics and Quality of Power 1998; 1: 244–248.
[27] Perunicic B., Mallini M., Wang Z., Lui Y. Power Quality Disturbance Detection and Classification Using Wavelets and Artificial Neural Networks. 8th Int. Conf. On Harmonics and Quality of Power 1998; 1: 77–82.
[28] Angrisani L., Daponte, P., D'Apuzzo, M. A method for the automatic detection and measurement of transients. Part II: Applications. Measurement: Journal of the International Measurement Confederation 1999; 25(1): 31–40.
[29] Parihar P., Liu E. Identification, Classification and Correlation of Monitored Power Quality Events. Power Engineering Society Winter Meeting 1999; 1: 437–441.
[30] Parsons A., Grady M., Powers, E. A Wavelet-Based Procedure for Automatically Determining the Beginning and End of Transmission System Voltage Sags. Power Engineering Society Winter Meeting 1999; 2: 1310–1313.
[31] Morcos M., Ibrahim, W. Electric Power Quality and Artificial Intelligence: Overview and Applicability. IEEE Power Engineering Review 1999; 6: 5–10.
[32] Santoso S, Lamoree, J, Grady, W M, Powers E J, Bhatt S C. Scalable PQ event identification system. IEEE Transactions on Power Delivery 2000; 15(2): 738–743.
[33] Eguiluz L.I., Mañana M., Lavandero J.C. Disturbance classification based on the geometrical properties of signal phase-space representation. PowerCon 2000. Perth. International Conference on Power System Technology. Proceedings 2000; 3: 1601–1604.
[34] Kezunovic M., Liao Y. Automated analysis of power quality disturbances. IEE Conference Publication 2001; 2(482).
[35] IEEE C37.111. Standard Common Format for Transient Data Exchange (COMTRADE) for Power Systems (1991).
[36] IEEE 1159.3. IEEE Recommended Practice for the Transfer of Power Quality Data (2003).

3

Joint Time–Frequency Analysis of the Electrical Signal

Juan-Carlos Montaño, Juan-Carlos Bravo[*] and María-Dolores Borrás[*]

Laboratorio de Electrónica,
Institute for Natural Resources and Agricultural Research (IRNAS)
Spanish Research Council (CSIC)
Reina Mercedes Campus, POB 1052,
41080–Sevilla, Spain
Email: montano@irnase.csic.es

Dpto. Ingeniería eléctrica [*]
Escuela Universitaria Politécnica
Universidad de Sevilla
Virgen de África 7
41011–Sevilla, Spain
Email: carlos_bravo@us.es, borras@us.es

3.1 Introduction

As is well known, a signal may be described in numerous ways. Classically, signals have been studied as a function of time or as a function of frequency separately. In these separate fields, time-domain functions indicate the evolution of the signal amplitude over time, while a function in the frequency domain shows how quickly such changes takes place.

The connection between time and frequency domains is made by the Fourier transform that uses orthogonal sinusoidal basis functions (BFs) showing the signal as an expansion [1, 2]. Since all the BFs are orthogonal (see Figure 3.1) the transform coefficients of such a representation are easily computed using inner products to project the signal onto the BFs. For example, stationary signals have frequency representations, such as the power spectrum, wich are simpler than time waveforms. So, the complex sinusoidal function corresponds to only one pulse in the frequency domain.

However, these BFs are global in nature and cannot represent the structure of the signal locally, so the advantage of the simple interpretation of a signal as a pure frequency is not an efficient way applicable to all signals, in particular to those whose spectra evolves with time.

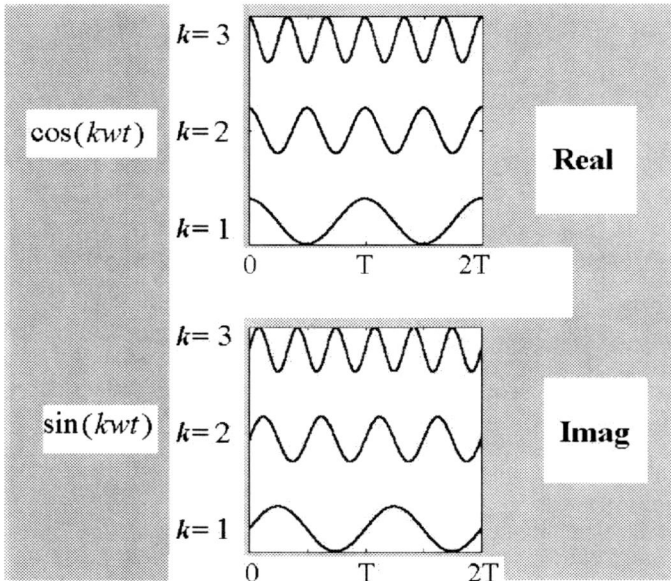

Figure 3.1. Basis functions of the Fourier transform

Common examples to show that the Fourier transform is unable to analyse "real-life signals" that are usually of relatively short duration include biomedical signals, such as the well-known ECG, or the disturbance generated by the capacitor-switching restrike on opening (see Figure 3.2). A transient oscillation with natural frequency determined by the capacitance and inductance of the system can be observed.

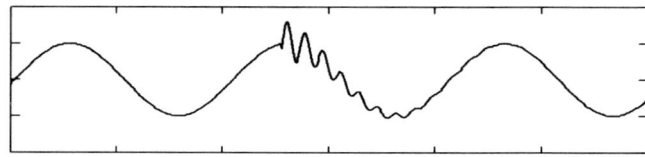

Figure 3.2. Transient oscillation due to a capacitor-switching disturbance

In daily life, there are more transient signals than stationary ones which are simpler. For non stationary signals, a representation in a similar way like Fourier basis was the goal of many previous research studies. Instead of this, models capable to combine effectively time and frequency domains are the most suitable. This is principally motivated by the impossibility of the classical power spectrum to reveal important information about the evolution over the time of a real signal frequency content. In this way, a large variety of solutions have been proposed and extensively studied: the Gabor expansion, time-dependent spectra and wavelets. In what follows we entitle all these new techniques able to complement the classical separate time analysis and frequency analysis *joint time-frequency analysis* (JTFA).

The key question is how to select properly the basis for processing a particular kind of signal. As is desirable, the decomposition coefficients of a signal in a basis should characterise a representation capable of distinguishing some particular signal features. For example, wavelet coefficients give explicit information on the location and type of signal singularities. The problem is to find a criterion for selecting a basis that is naturally well adapted to represent a particular class of signals.

Mathematical approximation theory to signal processing suggests choosing a basis that can construct precise signal aproximations with a linear combination of a small number of vectors selected inside the basis. A good measure of the efficiency of a basis will be given by applications like, for example, compact coding, noise estimation of signals, of course, optimal detection and classification of disturbances in electrical signals.

Nowadays, conventional approaches to signal representation have been improved, such as the Fourier transform signal representation using nonorthogonal BFs [3]. This study is more complex and generally involves choosing a small set of BFs for the representation from a set of possible nonorthogonal BFs.

The main consideration in choosing a BF for representing signals in a data-efficient manner, implies that the selected BFs should be well localized both in spatial and frequency domains, and should also be adaptable to represent the local structure of the signal. To clarify this point we make a brief approximation to the JTFA intrinsic limits imposed by the uncertainty principle. As we will see, the optimum joint localization, as specified by the Heisenberg's inequality, is achieved when the analysis functions are complex modulated Gaussians, which are also referred to as Gabor functions.

In this chapter, we also introduce the basic concepts and well-tested algorithms for JTFA. Analogous to the classical Fourier analysis, linear and non linear procedures are studied and compared: the linear (*e.g.*, short-time Fourier transform, Gabor expansion, wavelet transform) and the quadratic transforms (*e.g.*, Wigner–Ville distribution). This study offers the opportunity to show that non linear is not always synonymous with difficult.

Numerical experiments are performed over a variety of simulated electrical signals with a given harmonic content, classified according to the kind of low-frequency disturbance (voltage sag, swell and momentary interruption) and high-frequency disturbance (oscillatory transient) present.

3.2 Application of JTFA to Electrical Signals

As we have already commented, one of the key features in signal processing is the choice of a suitable basis to represent in an efficient way the kind of considered signals. In particular, for electrical signals, a first approach in power quality is to apply different JTFA methods to the entire original signal. In this manner we try to study its most relevant features.

Alternatively, other mixed power-quality analysis strategies [4–7] have usually been divided into those that address steady-state concerns, such as harmonic distortion, and transient concerns, like those resulting from faults or switching transients. Techniques such as Fourier spectral analysis are often applied to steady-state events while wavelets, classical transient analysis, computer modelling are traditionally used for transient events.

In this environment, in order to determine the sources and causes of harmonic distortion, one can detect and localize those disturbances for further classification. Software procedures have been developed for this purpose, such as the FFT [8]; however, due to the amount of stored data and the time of processing required, such a procedure is slow and not very efficient.

Continuous and discrete wavelet transforms (DWT) have been used in the analysis of non stationary signals and many papers such as [9,10], have been presented proposing the use of wavelets for power-system analysis.

Steady-state Events
While there are a few cases where the distortion is randomised, most distortion is periodic, or harmonic. That is, it is nearly the same cycle after cycle, changing very slowly, if at all. The advantage of using a Fourier series to represent distorted waveforms is that it is much easier to find the system response to an input that is sinusoidal. Conventional steady-state analysis techniques can be used. The system is analysed separately at each harmonic. Then the outputs at each frequency are combined to form a new Fourier series, from which the output waveform may be computed, if desired.

Harmonics, by definition, occur in the steady state, and are integer multiples of the fundamental frequency. The waveform distortion that produces the harmonics is present continually or at least for several seconds. Transients are usually dissipated within a few cycles. Transients are associated with changes in the system such as switching a capacitor bank. Harmonics are associated with the continuing operation of a load.

Usually, the higher-order harmonics (above the range of the 25th to 50th, depending on the system) are negligible for power-system analysis. While they may cause interference with low-power electronic devices, they are usually not damaging to the power system. It is also difficult to collect sufficiently accurate data to model power systems at these frequencies.

Transient Events

The different simulated transient events we have considered are depicted in Figure 3.3. All of them act over a fundamental signal with 1024 samples and amplitude equal to $220\sqrt{2}$ V.

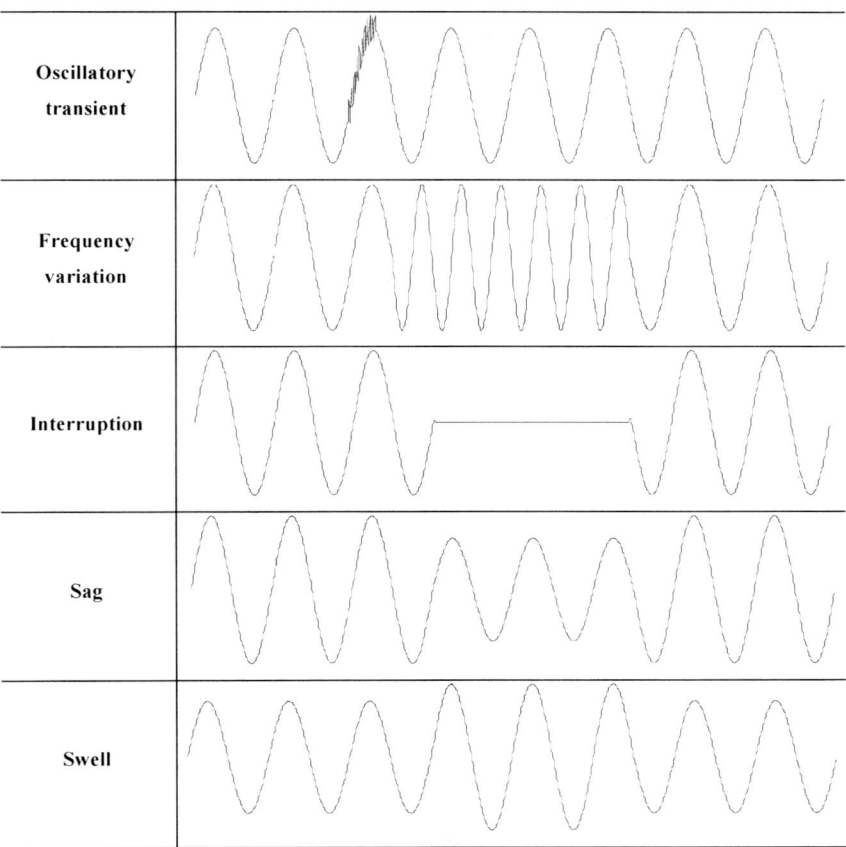

Figure 3.3. The simulated transient events

Harmonic distortion is blamed for many power-quality disturbances that are actually transient. A measurement of the event may show a distorted waveform with obvious high-frequency components. Although transient disturbances contain high-frequency components, transients and harmonics are distinctly different phenomena and are analysed differently. Transient waveforms exhibit the high frequencies only briefly after there has been an abrupt change in the power system. The frequencies are not necessarily harmonics; they are whatever the natural frequencies of the system are at the time of the switching operation. These frequencies have no relation to the system fundamental frequency.

Continuous and discrete wavelet transform (CWT and DWT) have been used in the analysis of non stationary signals and, recently, several papers [11–14] and books [15–17] have been presented proposing the use of wavelets for identifying various categories of power-system disturbances. They are able to remove noise and achieve high compression ratios because of the "concentrating" ability of the wavelet transform. It has proven a powerful signal processing tool in communications in such areas as, data compression, denoising, reconstruction of high-resolution images, and high-quality speech.

3.3 Review of Fundamental Mathematical Tools

A signal or function $f(t)$ can often be better analysed or processed if expressed as a linear combination of a set of orthonormal functions $\Psi_l(t)$

$$f(t) = \sum_{l \in Z} c_l \Psi_l(t), \qquad (3.1)$$

where l is an integer index, and c_l the expansion coefficients that may be calculated, for continuous-time signals, by the inner product (or scalar product)

$$c_l = \langle f, \tilde{\Psi}_l \rangle = \int_{-\infty}^{\infty} f(t) \tilde{\Psi}_l^*(t) dt, \qquad (3.2)$$

or for discrete-time signals,

$$c_l = \langle f, \tilde{\Psi}_l \rangle = \sum_{k=-\infty}^{\infty} f(k) \tilde{\Psi}_l^*(k). \qquad (3.3)$$

In Equations 3.2 and 3.3 the complex conjugate operation is depicted by "*" and $\{\tilde{\Psi}_l\}_{l \in Z}$ is a dual set obtained from the original frame $\{\Psi_l\}_{l \in Z}$ [18]. This dual frame is not generally unique and then, the expansion result is redundant. If the dual frame $\{\tilde{\Psi}_l\}_{l \in Z}$ is unique, the redundancy is avoided and the set of $\{\Psi_l\}_{l \in Z}$ becomes an orthogonal basis. In this case:

$$\langle \Psi_l, \tilde{\Psi}_{l'} \rangle = k\delta(l - l'), \qquad (3.4)$$

where

$$\delta(l) = \begin{cases} 1 & l = 0 \\ 0 & \text{otherwise} \end{cases}$$

and k denotes a constant number. In an orthogonal expansion, expansion coefficients c_l correspond with the projection of signals onto the elementary functions Ψ_l. Moreover, when $\{\Psi_l\}_{l \in Z}$ forms a basis, $\{\Psi_l\}_{l \in Z}$ and $\{\widetilde{\Psi}_l\}_{l \in Z}$ are called biorthogonal functions [21].

The relations defined in Equations 3.1, 3.2 and 3.3 may be seen as transformations, direct and inverse respectively. Is for what Ψ_l and $\widetilde{\Psi}_l$ are called synthesis and analysis functions respectively.

From another perspective, the inner product may be understood as the affinity between the signal f(t) and the dual function $\widetilde{\Psi}_l$ in the sense that the larger the inner product c_l, the closer the signal f(t) is to the dual function $\{\widetilde{\Psi}_l\}_{l \in Z}$.

3.3.1 Fourier Theory

The Fourier transform is the main tool for signal spectral decomposition. It represents a signal f(t) as a superposition of a complex exponential of definite frequency f and infinite time duration, computing the inner products of the signal to be analysed with the complex exponential, *i.e.*

$$\hat{F}(f) = \int_{-\infty}^{+\infty} f(t) e^{-j2\pi ft} dt, \qquad (3.5)$$

so the original signal can be recovered by means of the inverse formula

$$f(t) = \frac{1}{2\pi} \int_{-\infty}^{+\infty} \hat{F}(f) e^{j2\pi ft} df. \qquad (3.6)$$

Therefore, the Fourier transform (FT) uses complex sinusoidal functions as BFs to analyse and reconstruct a given signal.

If the signal is discrete and periodic

$$f(n + mN) = f(n) \quad m \in Z,$$

the orthogonal basis for the synthesis functions are

$$\{e_\kappa(n)\} = \{e^{\left(\frac{j2\pi}{N}\right)\kappa n}\} \quad \kappa = 0, 1..., N-1,$$

and the Fourier expansion

$$f(n) = \sum_{\kappa=0}^{N-1} \alpha_\kappa e_\kappa(n) = \sum_{\kappa=0}^{N-1} \alpha_\kappa e^{\left(\frac{j2\pi}{N}\right)\kappa n},$$

where the expansion coefficients α_κ or discrete Fourier transform are derived from the inner product as follows

$$\alpha_\kappa = \left\langle f(n), e^{j2\pi\frac{\kappa n}{N}} \right\rangle = \frac{1}{N} \sum_{n=0}^{N-1} f(n) e^{-j2\pi\frac{\kappa n}{N}}. \tag{3.7}$$

Since the complex sinusoidal functions $e^{j2\pi ft}$ that configure the elementary BFs are orthonormal with respect to the scalar product, the dual function and the elementary function have the same form. So both can be used for analysis or synthesis and their roles are interchangeable.

In practice, we never work with the pure mathematical function f(t), but we have partial information concerning it, that is, samples of the signal at some regular time interval T_s. Therefore, the basic input data will be the array

$$f(n) \equiv f(tn) \equiv f(nT_s), \quad n \in 0, 1, ..., N-1, \tag{3.8}$$

where N is the total number of samples of the signal. The quantities T_s and N determine the maximum and minimum frequency we are able to resolve. On the one hand, according to Shannon's Theorem, we can not go beyond larger frequencies than $\omega_{max} = \omega_s/2$, where $\omega_s = 1/T_s$ is the sample frequency. On the other hand, the minimum frequency will be given by the inverse of the time interval in which we have samples of the signal, that is, $\omega_{min} = 1/NT_s$.

To extract the harmonic content of a given signal by using the standard FFT algorithm, we compute the amplitudes F_k for the definition of the vector f(n) as a superposition of complex exponential vectors $e_k(n) = \exp\{j2\pi nk/N\}$, where $k \in 0 ... N-1$, and j is the imaginary unit. Each of these vectors has a definite frequency $\omega_k = k\omega_{min}$, so the mapping from k to ω is given by

$$k \rightarrow \omega_k = \frac{k}{NT_s}. \tag{3.9}$$

In this context, we have

$$f(n) = \frac{1}{\sqrt{N}} \Re \left(\sum_{k=0}^{N-1} F_k e^{-j2\pi\frac{nk}{N}} \right), \tag{3.10}$$

being the expression for each one of the phasors (FFT)

$$F_k = \frac{1}{\sqrt{N}} \sum_{n=0}^{N/2-1} f(n) e^{j2\pi\frac{nk}{N}}. \tag{3.11}$$

The component of the signal f(n) at the frequency ω_k is then given by

$$c_k(n) = |F_k|\cos(2\pi o_k t_n + \arg(F_k)). \tag{3.12}$$

The fundamental component of f(n) is the component corresponding to the value of $k=k_{fun}$ such that $\omega_{min} k_{fun} = \omega_0$, where ω_0 is some prefixed frequency. We refer to the harmonic *components* of f(n) as the components of the Fourier transform corresponding to integer multiples of k_{fun}, i.e. those values of k belonging to the subset $\Delta = (k_{fun}, 2k_{fun}, 3k_{fun}, \cdots, Mk_{fun})$, where M is the highest order for the last harmonic component considered, and it has to be set according to some convention. The harmonic content of the analysed signal is the superposition of all the previous components and defined by

$$h(n) = \sum_{k \in \Delta} |F_k|\cos(2\pi o_k t_n + \arg(F_k)). \tag{3.13}$$

Another relevant magnitude in signal processing is the power spectrum. This quadratic magnitude is more useful in many applications than the form of the time function because it normally displays an easier pattern to characterise the signal. The graph of conventional power shows the frequency contents. Nevertheless, the power spectrum itself does not indicate how these frequencies evolve with time.

The power spectra is simply the square of the Fourier transform and can also be viewed, by means of the Wiener–Khinchin theorem, as the Fourier transform of a certain correlation function, i.e.

$$P(\omega) = |S(\omega)|^2 = \int_{-\infty}^{\infty} F(\tau) e^{-j\omega t} d\tau, \tag{3.14}$$

where the correlation function $F(\tau)$ is defined as the average of the instantaneous correlation, i.e.

$$F(\tau) = \int_{-\infty}^{\infty} f(t) f^*(t-\tau) dt = \int_{-\infty}^{\infty} f(t + \tau/2) f^*(t - \tau/2) dt. \tag{3.15}$$

3.4 Time-Frequency Analysis Limits: Uncertainty Principle

Important information often appears through a simultaneous analysis of the signal's time and frequency properties. But it is not a trivial task to decompose a one-dimensional signal that exhibits two important properties, localization in time of transient phenomena and the presence of specific frequencies. It is necessary to understand how the uncertainty principle limits the flexibility of time-frequency transforms. [2, 17, 19–20].

The uncertainty principle states that the energy spread of a signal or function and its Fourier transform cannot be simultaneously arbitrarily small. Heisenberg inequality limits how precisely the localization in time and the presence of specific frequencies in a signal can be computed. This imposes a lower bound on the product of the temporal variance and the frequency variance of the signal:

$$\Delta f \cdot \Delta t \geq \frac{1}{4\pi}. \qquad (3.16)$$

The main tool used to decompose a signal is the expansion in orthonormal bases whose elements are well concentrated in time and in frequency. These bases are a family of waveformes called *time-frequency atoms*. Idealized representations of these atoms are drawn in the time-frequency plane. This plane is useful for idealizing measurable quantities associated to transient signals. In Figure 3.4, a waveform is represented by a rectangle in this plane with its sides parallel to the time and frequency axes.

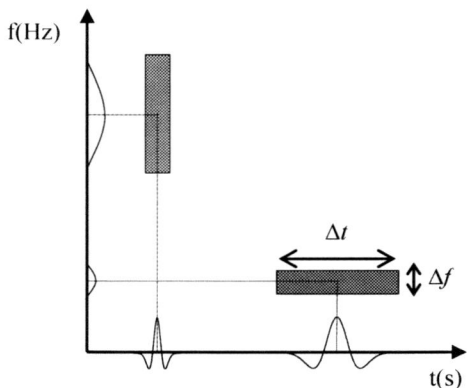

Figure 3.4. Time-frequency boxes

The amplitude of a waveform can be encoded by shading the rectangle in proportion to its waveform's energy. The uncertainty in time and the uncertainty in frequency are given by the width and height of the rectangle, respectively. The time-frequency plane must be manipulated carefully because only rectangles with areas at least $1/4\pi$ may correspond to time-frequency atoms.

A basis of time-frequency atoms correspond to a cover of the plane by rectangles; an orthonormal basis may be depicted as a cover by disjoint rectangles. By this procedure, the standard basis or Dirac basis consists of the cover by the tallest, thinnest patches allowed by the sampling interval and the underlying synthesis function: it has optimal time localization and no frequency localization. On the contrary, the Fourier basis may be represented as a rotation by 90° of the Dirac basis and it has optimal frequency localization but no time localization. Both bases are depicted in Figure 3.5.

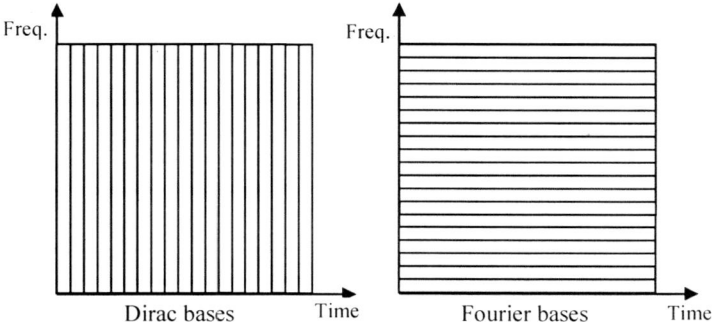

Figure 3.5. Dirac and Fourier bases

Wavelets and windowed Fourier functions have their energy well localized in time, while their Fourier transform is mostly concentrated in a limited frequency band. Windowed Fourier or trigonometrics transform have a fixed window size, the ratio of frequency uncertainty to time uncertainty is the aspect ratio of the information cell, as seen in Figure 3.6. The wavelet basis is an octave-band decomposition of the time-frequency plane, a wavelet packet basis gives a more general covering, as seen in Figure 3.7.

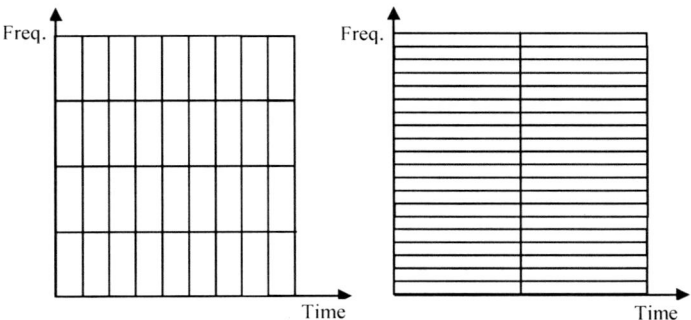

Figure 3.6. Time-frequency plane of different windowed Fourier bases

As we can see in Figure 3.6, the windowed Fourier transform has the same resolution in the size of its cell across the time-frequency plane. The resolution in time and frequency of the windowed Fourier transform depends on the spread of the window in time and frequency, this can be measured from the area of the Heisenberg cell. The choice of a particular scale depends on the desired resolution tradeoff between time and frequency.

The wavelet transform uses time-frequency atoms with different time supports; see Figure 3.7. This transform presents an exceptional time resolution and poor frequency resolution in the higher frequency band and *vice versa*. This is a direct consequence of the application of wideband windows at the higher-frequency band and narrowband windows at the lower-frequency band.

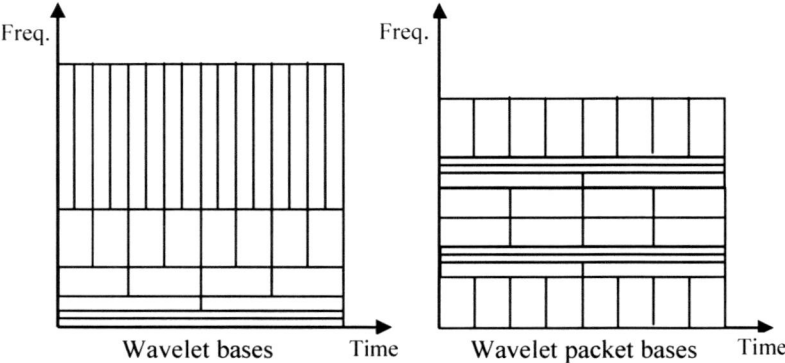

Figure 3.7. Different time-frequency wavelet cells

Hence, it is desirable to obtain a good frequency resolution at any frequency band, and the same for the time resolution. This is possible by means of the adaptive spectrogram. As we will show, this method not only offers good time-frequency resolution but does not suffer crossterm interference and is non-negative. Although the adaptive spectrogram does not do well for chirp-type signals, it does well for most of the signals of interest, especially to quasistationary signals or very short transient signals [21].

3.5 JTFA Linear Methods

In the next sections the most relevant JTFA algorithms are described. They have all been grouped into two categories: linear and quadratic algorithms.

The studied linear methods are: Gabor expansion (as inverse STFT), STFT (short-time Fourier transform used to compute the Gabor coefficients), adaptive representation (considered as the inverse adaptive transform) and the adaptive transform.

3.5.1 Windowed Fourier Transform (STFT) and Gabor Expansion

To improve the analysis for signals whose spectrum changes with time we introduce the widely used and well-defined *short-time Fourier transform (STFT)*, i.e.

$$\text{STFT}(t,\omega) = \int f(\tau) w^*_{t,\omega}(\tau) d\tau = \int f(\tau) w^*(\tau - t) e^{-j\omega\tau} d\tau . \tag{3.17}$$

As the function w(*t*) has typically a short time extent, it is named a *window function*, and when w(*t*) =1 for all *t*, one obtains the classic Fourier transform.

In the STFT, the function f(*t*) is multiplied by this window function w(*t*) and the Fourier transform is computed. Then, w(*t*) is shifted in time and the Fourier transform of the product is calculated again, and so on. By this procedure, for a prefixed shift τ of the window w(*t*), the window captures the features of the signal f(*t*) around τ. In others words, the window helps to localize the time-domain data before obtaining the frequency-domain information.

By observing Equation 3.17, the original signal f(*t*) is compared with a set of elementary functions $w(\tau - t)e^{j\omega\tau}$ that are simultaneously concentrated in both the time and frequency domains. Supposing w(*t*) and its corresponding Fourier transform centred at $(t, \omega)=(0,0)$ with a time duration Δt and frequency bandwidth $\Delta \omega$ respectively, then STFT(t, ω) shows the signal behaviour near $[t - \Delta t, t + \Delta t] \times [\omega - \Delta \omega, \omega + \Delta \omega]$.

The STFT can be also viewed as the convolution of the signal f(*t*) with a filter having an impulse response of the type $h(t) = w(-t)e^{j\omega t}$

$$\text{STFT}(t, \omega) = e^{-j\omega t} \int f(\tau) w^*(\tau - t) e^{j\omega(t-\tau)} d\tau .$$

The original signal f(*t*) can be totally recovered from

$$f(t) = \frac{1}{2\pi\pi w(0} \int \text{STFT}(t,\omega) e^{jt\omega} d\omega \tag{3.18}$$

obtained by taking the inverse Fourier transform with respect to STFT(t, ω)

$$\frac{1}{2\pi} \int \text{STFT}(t,\omega) e^{j\mu\omega} d\omega = \frac{1}{2\pi} \iint f(\tau) w(\tau - t) e^{j(\mu-\tau)\omega} d\tau d\omega$$
$$= \int f(\tau) w(\tau - t) \delta(\mu - \tau) d\tau = f(\mu) w(\mu - t)$$

when $\mu = t$. So, f(*t*) can be recovered for all *t* as long as w(0)≠ 0. If w(*t*)=0, another value of τ can be chosen to guarantee w(0)≠ 0.

To improve the degree of aproximation to the original signal f(*t*) at a particular time and frequency (t, ω), it is desirable that the window's standard deviations Δt and $\Delta \omega$ should be as narrow as possible. But unfortunately, the simultaneous choice of Δt and $\Delta \omega$ can not be arbitrary. As we explained in the previous section, the resolution in the time-frequency domain is limited by the Heisenberg inequality $\Delta \omega \cdot \Delta t \geq 0.5$.

Therefore, for a chosen window $w(t)$, a good time resolution (narrower Δt) compromises the frequency resolution (wider $\Delta\omega$) and *vice versa*. We get the equality only when $w(t)$ is a Gaussian-type function [21].

A Gaussian function was precisely selected by Dennis Gabor in 1946 as the elementary synthesis function, *i.e.*

$$g_\alpha(t) = \frac{1}{2\sqrt{\pi\alpha}} e^{-\frac{t^2}{4\alpha}},$$

where α is the control parameter for the degree of resolution in the time-frequency plane. This author introduced the Gabor expansion for a signal $f(t)$ as

$$f(t) = \sum_{m=-\infty}^{\infty}\sum_{n=-\infty}^{\infty} c_{m,n} g_{m,n}(t) = \sum_{m=-\infty}^{\infty}\sum_{n=-\infty}^{\infty} c_{m,n} g(t-mT) e^{jn\Omega t}, \qquad (3.19)$$

where T and Ω are the time and frequency steps.

The necessary condition of the existence of the Gabor expansion is that the sampling cell $T\Omega$ must be small enough to satisfy

$$T\Omega \leq 2\pi.$$

Also, Gabor's elementary synthesis functions reach the minimum sampling cell ($T\Omega = 2\pi$), called the critical sampling cell, his representation is the most compact expression because of an optimal concentration in the joint time-frequency domain according to the uncertainty principle.

Based upon the expansion theorem [22], for a complete set of Gabor elementary functions $\{g_{m,n}(t)\}$ there will be a dual function (or auxiliary function) $w(t)$ such that the Gabor coefficients can be computed by the regular inner product operation, *i.e.*

$$c_{m,n} = \int f(t) w^*{}_{m,n}(t) dt = \int f(t) w^*(t-mT) e^{-jn\Omega t} dt = \text{STFT}(mT, n\Omega),$$

which is the sampled STFT and is also known as the Gabor transform. Then, the Gabor expansion turns out to be the most elegant algorithm for computing the inverse of the sampled STFT.

It is worth noting that the Gabor expansion is not only restricted to be based on Gaussian elementary functions. It holds for almost any time-shifted and frequency-modulated signal $g_{m,n}(t)$. It can be shown [21] that for critical sampling, the Gabor elementary functions $\{g_{m,n}(t)\}$ are linearly independent. In this case, the dual function is unique and biorthogonal to $g(t)$. On oversampling ($T\Omega < 2\pi$), due to the use of a non-Gaussian Gabor elementary function, the dual function is not unique and its optimal selection is an essential issue.

It is also remarkable that the biorthogonality between the synthesis function $g_{m,n}(t)$. and the analysis function $w(t)$ is not a suficient condition but a necessary one for an efficient representaton of the signal behaviour near $[t - \Delta t, t + \Delta t] \times [\omega - \Delta\omega, \omega + \Delta\omega]$. The Gabor elementary functions can be optimally concentrated in the joint time frequency domain but the dual function $w(t)$ may not be localized. This implies that, although the Gabor coefficients $c_{m,n}$ can be computed by the inner product of the signal and dual function, they may be unable to describe properly the signal's local behaviour. A suitable description is obtained only if the dual function is well concentrated in the joint time-frequency domain.

The appropriate choice of dual functions, as well as the computation of the dual function itself, are topics widely studied bymany authors [21, 23 25]. Figure 3.8 shows our choice for the analysis function and the synthesis function to compute the Gabor coefficients. We have applied these dual functions to all disturbed simulated signals with 1024 samples length. The selected dual signals are both concentrated enough to work properly in the re-establishment of the original signal. They both are formed from 96 samples, with 16 samples per sampling interval and a total of 32 frequency bins.

Figure 3.8. Gaussian synthesis function (solid) and its dual (dashed)

As we have seen, from the continuous-time STFT it may be easy to re-establish the original signal. But this way is expensive in data storage because this representation is highly redundant by oversampling. In fact, we can recover the original signal in a more efficient way by using the sampled version of the STFT or discrete STFT. Then, a more compact representation is possible by replacing each Fourier transform in the STFT by the discrete Fourier transform. This results in a double-domain-sampled version of the STFT that is very suitable for practical implementation in digital signal processing:

$$\text{STFT}[mT, n\Omega] = c_{m,n} = \text{STFT}(t,\omega)\Big|_{t=mT, \omega=\frac{2\pi n}{NT}} = \sum_{i=0}^{N-1} f[i] \, w^*[i - mT] \, e^{-j\frac{2\pi n i}{N}},$$

where T and Ω are the time and frequency sampling intervals, N is the number of frequency bins and $w[k] \equiv w[k + T]$ is the N-point window.

The procedure of digitizing the continuous-time Gabor expansion essentially is a standard sampling process. The discrete Gabor expansion represents a discrete signal as a weighted sum of the time-shifted and frequency-modulated function $g_{m,n}[i] = g[i - mT] \, e^{j2\pi ni/N}$:

$$f[i] = \sum_{m} \sum_{n=0}^{N-1} c_{m,n} g[i - mT] \, e^{j2\pi ni/N}$$

and the condition for a perfect reconstruction implies that the oversampling rate, N/T must be greater than or equal to one.

3.5.2 Adaptive Representation and Adaptive Transform

The adaptive representation tries to better match the analysed signal by decomposing the signal f[i] as a sum of weighted linear parameter adaptive chirp-modulated Gaussian functions:

$$f[i] = \sum_{k=0}^{F-1} A_k g_k[i], \qquad (3.20)$$

where the linear chirp-modulated Gaussian function is defined by

$$g_k[i] = (\alpha_k \pi)^{-0.25} \exp\left\{ -\frac{[i - i_k]^2}{2\alpha_k} + j\left(2\pi f_k [i - i_k] + \frac{\beta_k}{2}[i - i_k]^2 \right) \right\}$$

which has four-tuple parameters (α_k, i_k, f_k, β_k). Hence, the adaptive representatioon is more flexible than the elementary function used in the Gabor expansion.

The parameter F in Equation 3.20 indicates the total number of elementary functions used by $g_k[i]$. A_k is the weight of each individual $g_k[i]$, as computed by the adaptive transform.

Scientists at National Instruments and Mallat and Zhang independently developed the adaptive representation. The adaptive methods discussed in this algorithm were implemented with the adaptive oriented orthogonal projective decomposition algorithm. The source code for this algorithm was developed by Qinye and Zhifang Ni at Xi'an Jiaotong University, China [31]. We have used the JTFA toolkit of the National Instruments[TM] LabView[TM] software to compute dual functions and represent this algorithm.

3.5.3 Wavelet Theory

The wavelet approach is more suitable than the Fourier one, especially when signals are nonstationary. Wavelet algorithms process data on a different scale or resolution. In wavelet analysis, the scale that we use to look at data plays a special role. A basis function varies in scale by chopping up the same function or data space using different scale sizes. Various wavelets are obtained from a single wavelet $\Psi(t)$ (mother wavelet) by scaling and shifting operations [2, 26].

The continuous wavelet transform was developed as an alternative approach to the short-time Fourier transform to overcome the resolution problem. The wavelet analysis is done in a similar way to the STFT analysis, in the sense that the signal is expanded with a function, $\Psi(t)$, similar to the window function in the STFT, and the transform is computed separately for different segments of the time-domain signal. However, there are two main differences between the STFT and the CWT:

1. The Fourier transforms of the windowed signals are not taken, and therefore a single peak will be seen corresponding to a sinusoid, *i.e.* negative frequencies are not computed.

2. The width of the window is changed as the transform is computed for every spectral component, which is probably the most significant characteristic of the wavelet transform.

The continuous wavelet transform is defined as follows

$$\text{CWT}_X^\psi(\tau,s) = \Psi_X^\psi(\tau,s) = \frac{1}{\sqrt{|s|}} \int x(t)\ \psi^*(\frac{t-\tau}{s})dt .$$

As seen in the above equation, the transformed signal is a function of two variables, τ and s, the translation and scale parameters, respectively. $\Psi(t)$ is the mother wavelet.

In today's world, computers are used to do most computations. It is apparent that neither the FT, the STFT, nor the CWT can be practically computed by using analytical equations, integrals, *etc.* It is therefore necessary to discretize the transforms. As in the FT and STFT, the most intuitive way of doing this is simply sampling the time-frequency (scale) plane. Again intuitively, sampling the plane with a uniform sampling rate sounds like the most natural choice. In the case of WT, a fast algorithm is actually available to compute the wavelet transform of a signal, the discrete wavelet transform (DWT).

For the wavelet expansion, Equation 3.1 is expressed as

$$f(t) = \sum_j \sum_k c_{j,k} \Psi_{j,k}(t), \qquad (3.21)$$

where j and k are integer indices and $\Psi_{j,k}(t)$ are the wavelet functions that form an orthogonal basis. The set of coefficients $c_{j,k}$ is called the discrete wavelet transform (DWT) of $f(t)$ function, and can be calculated by

$$c_{j,k} = \int f(t)\Psi_{j,k}(t)dt = \langle f(t), \Psi_{j,k}(t)\rangle.$$

The DWT analyses the signal at different frequency bands with different resolutions by decomposing the signal into a coarse approximation and detailed information. DWT employs two sets of functions, called scaling functions and wavelet functions, which are associated with lowpass and highpass filters, respectively. The decomposition of the signal into different frequency bands is simply obtained by successive highpass and lowpass filtering of the time-domain signal.

The translation and dilation operations applied to the mother wavelet are performed to calculate the wavelet coefficients, which represent the correlation between the wavelet and a localised section of the signal. The wavelet coefficients are calculated for each wavelet segment, giving a time-scale function relating the wavelets correlation to the signal.

From Equation 3.21 the wavelet expansion has the form

$$f(t) = \sum_k \sum_j c_{j,k} \Psi_{j,k}(t) = \sum_k a_{J_0,k} \varphi_{J_0,k}(t) + \sum_{j=J_0}^{\infty} \sum_k d_{j,k} \Psi_{j,k}(t), \qquad t \in \Re,$$

where J_0 is a non-negative integer. This expansion is similar to that by Fourier analysis showing a linear combination of wavelet coefficients, ($a_{J_0,k}$, $d_{j,k}$), a set of basis functions $\varphi_{J_0,k}(t)$, called scaling functions, and $\Psi_{j,k}(t)$. Sets $a_{J_0,k}$ and $d_{j,k}$ are the discrete wavelet transform (DWT) of $f(t)$ and can be calculated by

$$a_{J_0,k} = \langle f(t), \varphi_{J_0,k}(t)\rangle, \qquad (3.22)$$

$$d_{j,k} = \langle f(t), \Psi_{j,k}(t)\rangle. \qquad (3.23)$$

In practice, the wavelet expansion must be truncated at $j = J-1$ such that

$$f(t) = \sum_{k=0}^{2^{J_0}-1} a_{J_0,k} \varphi_{J_0,k}(t) + \sum_{j=J_0}^{J-1} \sum_{k=0}^{2^j-1} d_{j,k} \Psi_{j,k}(t), \quad t \in R. \qquad (3.24)$$

In this expression, the first sum is a coarse representation of $f(t)$, where $f(t)$ has been replaced by a linear combination of 2^{J_0} translations of the scaling function $\varphi_{J_0,0}$. The remaining terms are the detailed representation. For each j level, 2^j translations of the wavelet $\psi_{j,0}$ is added to obtain a more detailed approximation of $f(t)$.

The DWT is implemented using multiresolution analysis (MRA) [26–28] to decompose a given signal into scales with different time and frequency resolution. MRA, as implied by its name, analyses the signal at different frequencies with different resolutions. Every spectral component is not resolved equally as was the case in the STFT.

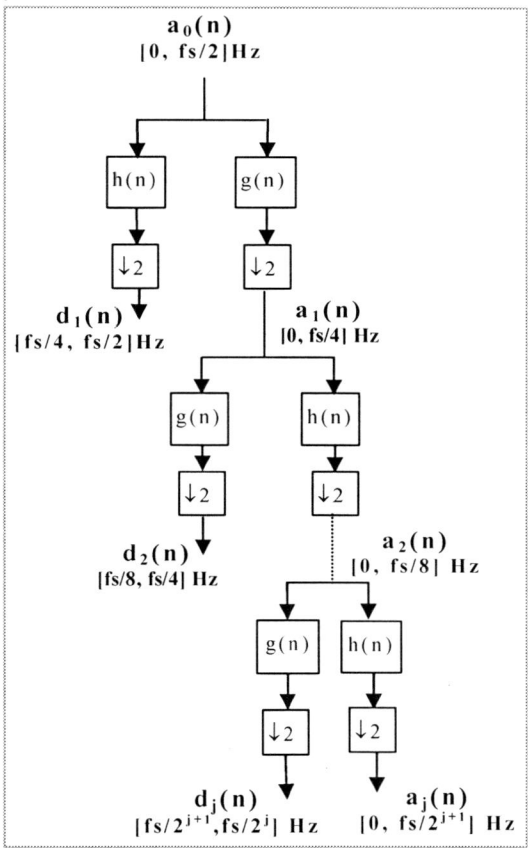

Figure 3.9. Multiresolution signal decomposition

MRA (see Figure 3.9) is designed to give good time resolution and poor frequency resolution at high frequencies and good frequency resolution and poor time resolution at low frequencies. This approach makes sense, especially when the signal at hand has high-frequency components for short durations and low-frequency components for long durations. Fortunately, the electrical signals that are encountered in practical applications are often of this type. The analysis filter bank divides the spectrum into octave bands. The cutoff frequency for a given level j is found by

$$fc = \frac{fs}{2^{j+1}}. \tag{3.25}$$

To construct the scale analysis of f(t), it is necessary to express the scaling function, $\varphi(t)$, and the wavelet function, $\psi(t)$, in terms of a weighted sum of shifted $\varphi(2t)$ and $\psi(2t)$ respectively:

$$\varphi(t) = \sqrt{2}\sum_{n} h(n)\varphi(2t-n) \quad n \in Z, \qquad (3.26)$$

$$\psi(t) = \sqrt{2}\sum_{n} g(n)\varphi(2t-n) \quad n \in Z, \qquad (3.27)$$

where h(n) and $g(n) = (-1)^n h(1-n)$ are sequences that represent discrete filters and Z is the set of all integers. Both functions are used to obtain the WTC of the signal. The DWT synthesis of f(t) [29] is given, in general, by:

$$a_{j,k} = \sum_{m} h(m-2k)a_{j+1,m}, \qquad (3.28)$$

$$d_{j,k} = \sum_{m} g(m-2k)d_{j+1,m}. \qquad (3.29)$$

Equations 3.28 and 3.29 represent the DWT algorithm. Sequence f(n), the digitalized version of f(t), is decomposed by the MRA into the smoothed version $a_1(n)$ (it contains low-frequency components), and the detailed version $d_1(n)$ (it contains higher-frequency components). This is a first-scale decomposition, the next higher scale decomposition is now based on signal $a_1(n)$ and so on. To eliminate the redundant information, a decimation operator is applied at each iterative step given by Equations 3.28 and 3.29. The implementation of the MRA is represented in Figure 3.9.

A reconstruction of the original fine-scale coefficients of the signal can be made from a combination of both the scaling function and the wavelet coefficients of coarse resolution.

Practical Application

The DWT used in this example (Daubechies family Db4) is applied to a digitized function with N samples, getting signals $a_j(n)$ and $d_j(n)$, where j is the index level. This family Db4 is particularly appropriate to detect disturbances of high frequency (transients) as it is more localized in time than other members of the same family, for example family Db20.

In Figure 3.10 we have shown a sinusoidal signal with 1024 sample points and an added oscillatory transient. The outputs of the DWT implementation are different signals obtained from each highpass filter h(n), corresponding to a particular level j. Assuming a 12.8–kHz sample rate, Figure 3.11 shows the relation of each level 2j with a frequency band containing components of the decomposed signal.

Joint Time–Frequency Analysis of the Electrical Signal   61

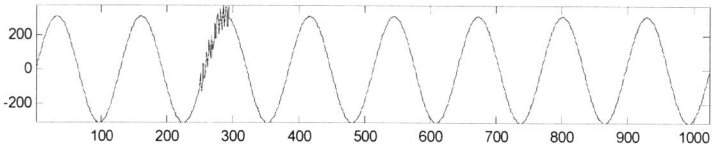

Figure 3.10. The oscillatory transient

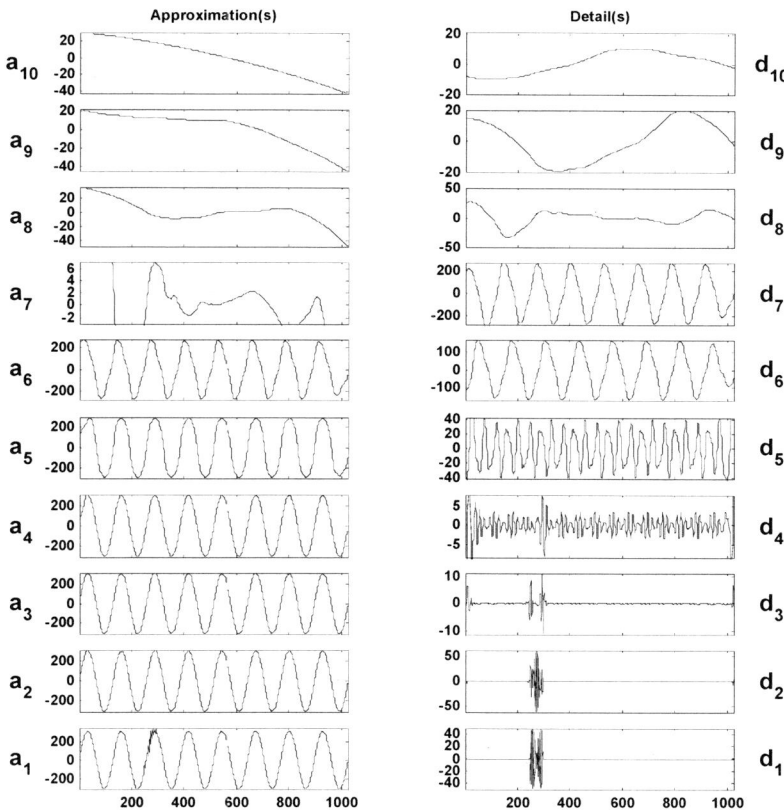

Figure 3.11. The scale analysis using the MRA technique

Figure 3.11 shows the decomposition of the input signal into approximation signals $a_j(n)$ and detail signals $d_j(n)$. In our case, $j= 1...9$ and the number of coefficients and frequency bands, at each level j, are shown in Table 3.1. The MatLab® program have been used to calculate the *DWT* for the input signal.

Table 3.1. Number of coefficient and frequency band

	a_{jk}	Number of Coefficients	d_{jk}
level 1	0–3200 Hz	515	3200–6400 Hz
level 2	0–1600 Hz	261	1600–3200 Hz
level 3	0–800 Hz	134	800–1600 Hz
level 4	0–400 Hz	70	400–800 Hz
level 5	0–200 Hz	38	200–400 Hz
level 6	0–100 Hz	22	100–200 Hz
level 7	0–50 Hz	14	50–100 Hz
level 8	0–25 Hz	10	25–50 Hz
level 9	0–12.5 Hz	8	12,5–25 Hz
level 10	0–6.25 Hz	7	6.25–12.5 Hz

3.6 Time-dependent Spectrum: Quadratic Transforms

The quadratic JTFA methods include the following: STFT spectrogram, Wigner–Ville distribution (WVD), Cohen's class, conic distribution, Choi–Williams distribution, Gabor Spectrogram and adaptive spectrogram.

3.6.1 STFT Spectrogram

In adition to the linear transforms, like the Gabor expansion and wavelets, another important method for the joint time-frequency analysis is the time-dependent spectrum. The objective is to find a representation capable of describing the signal's power spectrum evolving in time. A similar procedure to the conventional power spectrum gives a first approach of computing the time-dependent spectrum by taking the square of the STFT:

$$SP[t,\omega] = \left| \int_{-\infty}^{\infty} f[\tau] w^*[\tau - t] \, e^{-j\omega\tau} d\tau \right|^2 ,$$

and the sampling version yields to

$$SP[mT, n] = \left| \sum_{i=0} f[i] w^*[i - mT] \, e^{-j2\pi \frac{ni}{N}} \right|^2 .$$

Joint Time–Frequency Analysis of the Electrical Signal   63

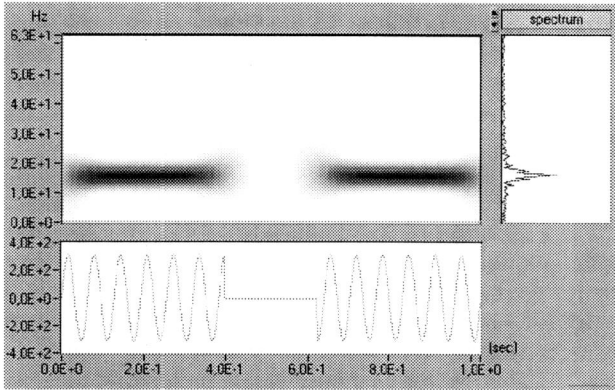

Figure 3.12. Interruption: STFT spectrogram with a Hamming window (32 samples)

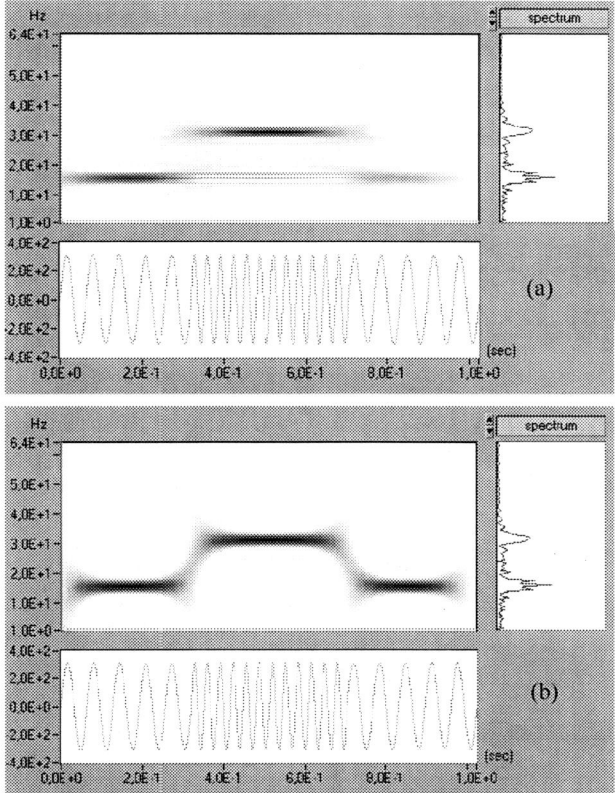

Figure 3.13. Frequency variation analysed by STFT spectrogram with a wide window of 128 samples (a), and also with a narrow window of 32 samples (b). Different time-frequency resolutions are achieved

As an example, we show in Figure 3.12 the conventional spectrum of the signal as well as its time waveform and the joint time-frequency representation for a signal with a momentary interruption. Virtual instruments developed by scientists at the National Instruments Co. have been used for computing the examples of the JTFA methods.

The STFT spectrogram is simple and fast but suffers from the so called window effect. This is where the dimension of the window seriously compromises time and frequency resolutions. So, with a wide window, the time-dependent spectrum has high frequency resolution but poor time resolution (Figure 3.13 a). On the other hand, with a narrow window, the time-dependent spectrum has poor frequency resolution but high time resolution (Figure 3.13 b). This fact makes the STFT spectrogram inadequate for applications where both high time and frequency resolutions are required.

3.6.2 Wigner–Ville Distribution and Wigner–Ville Pseudo-Distribution

As mentioned at the end of Sect. 3.3, the conventional correlation is the average of instantaneous correlations $f(t+\tau/2)f^*(t-\tau/2)$:

$$R(\tau) = \int_{-\infty}^{\infty} f(t+\tau/2)f^*(t-\tau/2)dt = \int_{-\infty}^{\infty} R(t,\tau)dt.$$

In the instantaneous correlation $R(t,\tau)$, the time information has still not been removed by averaging. So, this function is suitable to characterise the time-dependent spectrum. For a signal $f(t)$ the Wigner–Ville distribution is defined as

$$WVD(t,\omega) = \int_{-\infty}^{\infty} R(t,\tau)e^{-j\omega t}dt.$$

This distribution does not have any window effect, hence it can characterise a signal's properties in the joint time-frequency domain better than STFT. The WVD is also simple and fast but, if the analysed signal contains more than one component, the WVD method suffers from crossterm interference.

In support of algorithm implementation, the discrete version for a signal $f[i]$ is

$$WVD[i,k] = \sum_{m=-N/2}^{N/2} R[i,m]e^{-j2\pi\frac{km}{N}},$$

where $R[i,m]$ is the instantaneous correlation function given by

$$R[i,m] = z[i+m]z^*[i-m]$$

$z[i]$ being the analytical or interpolated form of $f[i]$ [21].

The crossterm reflects the correlation between a pair of corresponding autoterms, always sits halfway between two corresponding autoterms, and oscillates frequently. Although its magnitude can be very large, its average usually is limited.

To improve the effect of the crossterm interference, we can assign different weights to the instantaneous correlation R[i,k] to suppress the less important parts and enhance the fundamental parts.

Usually, two methods have been used to apply the weighting function to the instantaneous correlation R[i,m]. The first method is developed in the time domain:

$$\text{PWVD}[i,k] = \sum_{m=-N/2}^{N/2} w[m] R[i,m] \, e^{-j2\pi \frac{km}{N}}, \tag{3.30}$$

which is named the pseudo Wigner–Ville distribution (PWVD) and uses a Gaussian window function w[m]. The second method is developed in the frequency domain, where different weights are assigned to the instantaneous correlation R[i, m]:

$$\text{WVD}[i,k] = \sum_{m=-N/2}^{N/2} G[m] R[i,m] \, e^{-j2\pi \frac{km}{N}}, \tag{3.31}$$

where G·[m] is also a Gaussian window function. This equation is equivalent to

$$\text{PWVD}[i,k] = \sum_{m=-N/2}^{N/2} \left(\sum_n g[n] R[i-n,m] \right) e^{-j2\pi \frac{km}{N}}, \tag{3.32}$$

where g[n] is the inverse Fourier transform of G·[m] in Equation 3.31.

3.6.3 Cohen Class

Joining the two aproaches from Equations 3.30 and 3.32

$$C[i,k] = \sum_{m=-N/2}^{N/2} \sum_n \Phi[n,m] R[i-n,m] \, e^{-j2\pi \frac{km}{N}}, \tag{3.33}$$

where Φ[i,m] denotes the kernel function. The window functions w[m] in Equation 3.30 and g[m] in Equation 3.32 are special cases of Φ[i,m] in Equation 3.33.

In 1966, Leon Cohen developed the representation C[i,k] in Equation 3.33, so it is traditionally known as Cohen's class [30]. Compared with the PWVD in Equation 3.30 or Equation 3.32, the Cohen's class method is more general and flexible. Most quadratic equations known so far, such as the STFT spectrogram, WVD, PWVD, CWD (Choi–Williams distribution), and the cone-shaped distribution, belong to the Cohen's class.

3.6.4 Choi–Williams Distribution

If the kernel function in Equation 3.33 is defined by

$$\Phi[i,m] = \sqrt{\frac{\alpha}{4\pi m^2}} e^{-\alpha\, i^2/(4m^2)}, \qquad (3.34)$$

the Choi–Williams (CWD) distribution is obtained. Crossterm interference and time frequency resolution can be improved by adjusting the parameter α in Equation 3.34. However, the CWD method cannot reduce all the crossterms. Furthermore, the computation speed of CWD is very low.

3.6.5 Conic Distribution

The conic distribution is obtained when the kernel function in Equation 3.33 is defined by

$$\Phi[i,m] = \begin{cases} e^{-\frac{\alpha\, m^2}{c}} & \forall\ i < m \\ 0 & \text{otherwise} \end{cases} \qquad (3.35)$$

In this application the constant c is set to 500. Here crossterm interference and time-frequency resolution can be balanced again by adjusting the parameter α in Equation 3.34. The conic distribution method is faster than the CWD method.

3.6.6 Gabor Spectrogram

By applying the Gabor expansion as well as the Wigner–Ville pseudo-distribution's window method to a signal, the significance of each term corresponding to the signal's energy at point $[i,k]$ can be identified. The idea is to preserve those terms that have the most important contributions at point $[i,k]$ and remove those terms that have insignificant influence on the signal's energy. The Gabor spectrogram is defined by

$$GS_D[i,k] = \sum_{|m-m'|+|n-n'| \leq D} C_{m,n} C_{m',n'} WVD_{h,h'}[i,k], \qquad (3.36)$$

where $WVD_{h,h'}[i,k]$ denotes the cross WVD of frequency-modulated Gaussian functions. The degree of smoothing is controlled by the order of the Gabor spectrogram, D. If $D = 0$, $GS_0[i,k]$ is non-negative and similar to the STFT spectrogram. When D goes to infinity, the Gabor spectrogram converges to the WVD.

A higher-order Gabor spectrogram has better resolution but more crossterm interference and *vice versa*. Additionally, the higher the order, the longer the computation time. The Gabor spectrogram has better resolution than the STFT spectrogram and much less crossterm interference than the cone-shaped, Choi–Williams or Wigner–Ville distributions.

Joint Time-Frequency Analysis of the Electrical Signal   67

The Gabor spectrogram is usually less sensitive to window length than the STFT spectrogram. Figure 3.14 represents a comparison between different window sizes for a signal with a sag with a 50% of amplitude decrease. The time resolution is best with a narrow window but it offers a poor frequency resolution. Alternatively, a wide window offers high frequency resolution but poor time resolution.

The amplitude transient events (sag and swells) are successfully represented by this method. In this figure we can see clearly the decrease of brightness into the sag zone because of the energy reduction.

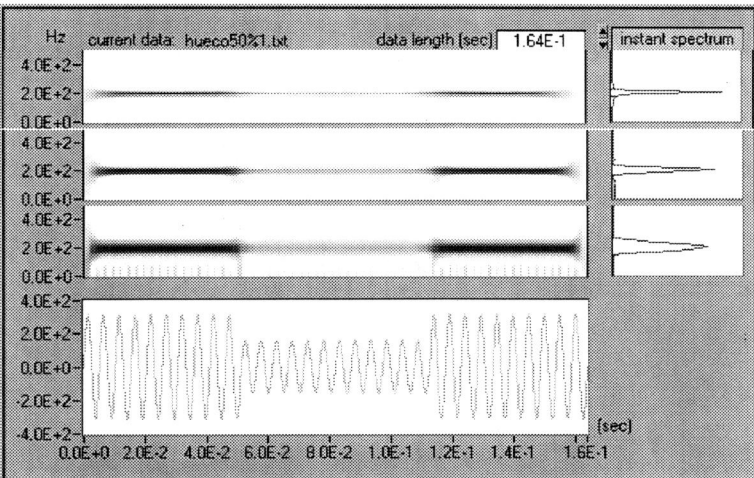

Figure 3.14. Gabor spectrogram of order 3 with a wideband window of 32 samples (bottom), mediumband of 64 samples (middle) and narrowband of 128 samples (top). Analysed signal: sag with a 50% decrease.

3.6.7 Adaptive Spectrogram

The adaptive spectrogram method is an adaptive representation based spectrogram (see Equation 3.20) computed by

$$AS[i,n] = 2\sum_{k=0}^{D-1} |A_K|^2 \exp\left\{-\frac{[i-i_k]^2}{\alpha_k} - (2\pi)^2 \alpha_k [n - f_k - \beta_k i]^2\right\}. \quad (3.37)$$

In general, the adaptive spectrogram gives the best time-frequency resolution especially when the analysed signal is a sum of linear chirp-modulated Gaussian functions. Unfortunately, the computation speed of the adaptive spectrogram increases exponentially with the analysed data size.

3.7 Algorithms Summary

If we consider the linear JTFA methods as the logical evolution of the conventional Fourier transform, the quadratic JTFA is the counterpart of the standard power spectrum. The main difference between linear and quadratic methods is that the linear transforms are invertible. As with the fast-Fourier transform, the original signal can be reconstructed based on the Gabor coefficients or on the wavelet coefficients for example. Thus, the linear transforms are suitable for signal processing. In contrast, the quadratic form is not reversible. The original time waveform can not be restored from the time-dependent spectrum. However, the quadratic JTFA describes the energy distribution of the signal in the joint time-frequency domain, which is useful for signal analysis.

Linear methods are fast and have an easy formulation and implementation but they do not have enough resolution for many applications. On the other hand, quadratic methods have better resolution in the joint time-frequency domain although they show, in general, crossterm interference that masks the autoterms. Moreover, no quadratic algorithm can significantly measure a signal's energy point to point in the joint time-frequency domain.

Table 3.2 shows the most important advantages and disadvantages of the quadratic JTFA methods.

Table 3.2 Quadratic algoritms of the JTFA

Method	Resolution and cross terms	Speed
Adaptive	Very high resolution, no crossterms and non-negative	low
CWD	Fewer crossterms than the PWVD	Very low
Conic distribution	Fewer crossterms interference than PWVD or CWD	low
Gabor	Good resolution, robust, minimum number of crossterms.	Medium
PWVD	Best resolution with various signal types. Important quantity of crossterms.	High
STFT	Poor resolution, robust and non-negative.	High

Joint Time-Frequency Analysis of the Electrical Signal   69

Table 3.3 summarizes the most relevant method for the analysis of transient events. We have made a scale zoom in the adaptive representation to show in detail the best energy concentration in the joint time-frequency domain.

Table 3.3. Oscillatory transient: quadratic methods comparison

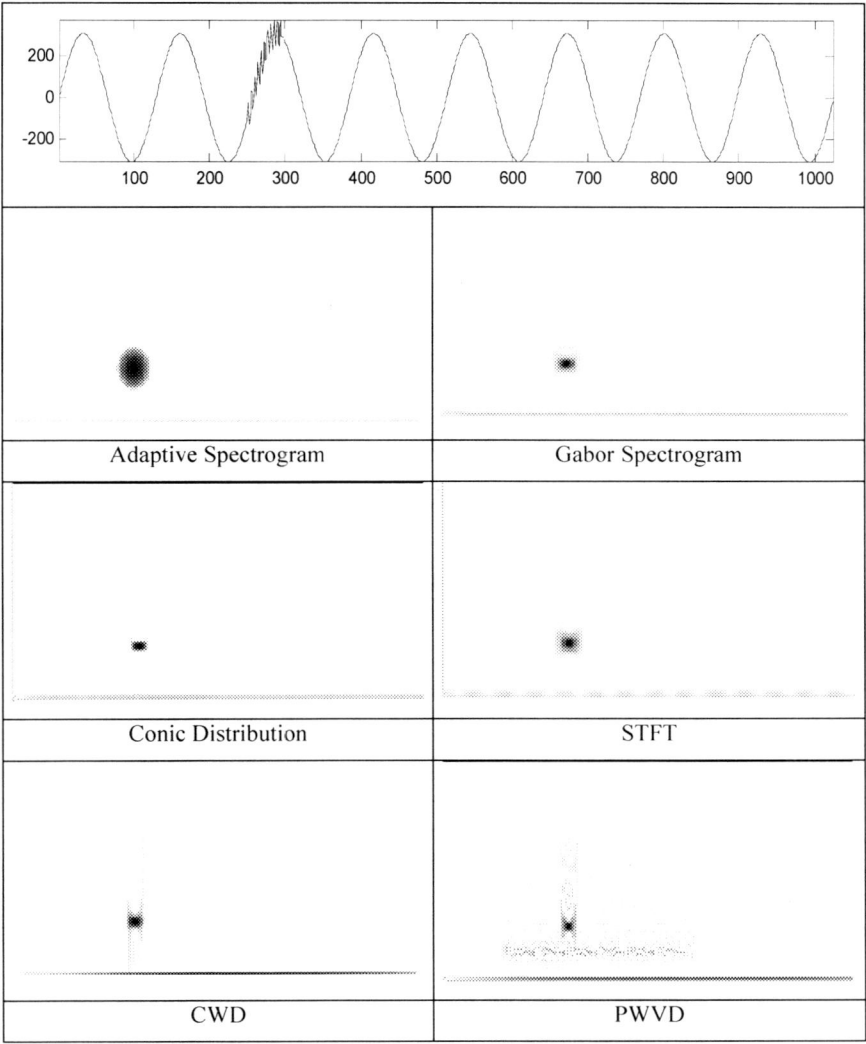

Finally, Table 3.4 shows the results of the application of several quadratic methods over a signal with frequency distortion. The best compromise between resolution and speed is given by the Gabor spectrogram.

Table 3.4. Frequency variation: quadratic methods comparison

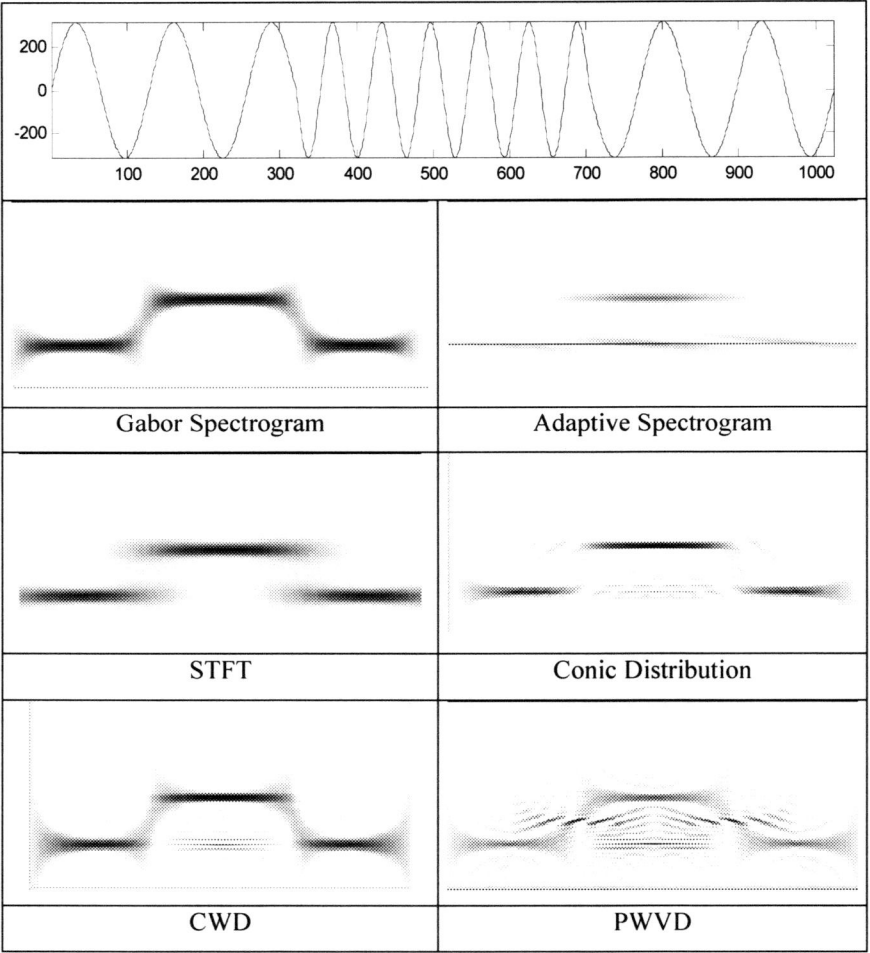

References

[1] Gröchenig K, (2001), Foundations of Time–Frequency Analysis, Birkhäuser, Boston.
[2] Hogan J, Lakey J, (2005), Time–Frequency and Time–Scale Methods, Birkhäuser, Boston.
[3] Ben–Arie K, Raghunath K, (1992), Image expansion by non–orthogonal wavelets for optimal template matching, Pattern Recognition, Vol.III. Conference C: Image, Speech and Signal Analysis, Proceedings., 650–654.
[4] Bravo JC, Borrás D, Castilla M, Montaño JC, López A, Gutiérrez J, (2005), Joint wavelet–Fourier analysis of power system disturbances, Proc. of the 9CHLIE05, Marbella, Spain.
[5] Montaño JC, Castilla M, Bravo JC, López A, Gutiérrez J, Borrás D, (2004), Analysis of Electrical Signal Disturbances. A New Strategy, Proc. of the ICREPQ04, Barcelona, Spain.
[6] Borrás D, Bravo JC, Montaño JC, Castilla M, López A, Gutiérrez J, (2005), A new advanced hybrid analysis method in power systems, IDAACS 2005, Sofia, Bulgaria.
[7] Borrás D, Castilla M, Moreno N, Montaño JC, (2001), Wavelet and neural structure: a new tool for diagnostic of power system disturbances, IEEE Trans. on Industry App. 37, 1, 184–190.
[8] Dugan RC, McGranaghan MF, Beaty HW, (1996), Electrical Power Systems Quality, McGraw–Hill, NY.
[9] Arrillaga J, Bradley DA, Bodger PS, (1985), Power Systems Harmonics, JohnWiley & Sons, New York.
[10] Derek AP, (1995), Power Electronic Converter Harmonics, IEEE Press, New York.
[11] Pillary P, Ribeiro P, Pan Q, (1996), Power quality modeling using wavelets, IEEE Proc. of the 7th International Conference on Harmonics and Quality of Power (ICHQP), Las Vegas, Nevada, USA, 625–631.
[12] Angrisani L, Daponte P, D'Apuzzo M, Testa A, (1998), A Measurement Method Based on the Wavelet Transform for Power Quality Analysis, IEEE Transactions on Power Delivery, 13, 4.
[13] Robertson DC, Camps OI, Mayer JS, Gish WB, (1996), Wavelets and electromagnetic power system transients, IEEE Trans. Power Delivery, 11, 2, 1050–1058.
[14] Poisson O, Rioual R, Meunier M, (2000), Detection and Measurement of Power Quality Disturbances Using Wavelet Transform, IEEE Trans. Power Delivery, 15, 3, 1039–1044.
[15] Bollen MHJ, (2000), Understanding Power Quality Problems, IEEE Press, New York.
[16] Martínez–Velasco J.A. (1997), Computer Analysis of Electric Power System Transients, IEEE Press, New York.
[17] Mallat S, (2001), A Wavelet Tour of Signal Processing, Academic Press, London.
[18] Daubechies I, (1990), The wavelet transform, time–frequency localization, and signal analysis, IEEE Trans. Infinn. Theory, 961–1005.
[19] Wickerhauser M.V, (1994), Adapted Wavelet Analysis from theory to software, IEEE Press, New York.
[20] Gröchenig K, (2003), Uncertainty principles for time–frequency representations. In Advances in Gabor Analysis , pp. 11–30, Birkhäuser, Boston.
[21] Qian S, Chen D, (1996), "Joint Time–Frequency analysis. Methods and applications", Englewood Cliffs, NJ: Prentice Hall.
[22] Qian S, Chen D, (1999), Understanding the nature of signals whose power spectra change with time. Joint analysis, IEEE Signal Processing Magazine.
[23] Qian S, Chen D, (1993), "Discrete Gabor transform," IEEE Trans. Signal Processing, vol. 41, no. 7, pp. 2429–2439.

[24] Farkash S, Raz S, (1990), Time–variant filtering via the Gabor expansion, Signal Processing I/: Theories and Applications, 509–512, New York: Elsevier.
[25] Hlawatsch F and Krattenthaler W, (1992), "Bilinear signal synthesis," IEEE Trans. Signal Processing, vol. 40, no. 2, pp. 352–363.
[26] Walnut DF, (2002), An Introduction to Wavelet Analysis, Birkhäuser, Boston.
[27] Mallat S, (1989), A theory for multiresolution signal decomposition: the wavelet representation, IEEE Trans. on Pattern Anal. and Mach. Intell., 11, 674–693.
[28] Burrus CS, Gopinath RA, Guo H, (1998), Introduction to Wavelets and Wavelet Transforms, Prentice Hall, New Jersey.
[29] Kumar PS, Satish DL, (1998), Multiresolution Signal Decomposition: A new tool for fault detection in power transformers during impulse test, IEEE Trans. on Power Delivery, 13, 4.
[30] Cohen L, (1995), Time–frequency Analysis, Englewood Cliffs, Nat. Inst., Prentice Hall.
[31] Yin Q, Ni Z, Qian S, Chen D, (1997), Adaptive oriented orthogonal projective decomposition, J. Electron. (Chinese), 25, 4, 52–58.

4

Measurement and Analysis of Voltage Events

Julio Barros, Enrique Pérez, Ramón I. Diego

Dept. Electronics and Computers,
University of Cantabria,
Av. de los Castros,
39005 Santander, Spain.
Emails: barrosj@unican.es, eperez25@boj.cnice.mecd.es, diegori@unican.es

This chapter describes the main issues related to the measurement and analysis of voltage events in power systems. Section 4.1 describes the definitions and the different magnitude and duration thresholds used in the main power-quality standards to distinguish between the different types of voltage events. Section 4.2 describes the performance of the standard method for detection and analysis of voltage events and other signal-processing methods used to overcome some of the drawbacks of the standard method. Section 4.3 details the main effects of voltage events on equipment and the use of voltage-tolerance curves to describe the performance equipment to voltage variations. Finally, Section 4.4 presents the results of several months of monitoring of the three–phase voltage supply in a low–voltage distribution network and the effects of the voltage events detected on different equipment connected to this network.

4.1 Introduction

The term voltage event is used to describe an abnormal and temporary variation of the magnitude of voltage supply. Voltage events are classified as voltage dips, interruptions and overvoltages and, depending on the power–quality standard employed; their limits are defined in different ways. At present, these voltage events are one of the most important power–quality phenomena owing to their frequency of occurrence and their effect on customers' equipment.

All of these voltage events are unpredictable and their frequency of occurrence varies greatly depending on the type of supply system and the point of observation, their distribution over the year being very irregular.

European Standard EN 50160 [1] and IEEE Standard 1159 [2] are the most important power–quality standards that consider these disturbances. According to EN 50160, a voltage dip is a sudden reduction of the voltage supply to a value between 90% and 1% of the nominal voltage, with duration between 10 ms and 1 min. A supply interruption is defined as a condition in which the voltage supply is lower than 1% of the nominal voltage. The interruption is classified as a short interruption if its duration is less than 3 min; otherwise the interruption is classified as a long interruption. Finally, the standard defines a temporary overvoltage as an overvoltage of relatively long duration, whereas a transient overvoltage is defined as a short–duration overvoltage with duration of a few milliseconds or less. These definitions are summarized in Table 4.1.

Table 4.1. **Definitions** of voltage events as used in EN 50160

Voltage event	Magnitude	Duration
Voltage dip	90%–1%	10 ms–1 min
Supply interruption	< 1%	< 3 min (short) > 3 min (long)
Temporary overvoltage	> 110%	From ms to ?
Transient overvoltage	> 110%	< few ms

On the other hand, IEEE Standard 1159 defines a voltage sag (dip) as a short–duration voltage decrease at the power–system frequency to a value between 90% and 10% of nominal voltage. An interruption is defined when the supply voltage decreases to less than 10% of the nominal voltage for a period not exceeding 1 min, and finally a swell is defined as an increase in the rms voltage for durations from 0.5 cycles to 1 min, with typical magnitudes between 110% and 180%.

The standard divides each type of voltage variation into three different categories depending on its duration: instantaneous (between 0.5 cycles and 30 cycles), momentary (between 30 cycles and 3 s) and temporary (between 3 s and 1 min).

If the rms decrease is longer than 1 min, then it is defined as a long–duration undervoltage. If the rms voltage increase is longer than 1 min, the event is defined as a long duration overvoltage, and finally if the interruption is longer than 1 min, it is defined as a sustained interruption. The IEEE definitions are summarized in Table 4.2.

Table 4.2. Definitions of voltage events as used in IEEE Standard 1159

Voltage event	Typical magnitude	Typical duration
Voltage sag (dip)	90%–10%	0.5 cycles – 1 min
Long–duration undervoltage	90%–80%	> 1 min
Interruption	< 10%	0.5 cycles – 1 min
Sustained interruption	< 10%	> 1 min
Swell	110%–180%	0.5 cycles – 1 min
Long–duration overvoltage	110%–120%	> 1 min

As can be seen from the definitions summarized in Table 4.1 and Table 4.2, EN 50160 and IEEE Standard 1159 use different magnitude and duration thresholds to distinguish between the different types of voltage events.

4.2 Monitoring of Voltage Events

The magnitude of a voltage event can be computed using the rms voltage, the fundamental voltage or the peak voltage. The use of any of these three methods will produce the same results in the case of a pure sinusoidal voltage waveform, but the presence of harmonics and other disturbances in voltage waveforms could lead to different results depending on the signal-processing method used.

The most common method used for the majority of power–quality instruments is the calculation of the rms voltage. In a digital system the rms magnitude of voltage supply is obtained using the following equation:

$$V_{rms} = \sqrt{\frac{1}{N}\sum_{i=1}^{N} v_i^2} ,$$

where v_i refers to the voltage samples and N is the number of samples taken in a window.

The rms method is simple and easy to implement but it has the drawback of its dependency on the window length and on the time interval for updating the values.

The window size can be selected from a half–cycle of the power–system frequency up to any multiple of half–cycles. On the other hand, the rms magnitude could be updated with each new sample taken, using an overlapping window, or it could be updated each sampling window when a non overlapping window is employed. The selection of the window length and the time for updating the values depends on the processing speed of the monitoring instrument and on the amount of memory available to store the results obtained. Until now, the majority of power–quality instruments available have used non overlapping windows with one-cycle size.

It is important to point out that a sudden change in the magnitude of a voltage supply is not immediately detected using the rms calculation, it being necessary that the new value of the voltage after the change is entirely within the sampling window to obtain its correct magnitude. Thus, depending on the window length used, the time interval for updating the rms values and the point on the wave where the voltage event begins, the magnitude and the duration of the voltage event can be very different [3–5]. Another limitation of the rms method is that no information about the phase-angle jump or the instant where the voltage event starts is given.

Recently, IEC Standard 61000-4-30 defined the $U_{rms(1/2)}$ magnitude as the basic measurement of a voltage event [6]. This magnitude is defined as the rms voltage measured over 1 cycle, commencing at a fundamental zero crossing, and refreshed each half-cycle.

According to this standard, in single-phase systems a voltage dip begins when the $U_{rms(1/2)}$ magnitude falls below the dip threshold and ends when this magnitude is equal to or above the dip threshold plus the hysteresis voltage. In polyphase systems a dip begins when the $U_{rms(1/2)}$ voltage of one or more channels is below the dip threshold and ends when the $U_{rms(1/2)}$ voltage of all measured channels is equal to or above the dip threshold plus the hysteresis voltage.

Voltage dips are characterized by a pair of data, the residual voltage or the depth and the duration. The residual voltage is the lowest $U_{rms(1/2)}$ value measured during the dip, and the depth is defined as the difference between the reference voltage and the residual voltage (generally expressed as a percentage of the reference voltage). The duration of a voltage dip is the time difference between the beginning and the end of the voltage dip.

Analogously in single-phase systems a swell begins when the $U_{rms(1/2)}$ voltage rises above the swell threshold and ends when the $U_{rms(1/2)}$ voltage is equal to or below the swell threshold minus the hysteresis voltage. In polyphase systems a swell begins when the $U_{rms(1/2)}$ voltage of one or more channels surpasses the swell threshold and ends when the $U_{rms(1/2)}$ voltage of all measured channels is equal to or below the swell threshold minus the hysteresis voltage.

A voltage swell is characterized by the maximum swell–voltage magnitude and the duration. The maximum swell–voltage magnitude is the largest $U_{rms(1/2)}$ value measured in any channel during the swell, and the duration of a voltage swell is the time difference between the beginning and the end of the swell.

Finally, in the case of voltage interruptions, IEC Standard 61000-4-30 defines that, in single-phase systems, a voltage interruption begins when the $U_{rms(1/2)}$ magnitude falls below the voltage–interruption threshold and ends when the $U_{rms(1/2)}$ value is equal to, or greater than, the voltage–interruption threshold plus the hysteresis. In polyphase systems, a voltage interruption begins when the $U_{rms(1/2)}$ voltages of all channels fall below the voltage–interruption threshold and ends when the $U_{rms(1/2)}$ voltage of any channel is equal to, or greater than, the voltage interruption threshold plus the hysteresis.

On the other hand, IEEE Standard 1159-1995 does not specify the window length or the time interval for updating the rms values that should be used to detect and to evaluate a voltage event. However, in the revision of this standard currently under discussion, the $U_{rms(1/2)}$ value, as defined by the IEC, is also proposed as the basic magnitude for voltage–dip detection and evaluation.

4.2.1 Performance of the IEC Standard Method in the Detection and Evaluation of Voltage Events

The $U_{rms(1/2)}$ method is simple and easy to implement, but shows a limited performance in the detection and in the estimation of voltage events, mainly in the case of short–duration and less severe events.

Using the $U_{rms(1/2)}$ value the magnitude of a voltage event is exactly computed only when the duration of the event is longer than the window width used (1 cycle). Furthermore, as the $U_{rms(1/2)}$ magnitude is computed each half–cycle of the fundamental component, the duration of a voltage event is given in integer multiples of half–cycles and thus, depending on the point on the wave where the voltage event begins and on its magnitude, the error in the duration could be very important, especially for short–duration and low–magnitude voltage dips. Finally, there is always a delay in the detection of the beginning of a voltage event that is a function of the magnitude and the point on the wave where the voltage event starts.

As an example of the performance of the $U_{rms(1/2)}$ method, Figure 4.1 shows the waveform of a 50% magnitude, 15 ms duration and 90° point-on-wave voltage dip and the corresponding ideal rms and $U_{rms(1/2)}$ magnitudes.

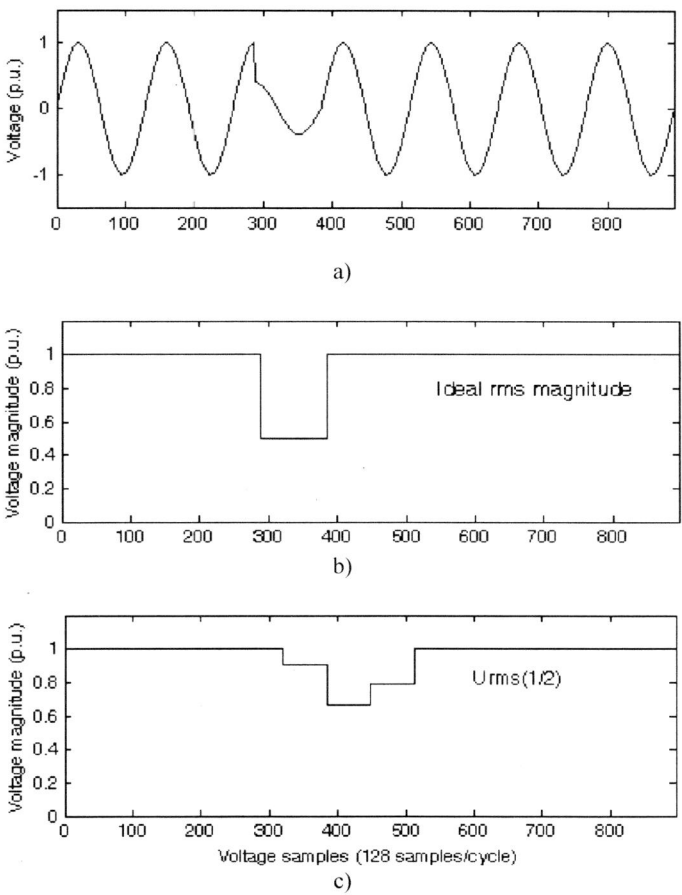

Figure 4.1. (a) Voltage waveform of a voltage dip, (b) ideal rms magnitude, (c) rms magnitude computed according to IEC 61000-4-30.

As can be seen, the shape of the $U_{rms(1/2)}$ magnitude is very different from the ideal rms value. The magnitude and duration of the voltage dip computed using the $U_{rms(1/2)}$ are 39.71% and 30 ms, respectively, and the delay in the detection of the beginning of the voltage event is 5 ms.

In order to evaluate the performance of the $U_{rms(1/2)}$ method in the detection and evaluation of voltage events, rectangular voltage dips of different magnitude, duration and instant of beginning are simulated in a 230V, 50Hz, pure sinusoidal voltage supply. For each simulation the $U_{rms(1/2)}$ magnitude is computed to detect the instant where the voltage dip begins and to obtain its magnitude and duration [7]. Neither phase jumps associated with voltage dips nor hysteresis voltages above the dip threshold are considered in the simulations. The sampling rate used is 6.4 kHz.

Table 4.3 and Figure 4.2 show the maximum error in the estimation of the magnitude of a voltage dip computed using the Urms(1/2) value, for voltage–dip magnitudes from 10% to 90%, durations of 0.5, 1 and 1.5 cycles, respectively, and different instants of beginning of the voltage dip (from 0° to 360° in steps of 5°).

Table 4.3. Maximum error in the depth for voltage dips of 0.5, 1 and 1.5 cycles

Voltage–dip magnitude (%)	Maximum error in magnitude (%)		
	0.5 cycle	1 cycle	1.5 cycles
10	51.30	25.52	0
20	52.75	26.40	0
30	54.36	27.76	0
40	56.15	29.40	0
50	58.12	31.40	0
60	60.26	33.81	0
70	62.60	36.73	0
80	65.14	40.25	0
90	67.84	44.42	0

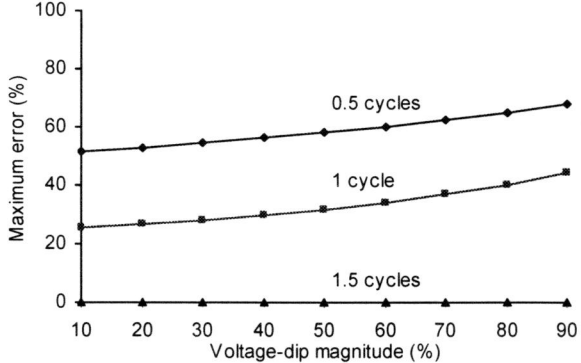

Figure 4.2. Maximum error in the depth for voltage dips of 0.5, 1 and 1.5 cycles

As can be seen, the error in the magnitude of the voltage dip is higher for shorter and more severe voltage dips. For voltage dips of 0.5 cycles the error in magnitude is in a range from 51.30% for 10% voltage dips and 0° or 180° of point on the wave of beginning, to 67.84% for 90% voltage dips and 0° or 180° of point on the wave of beginning. For voltage dips of 1 cycle duration, the error in magnitude is in a range from 25.52% to 44.42% for 10% and 90% voltage dips and 90° or 270° of point on the wave respectively. Finally, the error in magnitude is zero for voltage dips of 1.5 cycles or longer.

The large error observed in the computation of the magnitude of short-duration voltage dips, can lead to failures in the detection efficiency of the $U_{rms(1/2)}$ method for low-magnitude voltage dips, as the dip threshold might not be surpassed in this case.

Table 4.4 and Figure 4.3 show the minimum magnitude of a voltage dip, as a function of the duration, necessary to ensure 100% probability of detection of the voltage dip, independently of the point-on-wave where the voltage dip begins. All the voltage dips above the upper curve of Figure 4.3 are always detected using the $U_{rms(1/2)}$ method.

On the other hand, the lower curve in Figure 4.3 shows the maximum magnitude of a voltage dip as a function of the voltage–dip duration for 100% probability of non detection of voltage dips. Any voltage dip below the lower curve in Figure 4.3 is never detected using the $U_{rms(1/2)}$ method, independently of the point-on-wave where the voltage dip begins.

Voltage dips of magnitude-duration in a range between the two curves in Figure 4.3 could be detected using the $U_{rms(1/2)}$ method, with different detection probability depending on the instant of beginning of the voltage dip.

Table 4.4. Minimum/maximum voltage dip magnitude-duration for 100% probability of detection/non detection using the $U_{rms(1/2)}$ method

Minimum magnitude-duration for 100% detection		Maximum magnitude-duration for 100% non detection	
Duration (ms)	Magnitude (%)	Duration (ms)	Magnitude (%)
10	21.25	10	20.53
15	19.13	15	10.96
20	13.44	19.37	10.00
25	10.46	-	-
30.15	10	-	-

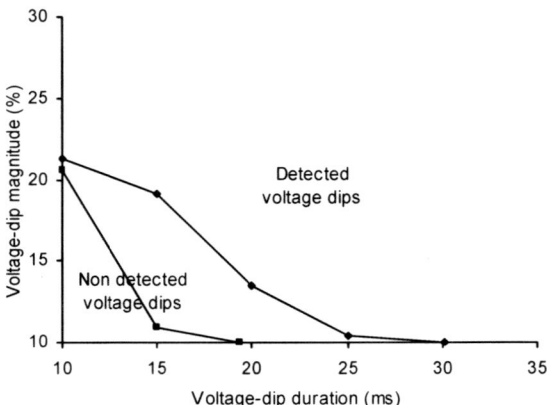

Figure 4.3. Minimum/maximum voltage–dip duration contour for 100% probability of detection/non detection using the $U_{rms(1/2)}$ method

4.2.2 Other Methods for Detection and Evaluation of Voltage Events

This section reviews the performance of other signal-processing tools used in the detection and analysis of voltage events. Discrete Fourier transform analysis, Kalman filtering and wavelet analysis have been used as alternative methods to overcome some of the drawbacks of the rms methods. An in depth review of the methods most commonly employed to detect and to classify voltage events can be found in [8].

4.2.2.1 Fourier Analysis
The traditional method to obtain the fundamental and harmonic components of a signal in a digital system is the application of the discrete Fourier transform (DFT) to the samples of the signal. As is well known, the results obtained using DFT are correct when the signal is stationary and periodic and the time window used is synchronized with the fundamental frequency of the signal. However, the results obtained are erroneous in the case of non stationary signals, as is the case of the voltage waveform in an event.

A way to overcome this problem is the use of the short–time Fourier transform (STFT). The STFT partitions the signal into time segments where the signal is considered stationary, applying the DFT within each segment. Once the size of the time window is selected, the time-frequency resolution obtained is fixed and it is the same for the whole frequency spectrum of the signal. To obtain different resolutions in different parts of the spectrum it would be necessary to apply STFT using different window sizes. The results obtained by applying the STFT give information about the time evolution of the harmonic components of the signal.

Different authors have studied the application of STFT for the detection and characterization of voltage events in power systems [9–11]. The results obtained show, as in the case of the rms methods, its dependence on the length of the time window selected. As in the rms method, a sudden change in the magnitude of voltage supply cannot be immediately detected, it being necessary again that the new value after the change is entirely within the sampling window to obtain its correct magnitude. The error in the estimation of the magnitude of a voltage event is 0% only when the duration of the event is longer than the time window used in the analysis, otherwise the error is a function of the magnitude, duration and point-on-wave where the voltage event begins.

An advantage of the STFT method is that it gives information about the magnitude and phase angle of the fundamental and harmonic components of voltage supply during the event.

Figure 4.4 shows an example of the performance of the STFT method in the detection and analysis of a voltage event. A 60% magnitude and 21.25–ms duration voltage dip is applied to a 230V, 50Hz pure sinusoidal voltage waveform. The voltage dip starts after two cycles and 5.15 ms of the voltage waveform and the sampling frequency used in the simulations is 6.4 kHz (128 samples/cycle in a 50–Hz power system). The STFT is applied using a sampling window of 1 cycle of the fundamental power-system frequency and the results are updated with each new sample taken, using the algorithm proposed in [12]. Figure 4.4 also shows the ideal rms and the $U_{rms(1/2)}$ magnitudes of the voltage waveform.

As can be seen, the results obtained using the STFT are better than those obtained using the $U_{rms(1/2)}$ magnitude, but the shape of the voltage supply during the event is still very different from the ideal rms value. Longer-duration voltage events are more accurately monitored using the STFT method.

4.2.2.2 Kalman Filtering
Kalman filters have been used as an alternative method for detection and analysis of voltage events in power systems. The change in magnitude of the fundamental component of voltage supply can be used as an efficient way to detect and analyse voltage events. Unlike the rms method, the Kalman filtering method gives information both about the magnitude and phase angle of the voltage supply during the event and the point-on-wave where the voltage event begins.

Kalman filtering uses a mathematical model of the signal in state-variable form. The fundamental component and different harmonic components in the voltage supply can be used as the state vector in order to obtain a higher-order model of the system to describe more accurately its behaviour and, thus, to obtain a better estimation of the rms magnitude of voltage supply. Each frequency component requires two state variables, the components in phase and in quadrature with respect to its respective rotating reference, $2n$ being the total number of state variables to represent n harmonic components.

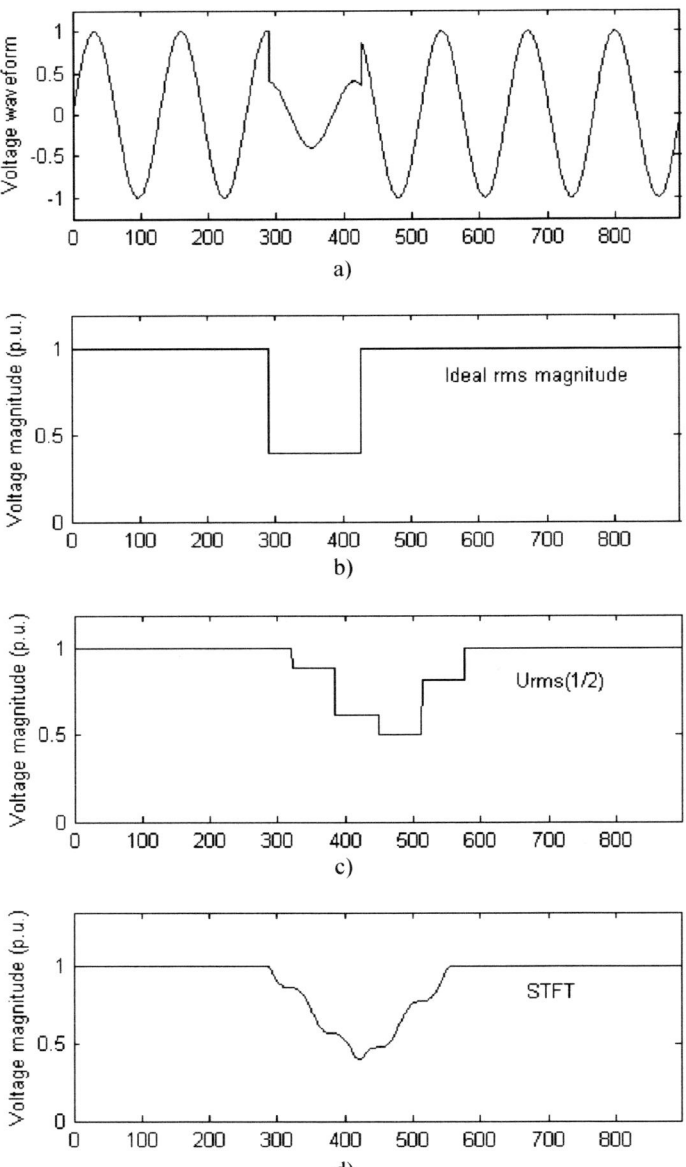

Figure 4.4. (a) Voltage waveform with a voltage dip, (b) ideal rms value, (c) rms magnitude computed using $U_{rms(1/2)}$, d) fundamental component using STFT

The state equation 4.1 describes the discrete system at instant k.

$$x_{k+1} = \Phi_k x_k + w_k , \qquad (4.1)$$

where x_k is the state vector at the instant k of size $2n$, Φ_k is the state-transition matrix and w_k is a random variable vector that represents the variation of the state variables due to an input white noise sequence. In expanded form, Equation 4.1 can be expressed as:

$$\begin{pmatrix} x_1 \\ x_2 \\ \vdots \\ x_{2n-1} \\ x_{2n} \end{pmatrix}_{k+1} = \begin{pmatrix} 1 & 0 & \cdots & & 0 \\ 0 & 1 & \cdots & & 0 \\ \vdots & & & & \vdots \\ 0 & 0 & \cdots & 1 & 0 \\ 0 & 0 & \cdots & 0 & 1 \end{pmatrix} \begin{pmatrix} x_1 \\ x_2 \\ \vdots \\ x_{2n-1} \\ x_{2n} \end{pmatrix}_k + w_k, \qquad (4.2)$$

The measurement equation 4.3 describes the relationship between the measurement and the state vector at the instant k.

$$z_k = H_k x_k + v_k = \begin{pmatrix} \cos(wk\Delta t) \\ -\sin(wk\Delta t) \\ \vdots \\ \cos(nwk\Delta t) \\ -\sin(nwk\Delta t) \end{pmatrix}^T \begin{pmatrix} x_1 \\ x_2 \\ \vdots \\ x_{2n-1} \\ x_{2n} \end{pmatrix}_k + v_k, \qquad (4.3)$$

where z_k is the measurement of the state vector at the instant k, H_k is the matrix that gives the ideal relationship between the measurement and the state vector and v_k is the measurement-error vector, assumed to be a white sequence uncorrelated with w_k, Δt is the sampling interval and T means the transpose. The covariance of the error measurement vector v_k is represented by R_k.

Starting from an initial estimation of the state vector and its associated error covariance, the Kalman filter uses the measurement to update the initial estimation. A linear combination of the initial estimation and the noise measurement is chosen in accordance with Equation 4.4

$$x_k = x'_k + K_k(z_k - H_k x'_k), \qquad (4.4)$$

where x'_k is the initial estimate, x_k is the estimate updated at the instant k and K_k is the filter coefficient. Making use of the state-transition matrix Φ_k we can project the filter ahead and use the measurement at the instant $(k+1)$ to obtain the new estimation for this instant x'_{k+1} and its error covariance P'_{k+1}, using Equations 4.5 and 4.6

$$x'_{k+1} = \Phi_k x_k + w_k, \qquad (4.5)$$

$$P'_{k+1} = \Phi_k P_k \Phi'_k + Q_k, \qquad (4.6)$$

where Φ'_k is the transpose of matrix Φ_k and Q is the covariance of vector w_k. Figure 4.5 shows the recursive algorithm of a Kalman filter.

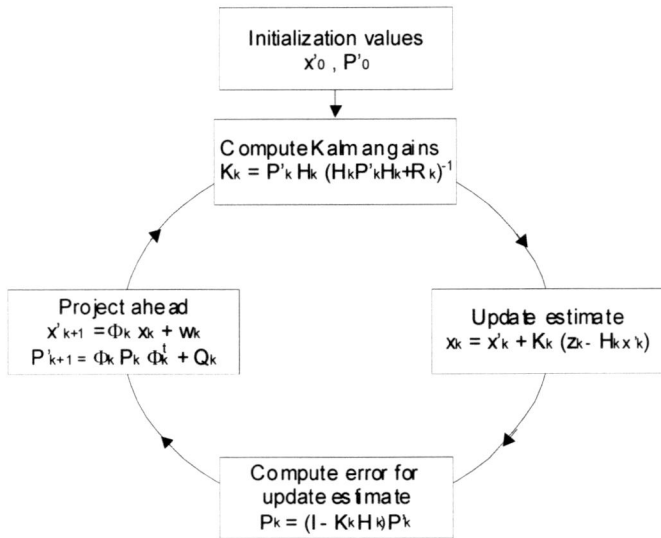

Figure 4.5. Recursive algorithm of a Kalman filter

After the initialization time of the filter, a new value of the state vector is obtained with each new sample of the input signal.

The detection properties of Kalman filtering and the accuracy in the estimation of the magnitude and duration of voltage events depend both on the model of the system used and on the magnitude, duration and point-on-wave where the voltage event begins.

References [13–15] discuss the Kalman filtering modelling issues and compare the speed of detection of voltage dips using linear Kalman filters of different order. References [16, 17] propose the use of a three 12–state Kalman filter (one per phase) for real-time detection and analysis of voltage events in power system. The fundamental component and odd harmonics from third to eleventh order are used to have an accurate representation of the voltage supply during the event. A new value of fundamental magnitude and phase angle of voltage supply is obtained with each new sample of the voltage waveform.

As an example of the performance of Kalman filters in the detection and analysis of voltage events, Figure 4.6 shows the waveform of a voltage dip measured in the low-voltage distribution network of our building and the fundamental component of voltage supply computed using a 12–state Kalman filter, with the fundamental and odd harmonic components from third to eleventh order of the voltage supply, as the state variables.

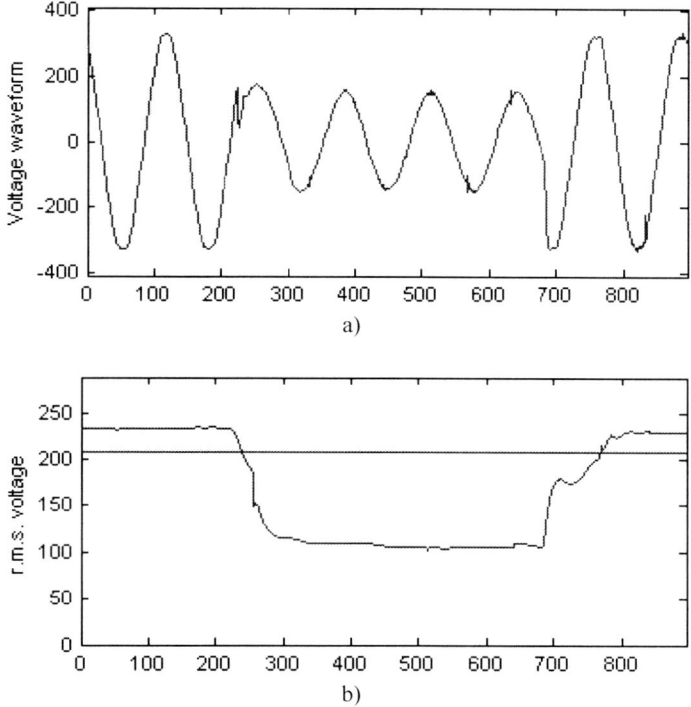

Figure 4.6. (a) Voltage waveform of a voltage dip measured in a low-voltage supply system, (b) magnitude of voltage supply computed using a 12-state Kalman filter model

The duration and the magnitude of the voltage dip shown in Figure 4.6 computed using the 12-state Kalman filter model is 63.4 ms and 49.1%, respectively.

As can be seen from the visual inspection of Figure 4.6, there seems to be a delay in the detection of the end of the voltage dip. In general, the detection of the end of a voltage event using Kalman filters could show worse performance than the detection of the beginning, mainly for short-duration or multiple-step voltage events, because, in such a case, the coefficients of the filter may not have converged to the new stationary values when the transition associated with the end of the voltage event arrives.

A possible solution to improve the performance of Kalman filtering in the detection and analysis of voltage events is the use of an extended Kalman filter to better estimate the non linear process associated with a voltage event. Reference [8] presents the performance of different extended Kalman filter models in the estimation of simulated and real voltage events measured in a low-voltage distribution system.

4.2.2.3 Wavelet Analysis

Wavelet analysis is a powerful signal-processing tool that is especially useful for the analysis of nonstationary signals. There is a lot of work in the technical literature dealing with the use and investigating the performance of wavelet-based algorithms for the analysis of time-varying disturbances in power systems [18–21].

Wavelets are short-duration oscillating waveforms with zero mean and fast decay to zero amplitude at both ends that are dilated and shifted to vary their time-frequency resolution. In wavelet analysis the wavelet function is compared with a section of the signal under study, obtaining a set of coefficients that represents how closely the wavelet function correlates with the signal in that section.

Analogously to the Fourier analysis, the discrete wavelet transform (DWT) is the digital representation of the continuous wavelet transform. The DWT decomposes a signal into different frequency components, but unlike the Fourier analysis, this decomposition provides a nonuniform division of the frequency domain instead of the uniform frequency decomposition of the DFT.

The DWT can be implemented using a multistage filter bank with the wavelet function as the lowpass filter (LP) and its dual as the highpass filter (HP), as is shown in Figure 4.7 for a three-level decomposition tree. Downsampling by two at the output of the low pass and high pass filters scales the wavelet by two for the next stage. $d(n)$ and $c(n)$ in Figure 4.7 are the outputs of the highpass and lowpass filters, respectively, and represent the detailed version of the high-frequency components of the signal and the approximation version of the low-frequency components.

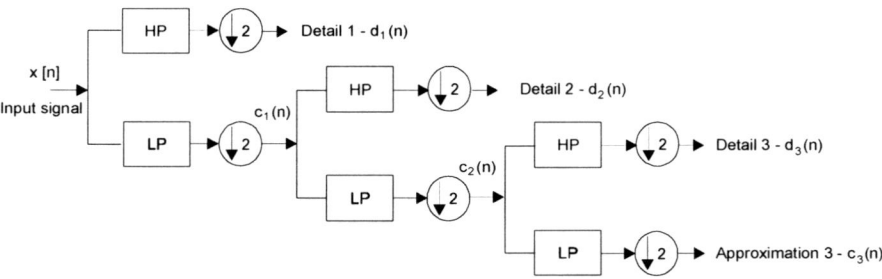

Figure 4.7. Multistage filter bank for a three-level decomposition tree

The detail and approximation coefficients of the DWT can be used to compute the rms magnitude of voltage and current waveforms and also of the output frequency bands of the wavelet decomposition tree, using the following equation [22]:

$$V_{rms} = \sqrt{\frac{1}{N}\sum_{i=1}^{N} v_i^2} = \sqrt{\frac{1}{N}\sum_{k} c_{j_0,k}^2 + \frac{1}{N}\sum_{j\geq j_0}\sum_{k} d_{j,k}^2} = \sqrt{V_{j_0}^2 + \sum_{j\geq j_0} V_j^2},$$

where v_i is the input signal (voltage samples), $c_{j,k}$ and $d_{j,k}$ are respectively, the approximation and detail coefficients, j is the decomposition level ($j = 1, 2, \ldots s$), with s the highest scaling level such that $2^s = N$, (N = number of samples taken in the sampling window) and k the sampling point. j_0 is the lowest level of the signal and includes the fundamental frequency component. V_{j0} is the rms value of the lowest-frequency subband and V_j is the set of rms values of each frequency subband higher than or equal to the scaling level j_0.

One of the key factors in the application of DWT is the selection of the most appropriate wavelet function depending on the type of disturbance to be detected and analysed. As a general rule shorter wavelets are best suited for detecting fast transients, while slow transients are better detected using longer wavelets [18].

References [23, 24] study the performance of different wavelet functions for the analysis of voltage dips, recommending the use of the Daubechies with six coefficients (db6) as the mother wavelet and the detail coefficients of the first decomposition level to detect the beginning and the end of a voltage dip. As can be seen in these references, these coefficients are insensitive to a steady-state signal but show a high variation in magnitude associated with the high-frequency components present at the beginning and the end of a voltage dip.

The use of these coefficients for detection of the beginning and the end of a voltage event requires the definition of an adequate threshold that should be surpassed only in the case of the transitions associated with the beginning and the end of a voltage event. Reference [23] proposes the use of $\mu \pm 3\sigma$ as an adequate threshold for detection of voltage dips in a transmission system, where μ is the mean value of the detail coefficients before the voltage dip and σ is the standard deviation.

As an example of the performance of wavelet analysis in the detection and analysis of voltage events, Figure 4.8 shows the results obtained for the same voltage dip in Figure 4.4. The sampling frequency used in the simulations is 6.4 kHz (128 samples/cycle in a 50Hz power system).

DWT is applied to the voltage samples using db6 as the mother wavelet function. The detail coefficients of the first decomposition level of the signal are used to determine the beginning, the end and the duration of the voltage dip and also to compute the rms magnitude of the voltage supply during the event.

The number of detail coefficients in Figure 4.8b is half the voltage samples due to the downsampling by two associated with the DWT implementation. As can be seen, these coefficients clearly point out the beginning and the end of the voltage dip. The duration of the voltage dip, computed as the time difference between the peaks of the detailed coefficients, is exactly 21.25 ms and the magnitude of the voltage event is exactly 60%.

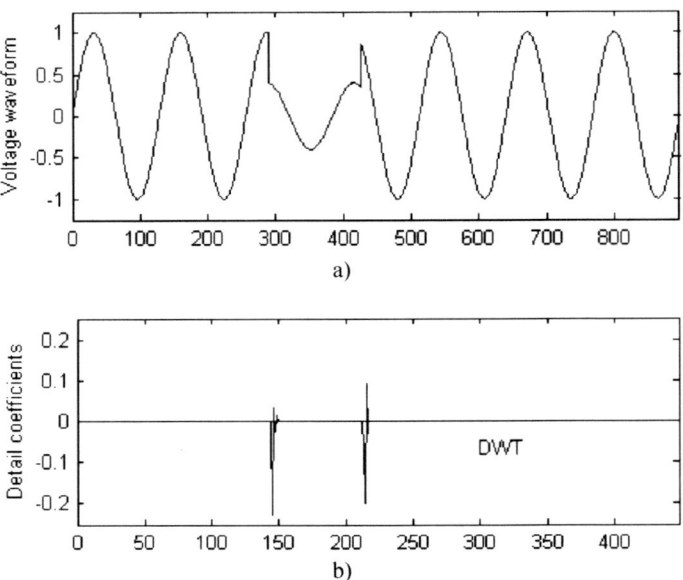

Figure 4.8. (a) Voltage waveform with a voltage dip, (b) magnitude of detail coefficients of the first-level decomposition of DWT

Unfortunately, these coefficients are so sensitive that other very short, high-frequency disturbances in voltage waveform, produce similar effects to the voltage events, making the use of DWT very difficult in an automatic voltage event analysis system [24].

Figure 4.9 shows the results obtained in the detection and analysis of a real voltage dip measured in the low-voltage distribution network in a building located in our campus. Figure 4.9a shows the waveform of the voltage supply prior to, during and after the voltage dip and Figure 4.9b shows the magnitude of the detail coefficients of the first level of the wavelet decomposition tree.

The beginning of the voltage dip can be detected from the peak values of these coefficients, although owing to the high-frequency noise present in the voltage supply during and after the voltage dip, the peak values are beyond the threshold level for voltage-dip detection ($\mu \pm 3\sigma$ as proposed in [23]) several times before and after the detection of the beginning and the end of the voltage dip.

The direct use of these peak values to signal the beginning and the end of the voltage dip without any other consideration may lead to erroneous results in the estimation of both the time-related parameters and the magnitude of the voltage dip.

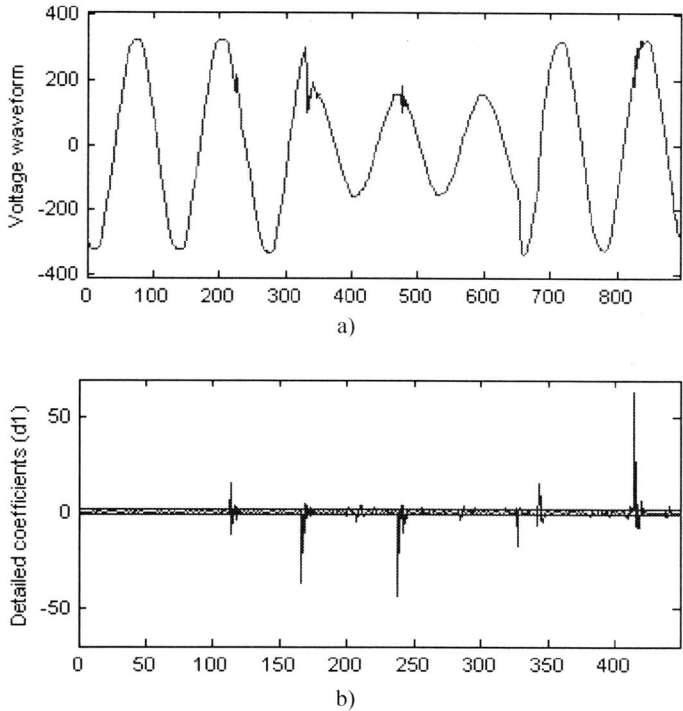

Figure 4.9. (a) Voltage waveform of a voltage dip, (b) magnitude of the detail coefficients of the first decomposition level

4.2.2.4 Hybrid Methods

Bearing in mind that from the results reported in this section, wavelet analysis shows the best performance in the detection and in the estimation of the time-related parameters of a voltage event, and Kalman filtering shows the best performance in the estimation of the magnitude and phase angle of a voltage supply during the event, a combination of wavelet analysis – Kalman filtering could be the best solution for the optimal characterization of voltage events in an automatic on line system.

References [8], [24] present the performance and the implementation issues of an automated real-time system for the detection and best estimation of voltage events in power systems. Wavelet analysis is used for the detection and estimation of the time-related parameters of the voltage event and Kalman filtering enables the confirmation of the beginning and the end of the voltage event, avoiding the erroneous detections that can be produced due to the very high sensitivity of the wavelet analysis, and on the other hand, the estimation of the magnitude and phase angle of the voltage supply during the event.

4.3 Effects of Voltage Events on Equipment

As was previously stated, voltage events may lead to malfunction or breakdown of equipment. Different types and different brands of equipment may have different sensitivity to voltage events. Among the most sensitive equipment to voltage events are personal computers, programmable logic controllers, adjustable-speed drives and discharge lamps [25, 26], furthermore, electrical and electronic equipment is becoming more sensitive to voltage events as its complexity increases.

Knowledge about the voltage tolerance of equipment is of the utmost importance in assessing its immunity to voltage events and in deciding the actions to be taken in order to mitigate their effects.

The recommended way of presenting the performance of equipment with respect to voltage variations is the use of voltage-tolerance curves. These curves, also known as power-acceptability curves, are plots of bus–voltage deviation versus time duration. They divide the voltage-deviation-time-duration plane into two regions: "acceptable power" and "unacceptable power".

The first voltage-tolerance curve, known as the CBEMA curve, developed by the Computer Business Equipment Manufacturers Association and subsequently taken up in IEEE Standard 446 [27], was applied primarily to data-processing equipment and shows that the sensitivity of equipment is very dependent on the duration of the voltage variation. Recently a revised CBEMA curve has been introduced by the Information Technology Industry Council. This curve, shown in Figure 4.10 and referred to as the ITIC curve [28], describes the AC input voltage envelope that typically can be tolerated (no interruption in function) by most information-technology equipment.

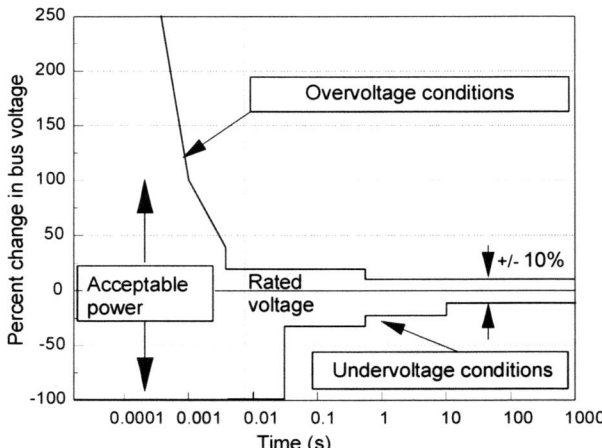

Figure 4.10. The ITIC curve

Although the ITIC curve is defined only for single-phase, 120volt, 60Hz, information-technology equipment, it has become a standard for measuring the performance of many other devices.

At present, IEC standard 61000-4-11 [29] is the power-quality standard that describes how to obtain the voltage tolerance of equipment, although the standard does not set any voltage-tolerance requirement. This standard defines the immunity test methods and range of preferred test levels and durations of voltage dips and short interruptions for which low-voltage electrical and electronic equipment has to be tested (Table 4.5).

This standard also describes two possible test configurations for mains-supply simulation, one using two variable transformers and two switches to generate voltage dips of different magnitude and duration or short interruptions of different duration, and a second test configuration using waveform generators and power amplifiers. This second configuration allows testing of equipment in the context of frequency variations and harmonics.

Table 4.5. Preferred test levels and durations of voltage dips to test the voltage tolerance of equipment

Test-level magnitude	Duration (in periods)
0%	0.5
	1
40%	5
	10
70%	25
	50

4.3.1 Voltage Tolerance of Equipment

The equipment investigated and reported in this section consists of personal computers and the results obtained are also valid for other information-technology equipment or consumer-electronic equipment that uses the same type of power supply. The measurement of the voltage tolerance of other equipment, such as programmable logic controllers, adjustable speed AC and DC drives, motor, lightning devices and contactors can be seen in references [30–40]. IEEE Standard 1346-1998 reports the voltage-tolerance ranges of several types of equipment currently in use [41].

Personal computers use switched-mode power supplies made up basically of a full-wave rectifier, a smoothing capacitor and a voltage controller. If the AC voltage drops, the voltage on the DC side of the rectifier also drops. If the non regulated DC voltage becomes too low the regulated DC voltage will also start to drop and finally the digital electronics in the personal computer will fail. The permissible duration of a voltage dip or an interruption depends on the size of the smoothing capacitor used in the switched-mode power supply of the computer.

Figure 4.11 shows the typical voltage-tolerance curve of a personal computer. It can be characterised by a pair of data (V_{min}, t_{max}). V_{min} is the minimum steady-state input voltage for which the personal computer can operate correctly and t_{max} is the maximum duration of an interruption (zero voltage) for which the computer still operate correctly. If the voltage magnitude during the voltage dip falls below the minimum for steady-state operation, V_{min}, and exceeds the permissible duration t_{max}, a malfunction will occur.

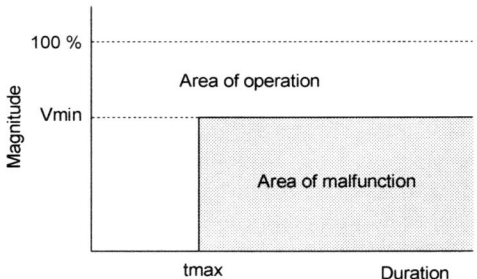

Figure 4.11. Voltage-tolerance curve of a personal computer

Different studies have been done on the voltage tolerance of personal computers, determining in each case the tolerance for zero voltage and the minimum steady-state magnitude for continuous operation of the computer. The results obtained show a wide range of tolerance to voltage fluctuation depending on the type of power supply employed in the computer. However, new studies are necessary because the use of new technologies is making equipment more sensitive to voltage dips.

An interesting approach can be seen in reference [42], where is shown an analytical process to develop voltage-tolerance curves for different types of loads.

4.3.2 Measurement System

An experimental test setup with a programmable AC power source has been used to generate voltage dips and short interruptions of different magnitude and duration under different supply conditions in order to investigate the behaviour of different types of equipment.

Figure 4.12 shows the structure of the experimental test setup. The AC power source is used to supply the equipment under test and produces arbitrary waveforms with programmable amplitude, frequency and waveshape. A personal computer controls the programmable AC power source through an HP-IB interface. The measurement system uses a data-acquisition board and voltage and current transducers to measure the voltage and currents in the device under test.

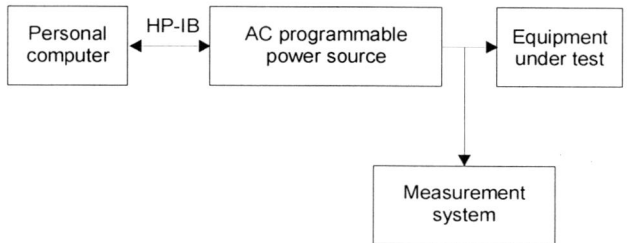

Figure 4.12. Experimental test setup for the determination of voltage tolerance of equipment

Software applications have been developed to control the AC programmable power source and to control the measurement system. The input AC voltage and the regulated DC voltage at the output of the power supply of the computer have been measured to characterise the response of the personal computer to voltage dips and short interruptions.

The methodology used in the test is the following:

- The AC power source is programmed to generate a pure sinusoidal 230V, 50Hz voltage waveform.
- Different interruptions (zero volts) in voltage supply are generated to determine t_{max} (the maximum duration of an interruption for which the personal computer would still operate correctly) starting from 0 ms in steps of 1 ms. The personal computer was considered to malfunction when the computer performed a restart of the operating system.
- Once t_{max} has been determined the magnitude of the voltage dip was incremented to determine V_{min} (minimum steady-state voltage for normal operation).

Figure 4.13 and Table 4.6 show the results obtained for two different personal computers in our laboratory.

Table 4.6. Voltage tolerance of two different personal computers

Equipment under test	t_{max} (ms)	V_{min} (%)
Computer 1	386	48.48
Computer 2	156	45.21

According to the results obtained, if the voltage magnitude during the voltage dip falls below the minimum for steady-state operation of the computers (48.48% of the nominal voltage for "Computer 1" and 45.21% for "Computer 2"), and exceeds the permissible duration of an interruption for which either of the two computers still operates correctly (386 ms for "Computer 1" and 156 ms for "Computer 2", respectively), a malfunction will occur.

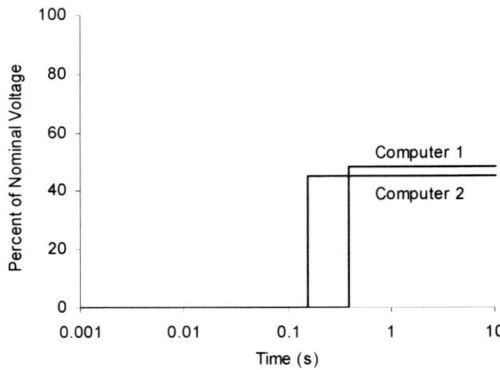

Figure 4.13. Voltage-tolerance curves of two different personal computers

4.3.3 Effect of Harmonic Distortion

Taking into account that the use of a test setup with a programmable AC power source enables testing of equipment in the context of harmonics, the purpose of this section is to investigate the effect of harmonic distortion in voltage supply on the voltage tolerance of equipment.

In recent papers [43, 44] the authors have shown that the voltage tolerance of personal computers depends on the harmonic distortion in voltage supply, with equipment being more sensitive to low-order harmonics and lower voltage peak magnitudes.

Harmonic distortion modifies the waveshape of the voltage supply, either flattening a perfect sinusoidal waveform or producing higher peak values than the pure sinusoidal, depending on the magnitude, phase difference and the harmonic order present in the voltage supply. To take this effect into account, a voltage crest factor (VCF) is defined as the relationship between the peak voltage and the rms voltage.

The effect of harmonic distortion on the VCF is different depending on the harmonic order and the phase difference with respect to the fundamental component. Thus, VCF is maximum when the phase difference is 180° for harmonics of third, seventh, eleventh, *etc.*, order and minimum when the phase difference is 0°. On the other hand, harmonics of fifth, ninth, thirteenth, *etc.*, order produce a maximum VCF in the voltage waveform when they are inphase with the fundamental component and minimum VCF when they are in opposite phase.

Lower peak values in voltage supply (and lower VCF) decrease the peak voltage in the smoothing capacitor in the nonregulated DC side of the power supply of a computer, making the computer more sensitive to voltage variations.

The experimental test setup of Figure 4.12 has also been used to generate voltage dips in a distorted voltage supply with harmonics of different order and different magnitude and phase difference with respect to the fundamental component, in order to check how the voltage tolerance of equipment depends on the harmonic distortion.

Table 4.7 and Figure 4.14 show the limits of the voltage tolerance measured for Computer 1, previously used in the last section, with a pure sinusoidal voltage supply. Curve "a" in the graph was obtained using a pure 230V, 50Hz sinusoidal voltage supply, whereas curves "b" an "c" were obtained when 5% of the 3rd order harmonic in phase with the fundamental component and 6% of the 5th order harmonic also in phase with the fundamental component, respectively, were present in the voltage supply before and during the voltage event. The individual harmonic voltages used in the tests are the compatibility levels defined in IEC standard 61000-2-2 for low-voltage networks [45].

Table 4.7. Voltage tolerance of Computer 1 in different supply conditions

Case	Voltage supply	Voltage peak magnitude	Voltage tolerance t_{max} V_{min}
a	50Hz pure sinusoidal	325.27 volts	386 ms, 48.48%
b	+ 5% 3rd harmonic	308.74 volts	337 ms, 55.00%
c	+ 6% 5th harmonic	344.06 volts	436 ms, 46.30%

According to the results obtained, the permitted duration of a short interruption in voltage supply for which Computer 1 still operated correctly varies from 337 ms to 436 ms, depending on the total harmonic distortion in voltage supply. On the other hand, the minimum steady-state voltage varies from 46.30% to 55.00% of the nominal voltage.

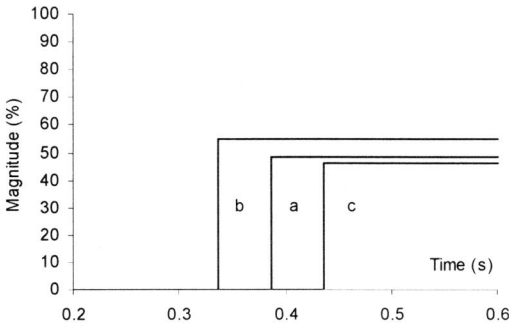

Figure 4.14. Voltage tolerance of a personal computer under different supply conditions

As can be seen from the results, the computer used in the test has a worse voltage-tolerance curve in case b (minimum peak voltage in voltage supply) than in case a (50Hz pure sinusoidal supply), whereas it has better voltage tolerance in case c (maximum peak voltage in voltage supply). Other phase differences between harmonics and higher-order harmonics have less effect on the voltage tolerance of the computer.

4.4 Voltage-event Surveys

The in depth knowledge of voltage events in a specific distribution network and the voltage tolerance of the equipment connected thereto, are both absolutely necessary to evaluate the effect of voltage variations on equipment and to design strategies to protect equipment from these power-quality disturbances.

This section presents the results of six months monitoring of the three-phase voltage supply in the low-voltage distribution network of a building on our campus. The statistical analysis of the voltage events recorded and the effect of these events on equipment are reported in this study [46].

A data-acquisition system based on the TMS320C31 digital signal processor, with three ADC input channels, 16 bits, 200 kHz throughput and three Hall-effect transducers has been developed to detect voltage events and to estimate the magnitude of the three-phase voltage supply during the event (Figure 4.15). The DAC converter is used to generate an alarm signal in the case of detection of a voltage event. The sampling rate used is 3.2 kHz/channel.

The system developed has been designed to detect and analyse voltage events that present a change in voltage supply greater than ±10% of the nominal voltage.

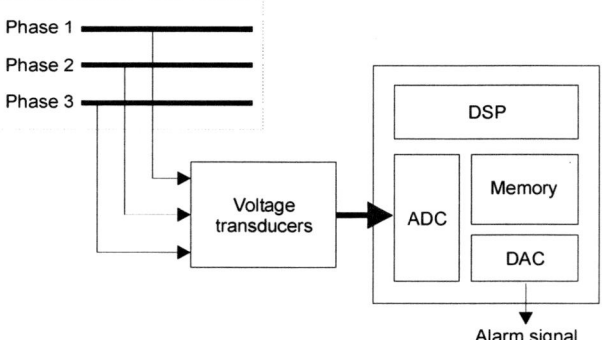

Figure 4.15. Hardware structure of the system for voltage-event detection and analysis

The voltage-event detection method used compares the magnitude of the fundamental component of voltage supply, as it is obtained in real time by three Kalman filters, one per phase, with the threshold of the accepted voltage supply variation as defined in the EN 50160 standard.

If the threshold of acceptable voltage-supply variation is surpassed for three consecutive voltage samples, an event record is generated. The event record contains the waveforms and the magnitude and phase angle of the three-phase voltage supply two cycles before the event, during the event and two cycles after the end of the event. The event finishes when the magnitude of the voltage supply returns to within the limits of acceptable voltage-supply variation. No hysteresis voltage above or below the event threshold is considered.

The monitoring was carried out in periods of approximately ten days each, distributed uniformly throughout the year. The building is supplied by a 12kV/380 V three-phase distribution transformer. A large percentage of the load consists of lighting, computers and other information-technology equipment such as printers, scanners, *etc.* There are also several laboratories in the building with rotating machines and electromechanical apparatus.

A voltage event is classified as a single event, irrespective of the shape and of the number of phases affected. In the case of a multiphase voltage event, the magnitude of the event is considered to be that of the most affected phase.

During the time of monitoring a total of 86 voltage events were recorded by the system. Of the events detected 45 (52.32%) were classified as voltage dips. Of this number of voltage dips, 18 (40%) produced an overvoltage in at least one of the phases not affected by the voltage dip, these voltage events were considered in the two categories simultaneously, as a voltage dip and as an overvoltage. Twenty-eight of the events recorded (32.55%) were classified as short interruptions and only one event was an overvoltage in the three phases simultaneously. Six events (6.97%) were unclassified and the rest of events (6.97%) correspond to short-duration transients (less than 10 ms) that were not included in any of the other three categories.

Table 4.8 shows the statistical classification of the voltage dips detected during the time of monitoring. The table is arranged in the same way as the UNIPEDE survey of voltage dips [47]. The evaluation of a voltage dip is defined by its duration and its depth.

Table 4.8. Statistical classification of voltage dips

Depth/ duration	10–100 (ms)	0.1–0.5 (s)	0.5–1 (s)	1–3 (s)	3–20 (s)	20–60 (s)	60–180 (s)
10%–15%	5	2	0	0	0	0	0
15%–30%	12	1	0	0	0	0	0
30%–60%	14	0	0	0	0	0	0
60%–99%	9	2	0	0	0	0	0

Table 4.9 summarizes the number and duration of the interruptions detected and recorded by the system. According to standard EN 50160, if the voltage supply value drops below 1% of the nominal voltage, the event is considered a short interruption. Of the short interruptions detected, 21 (75%) were less than 1 s and

the rest were between 1 s and 3 min. No interruptions longer than 3 min were detected during the time of observation.

Table 4.9. Statistical classification of short interruptions

Duration of interruptions	< 1 s	1 s – 3 min	> 3 min
Number of interruptions	21	7	0

Finally, Table 4.10 shows the evaluation of the overvoltages detected during the period of observation. Eighteen of the overvoltages (94.73%) were associated with a voltage dip in another voltage phase. The duration of an overvoltage, as is represented in Table 4.10, corresponds to the period during which the values measured remain more than 110% of the nominal voltage. The magnitude of an overvoltage is defined as the ratio expressed in % between the maximum value during the overvoltage and the nominal voltage. As can be seen in Table 4.10, no overvoltages longer than 1 s were detected during the time of monitoring. All the overvoltages detected were in the range of 15 ms to 30 ms.

Table 4.10. Statistical classification of overvoltages

Magnitude/ duration	< 1 s	1 s – 1 min	> 1 min
110%–120%	0	0	0
120%–140%	0	0	0
140%–160%	0	0	0
160%–200%	3	0	0
> 200 %	16	0	0

From the results reported it can be concluded that, as expected, overvoltages in voltage supply were much less common than voltage dips or short interruptions.

Finally, in order to evaluate the effect of the voltage events detected in the low-voltage distribution network monitored, Figure 4.16 shows a scatter plot of the voltage magnitude in % of the nominal voltage versus the duration time in seconds of all the voltage events detected, with the ITIC curve superimposed. This figure represents the sensitivity of standard information-technology equipment to the voltage events recorded. As can be seen in Figure 4.16, forty-four events were below the ITIC curve and nineteen events were above the curve.

On the other hand, Figure 4.17 shows the magnitude and duration of the voltage events recorded, superimposed on the voltage-tolerance curves of two personal computers in our laboratory. If the voltage magnitude during the voltage event falls below the minimum for steady-state operation of the computers (48.48% of the nominal voltage for "Computer 1" and 45.21% for "Computer 2" in Figure 4.17), and exceeds the permissible duration of an interruption for which either of the two computers still operates correctly (386 ms for "Computer 1" and 156 ms for "Computer 2", respectively), a malfunction will occur.

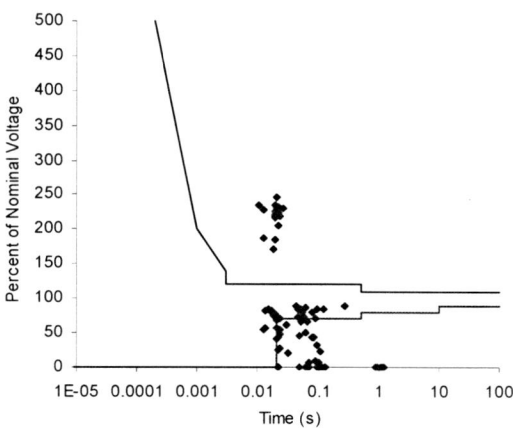

Figure 4.16. Scatter diagram of voltage events with the ITIC curve superimposed

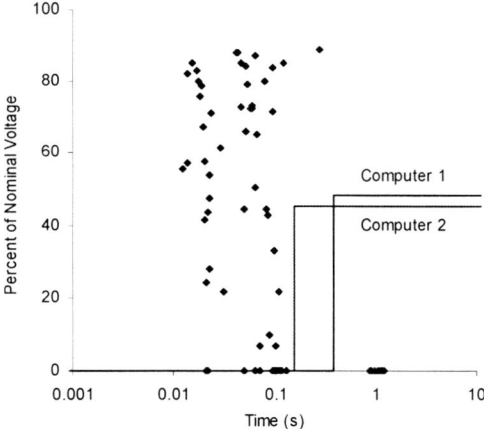

Figure 4.17. Scatter diagram of voltage events superimposed with the voltage-tolerance curve of two personal computers of our laboratory

From the results reported, it can be concluded that both "Computer 1" and "Computer 2" will fail on nine occasions, as a result of the voltage events undergone in the low-voltage distribution network during the time of monitoring.

References

[1] European Standard EN 50160: Voltage characteristics of electricity supplied by public distribution systems. CENELEC, 1994.
[2] IEEE Recommended Practice for Monitoring Electric Power Quality, IEEE Standard 1159 – 1995, New York: IEEE, 1995.
[3] Kagan N, Ferrari EL, Matsuo NM, Duarte SX, Sanommiya A, Cavaretti JL, *et al*. Influence of rms variation measurement protocols on electrical system performance indices for voltage sags and swells. 9th International Conference on Harmonics and Quality of Power 2000;3:790–795.
[4] Styvaktakis E, Bollen MHJ, Gu IYH. Automatic classification of power system events using rms voltage measurements. IEEE PES Summer Meeting 2002;2:824–829.
[5] Albu M, Heydt GT. On the use of rms value in power quality assessment. IEEE Transactions on Power Delivery 2003;18:1586–1587.
[6] International Electrotechnical Commission. IEC 61000-4-30. Electromagnetic compatibility (EMC) - Part 4: Testing and measurement techniques. Section 30: Power quality measurement methods, 2003.
[7] Barros J, Pérez E. Limitations in the use of rms value in power quality analysis. IEEE Instrumentation and Measurement Technology Conference, IMTC2006;2261–2264.
[8] Pérez E. New method for real-time detection and analysis of events in voltage supply using a combined wavelet - extended Kalman filtering model. PhD Thesis, University of Cantabria, Santander, Spain, 2006 (in Spanish).
[9] Heydt GT, Fjeld PS, Liu CC, Pierce D, Tu L, Hensley G. Applications of the windowed FFT to electric power quality assessment. IEEE Transactions on Power Delivery 1999;14:1411–1416.
[10] Styvaktakis E, Bollen MHJ, Gu IYH. Classification of power system events: voltage dips. 9th International Conference on Harmonics and Quality of Power 2000;2:745–750.
[11] Gu IYH, Bollen MHJ. Time-frequency and time-scale domain analysis of voltage disturbances. IEEE Transactions on Power Delivery 2000;15:1279–1284.
[12] Jacobsen E, Lyons R. The sliding DFT. IEEE Signal Processing Magazine 2003;20:74–80.
[13] Styvaktakis E, Gu IYH, Bollen MHJ. Voltage dip detection and power system transients. IEEE PES Summer Meeting 2001;683–688.
[14] Styvaktakis E, Bollen MHJ, Gu IYH. Expert system for voltage dip classification and analysis. IEEE PES Summer Meeting 2001;671–676.
[15] Styvaktakis E, Bollen MHJ, Gu IYH. Expert system for classification and analysis of power system events. IEEE Transactions on Power Delivery 2002;17:423–428.
[16] Barros J, Pérez E, Pigazo A. Real time system for identification of power quality disturbances. 17th International Conference and Exhibition on Electricity Distribution (CIRED 2003); Paper 29, Session 2.
[17] Barros J, Pérez E. Automatic detection and analysis of voltage events in power systems. IEEE Transactions on Instrumentation and Measurement 2006;55;1487–1493.
[18] Santoso S, Powers EJ, Grady WM. Electric power quality disturbance detection using wavelet transform analysis. IEEE – SP International Symposium on Time-Frequency and Time Scale Analysis, 1994;166–169.
[19] Santoso S, Powers EJ, Grady WM, Hofmann P. Power quality assessment via wavelet transform analysis. IEEE Transactions on Power Delivery 1996;11:924–930.
[20] Robertson DC, Camps OI, Mayer JS, Gish WB. Wavelets and electromagnetic power system transients. IEEE Transactions on Power Delivery 1996;11:1050–1058.
[21] Gaouda AM, Salama MM, Sultan MR, Chikhani AY. Power quality detection and classification using wavelet-multiresolution signal decomposition. IEEE Transactions

on Power Delivery 1999;14:1469–1476.
[22] Yoon W, Devaney M. Power measurement using the wavelet transform. IEEE Transactions on Power Delivery 1998;47:1205–1209.
[23] Parsons AC, Grady WM, Powers EJ. A wavelet-based procedure for automatically determining the beginning and end of transmission system voltage sags. IEEE PES Winter Meeting 1999;2:1310–1315.
[24] Barros J, Pérez E. A combined wavelet - Kalman filtering scheme for automatic detection and analysis of voltage dips in power systems. PowerTech2005;Paper 141:1–5.
[25] Bollen MHJ. Understanding Power Quality Problems. IEEE Press, 2000. Piscataway, NJ. USA
[26] McGranaghan MF, Mueller DR, Samotyj MJ. Voltage sags in industrial systems. IEEE Transactions on Industry Applications 1993;29:397–402.
[27] IEEE Standard 446. Recommended practice for emergency and standby power for industrial and commercial applications, IEEE, 1995. New York, USA.
[28] ITI (CBEMA) curve application note, Information Technology Industry Council (ITIC), Washington, USA, http://www.itic.org/technical/ iticurv.pdf
[29] International Electrotechnical Commission. IEC Standard. 61000-4-11. Electromagnetic compatibility (EMC), Part 4: Testing and measurement techniques. Section 11: Voltage dips, short interruptions and voltage variations immunity tets, 1994.
[30] Yalcinkaya G, Bollen MHJ, Crossley PA, Characterization of voltage sags in industrial distribution systems. IEEE Transactions on Industry Applications 1998;34:682–688.
[31] Sannino A, Bollen MHJ, Svensson J. Voltage Tolerance testing of three-phase voltage source converters. IEEE Transactions on Power Delivery 2005;20:1633–1639.
[32] Bollen MHJ, Zhang LD. Analysis of voltage tolerance of ac adjustable-speed drives for three-phase balanced and unbalanced sags. IEEE Transactions on Industry Application 2000;36:904–910.
[33] Pedra J, Córcoles F, Suelves FJ. Effects of balanced and unbalanced voltage sags on VSI-fed adjustable-speed drives. IEEE Transactions on Power Delivery 2005;20:224–233.
[34] Djokic SZ, Stockman K, Milanovic JV, Desmet JM, Belmans R. Sensitivity of ac adjustable speed drives to voltage sags and short interruptions. IEEE Transactions on Power Delivery 2005;20:494–505.
[35] Gómez JC, Morcos MM, Reineri CA, Campetelli GN. Behavior of induction motor due to voltage sags and short interruptions. IEEE Transactions on Power Delivery 2002;17:434–440.
[36] Das JC. Effects of momentary voltage dips on the operation of induction and synchronous motors. IEEE Transactions on Industry Applications 1990;26:711–718.
[37] Guasch L, Córcoles F, Pedra J. Effects of symmetrical and unsymmetrical voltage sags on induction machines. IEEE Transactions on Power Delivery 2004;19:774–782.
[38] Dorr DS, Mansoor A, Morinec AG, Worley JC. Effects of power line voltage variations on different types of 400-W high pressure sodium ballasts. IEEE Transactions on Industry Applications 1997;33:472–476.
[39] Djokic SZ, Milanovic JV, Kirschen DS. Sensitivity of ac coil contactors to voltage sags, short interruptions and undervoltage transients. IEEE Transactions on Power Delivery 2004;19:1299–1307.
[40] Pedra J, Córcoles F, Sainz L. Study of ac contactors during voltage sags. 10th International Conference on Harmonics and Quality of Power 2002;2:565–570.
[41] IEEE Standard 1346, Recommended practice for evaluating electric power system compatibility with electronic process equipment, IEEE, 1998. New York, USA.

[42] Kyei J, Ayyanar R, Heydt GT, Thallam R, Blevins J. The design of power acceptability curves. IEEE Transactions on Power Delivery 2002;17:828–833.
[43] Barros J, Diego RI. Effects of nonsinusoidal supply on voltage tolerance of equipment. IEEE Power Engineering Review 2002;22:46–47.
[44] Barros J, Diego RI, Pérez E. Measuring voltage tolerance of equipment under practical conditions. Second IASTED International Conference on Power and Energy Systems 2002;289–293.
[45] International Electrotechnical Commission. IEC Standard 61000-2-2: Electromagnetic Compatibility (EMC), Part 2-2: Environment - Compatibility levels for low-frequency conducted disturbances and signalling in public low-voltage power supply systems. 2nd Edition.
[46] Barros J, Pérez E. Measurement and analysis of voltage events in a low-voltage distribution network. 12th IEEE Mediterranean Electrotechnical Conference 2004;1083–1086.
[47] UNIPEDE Report 91. Voltage Dips and Short Interruptions in Electricity Supply Systems.

5

Transient Mitigation Methods on ASDs[*]

José Luis Durán-Gómez[1] and Prasad N. Enjeti[2]

[1]División de Estudios de Posgrado e Investigación
Instituto Tecnológico de Chihuahua
Ave. Tecnológico No. 2909
Chihuahua, Chih. México 31310
Email: jlduran@itchihuahua.edu.mx

[2]Deparment of Electrical and Computer Engineering
Texas A&M University
214 Zachry Engineering Center
College Station, TX, USA 77843-3128
Email: enjeti@ece.tamu.edu

5.1 Introduction

The application of adjustable-speed drives (ASDs) in commercial and industrial facilities is increasing due to their improved efficiency, energy savings and electronic process control. However, modern ASD equipment is more susceptible to utility transients such as: sags [1–3], swells, short-term power interruptions and utility capacitor-switching transient events, [1, 3–5]. Voltage sags generally cause an undervoltage trip in an ASD [1]. A short-term power interruption for a few cycles results in rapid decrease in ASD the DC-link voltage and requires an appropriate energy storage for ride-through [3]. On the other hand, a voltage swell or a capacitor-switching transient (CST), results in a momentary increase in the DC-link voltage (>1.3 pu) and most often results in an overvoltage trip [5, 7–10]. The utility CST is the most common cause of transient overvoltages, second only to lightning in frequency of occurrence in most systems [11, 12]. In ASD equipment, the DC-link capacitor, C_d, along with the DC-link inductance, L_d, along with utility line inductance, L_s, form an LC resonant circuit. During a CST event, the LC resonant circuit is excited and a current surge is produced, resulting in a rise of the DC-link voltage (>1.3 pu). In order to protect the IGBTs and diodes, overvoltage trip generally occurs, resulting in a nuisance tripping of the ASD.

---

[*]© 2002 IEEE. Reprinted, with permission, from IEEE Transactions on Power Electronics, vol. 17, no. 5, pp. 799-806, Sept. 2002.

Figure 5.1 illustrates a typical CST event and the momentary rise in the ASD DC-link voltage. Figure 5.1 shows a line-to-line voltage, $v_{ca,sec}$, in the secondary side of a distribution substation, as shown in Figure 5.2. In Figure 5.1, the term $v_{DC,link,1}$ is also shown which represents the ASD DC-link voltage on ASD1 in a typical power system (Figure 5.2).

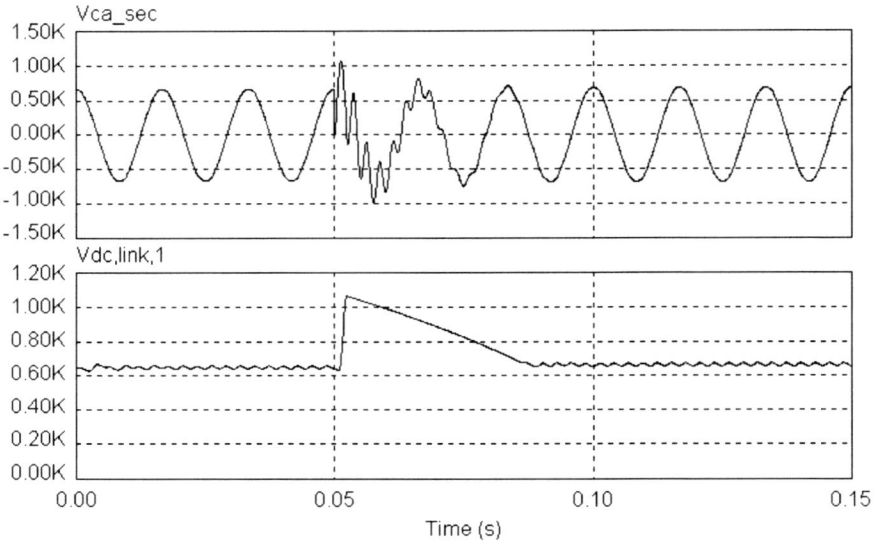

Figure 5.1. A typical capacitor-switching transient and the overvoltage occurrence in the ASD DC-link

Power-quality concerns due to CST events have been widely recognized in the literature [2, 5–8, 13]. Common mitigation methods to prevent nuisance tripping of ASDs so far suggested include [3–6, 10, 11, 13]:

i) Installing of 3–5% of additional line reactance in front of the ASD. This approach alters the LC resonance and reduces the overvoltage. However, additional inductance increases the susceptibility of ASDs tripping for common voltage sags [1].

ii) Suggested mitigation methods that can be implemented by the utility include [6, 9, 11, 13–15]: employ preinsertion resistors/inductors during capacitor switching [6], [11]; employ high-speed/static circuit breakers with zero-voltage closing control [6, 11, 13, 15]. These solutions have to be implemented by the electric utility and they are expensive.

In response to these concerns, this chapter explores new approaches to mitigate nuisance tripping of PWM ASDs due to utility capacitor-switching transient (CST) events. Modifications to ASD hardware are suggested to electronically damp the CST event, thus avoiding overvoltage in the DC-link and preventing nuisance tripping of ASDs. In this chapter an approach will be discussed that uses the soft-charge resistor available in most ASDs. The soft-charge resistor is momentarily introduced in the series path of the power flow to effectively damping the CST. It will be discussed that the soft-charge resistor can be introduced along with a DC-link (reactor), L_d, available in most ASD equipment. Further, the damping effect is electronically adjustable.

5.2 Transient Analysis

5.2.1 Analysis of CST Events

Utilities often use capacitor banks to maintain the distribution voltage level under varying load demands. The utility CST event is a relatively common power-system phenomena. Figure 5.2 illustrates an example of a one-line diagram of a typical utility CST event in a power-distribution system.

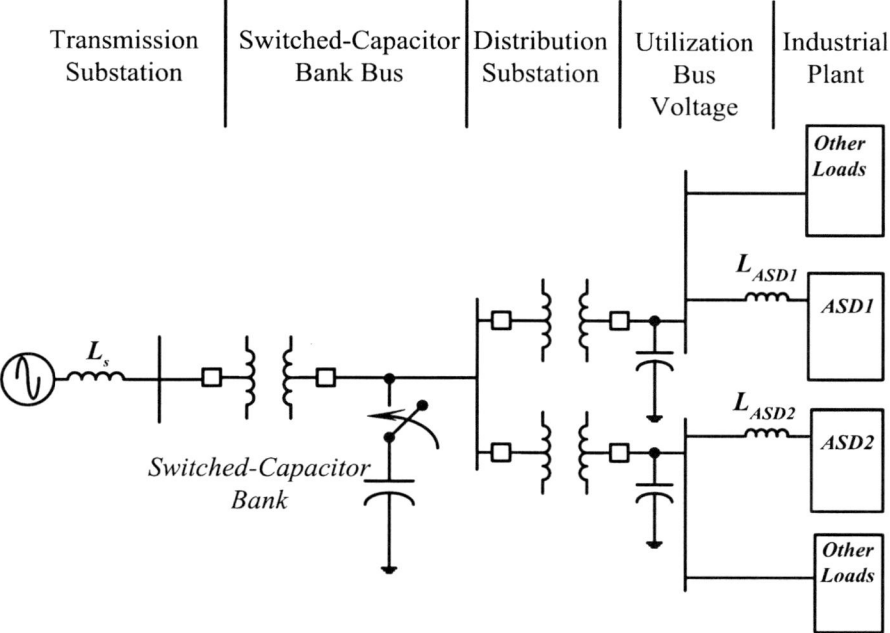

Figure 5.2. Example of one-line diagram of a typical power system

5.2.1.1 Analysis

In order to evaluate the effect of utility CST on ASDs, a simplified representation and an equivalent circuit of the power system is shown in Figure 5.3.

Figure 5.3a illustrates the simplified representation for a CST event in a typical power system. The analysis can be simplified by modeling the power system as an LC circuit, (Figure 5.3b), for the transient response [16].

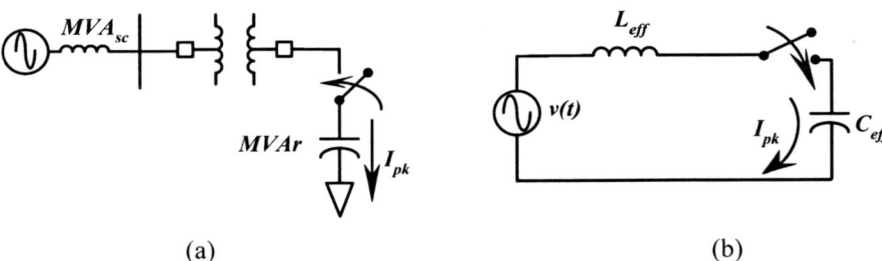

Figure 5.3. Equivalent circuit of the power system for the capacitor-switching transient event, (a) simplified representation, (b) equivalent circuit

From Figure 5.3b the following differential equation can be written,

$$L_{eff}\frac{d}{dt}i(t) + \frac{1}{C_{eff}}\int i(t)dt = v(t), \tag{5.1}$$

where L_{eff} is the effective line inductance and C_{eff} is the effective capacitance being switched. Defining, the characteristic impedance, Z_0,

$$Z_0 = \sqrt{\frac{L_{eff}}{C_{eff}}}, \tag{5.2}$$

and the resonant frequency,

$$\omega_n = \frac{1}{\sqrt{L_{eff}C_{eff}}}, \quad n = \frac{\omega_n}{\omega}, \tag{5.3}$$

where n is the per-unit natural frequency and ω is the fundamental power system frequency.

Also, defining the peak current,

$$I_{pk} = \frac{V_{pk}}{Z_0}, \tag{5.4}$$

where V_{pk} is the peak value of the fundamental frequency voltage. The peak value of the capacitor voltage V_{pk} can be expressed as,

$$V_{pk} = I_{pk} Z_0 \left(\frac{n^2}{n^2-1}\right). \tag{5.5}$$

The worst case of the voltage peak may occur on restriking the capacitor bank with a trapped charge of 1 pu In this case, if the restrike occurs when the source voltage reaches its peak, the voltage across the contacts of the switch will be twice the source peak system voltage as defined by,

$$V_{pk,crit} = V_{pk} - (-V_{Ceff}) = 2V_{pk}. \tag{5.6}$$

The instantaneous current in this case can be expressed as,

$$I_{pk} = \frac{2V_{pk}}{Z_0}. \tag{5.7}$$

5.2.1.2 Effect of Utility CST on ASDs

In this section the effect of a utility CST event on an ASD is examined. Figure 5.4a shows the topology of a typical ASD. Figure 5.4b shows the equivalent circuit for analysis. In Figure 5.4b, V_s and its symbol represent the line-to-line voltage applied to the ASD system, which is short-circuited at the event of a CST event and can reach a critical peak value given by Equation 5.6.

From the equivalent circuit (Figure 5.4b), we have

$$L_{eff} \frac{d}{dt} i_d(t) = -v_{DC}(t) - V_{DC}(0^+) + v_s(t), \tag{5.8}$$

$$v_{DC}(t) + V_{DC}(0^+) = R_o i_{d,av}(t), \tag{5.9}$$

$$C_d \frac{d}{dt} v_{DC}(t) = i_d(t) - i_{d,av}(t), \tag{5.10}$$

where, $L_{eff} = 2L_s + L_d$ represents the effective ASD input inductance. The utility input voltage (V_s) represents the source voltage with a magnitude equal to two times the maximum line-to-line voltage (for a worst case CST). Also, the steady-state V_{DC} is [17]

$$V_{DC} = 1.35 V_{LL}. \tag{5.11}$$

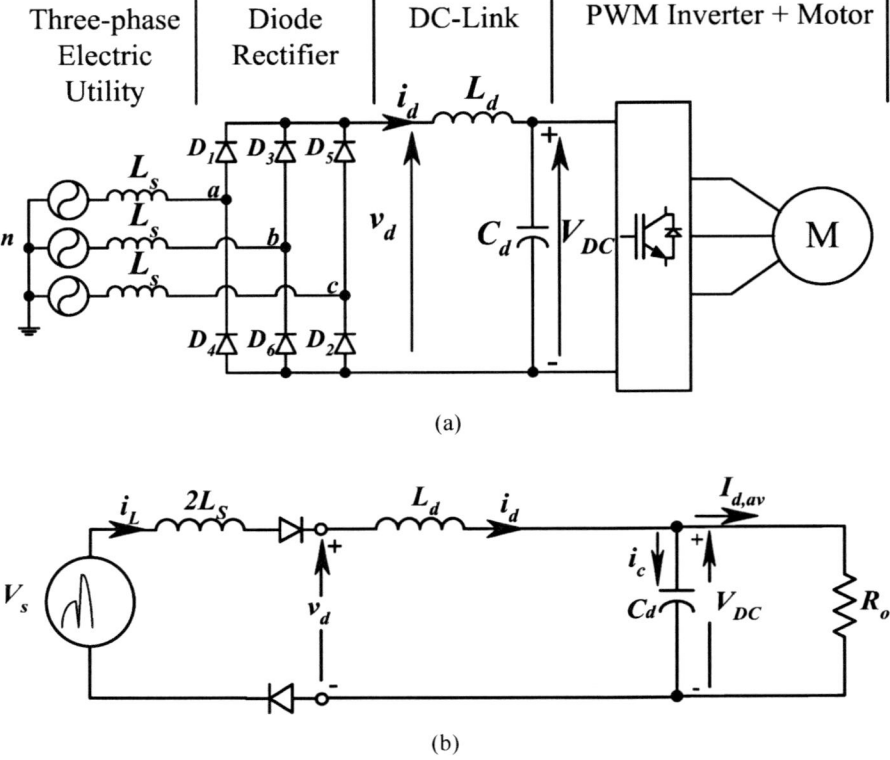

Figure 5.4. Adjustable-speed drive, (a) typical ASD topology, (b) equivalent circuit of an ASD under a CST event

From Equations 5.8, 5.9, and 5.10, a time-domain equation for the ASD DC-link voltage is derived as,

$$\frac{v_{DC}(t)}{(2\sqrt{2})V_{LL}} = \left[1 - \frac{e^{-\zeta\omega_n t}}{\sqrt{1-\zeta^2}} \sin(\omega_n \sqrt{1-\zeta^2}\, t + \theta) + \frac{1.35}{(2\sqrt{2})} e^{-\zeta\omega_n t} \cos(\omega_n \sqrt{1-\zeta^2}\, t)\right],$$

(5.12)

where, the resonant frequency is defined as,

$$\omega_n = \sqrt{\frac{1}{L_{eff} C_d}}.$$

(5.13)

The damping ratio is expressed as,

$$\zeta = \frac{1}{2R_o}\sqrt{\frac{L_{eff}}{C_d}}, \qquad (5.14)$$

$$\theta = \tan^{-1}\left(\frac{\sqrt{1-\zeta^2}}{\zeta}\right). \qquad (5.15)$$

Evaluating Equation 5.12, the peak time t_{pk} can be expressed as,

$$t_{pk} = \frac{\pi}{\omega_n\sqrt{1-\zeta^2}}. \qquad (5.16)$$

The ASD DC-link voltage peak can be computed from Equation 5.12 at the time defined by Equation 5.16, as expressed in per-unit quantity by,

$$V_{DC,pk} = 2\sqrt{2} + \sqrt{2}\left[\frac{2\pi-3}{\pi}\right]\frac{e^{\frac{-\zeta}{\sqrt{1-\zeta^2}}\pi}}{\sqrt{1-\zeta^2}}\sin(\theta). \qquad (5.17)$$

5.2.1.3 Effect of Line Inductance L_s and ASD Load (R_o) on $V_{DC,peak}$

In this section, the effect of line input inductance L_{eff} and ASD load, R_o (Figure 5.4b) on $V_{DC,pk}$ is explored. The following per-unit quantities are defined:

utility line-to-line voltage V_{LL} = 1 pu
input power ASD VA = 1 pu
DC-link voltage V_{DC} = 1.35 pu

Figure 5.5 shows the variation of $v_{DC}(t)$ as a function of time and for different L_s, R_o values in per-unit. Figure 5.5b shows the expanded version of Figure 5.5a around the peak of $v_{DC}(t)$. It is clear from Figure 5.5a and b that for lower values of L_{eff}, the rate of rise of $v_{DC}(t)$ from nominal (5.35 pu) to $V_{DC,pk}$ (4.31 pu) is shorter. Also, the ASD load (R_o) has a negligible effect on the $V_{DC,pk}$ (Figure 5.5a).

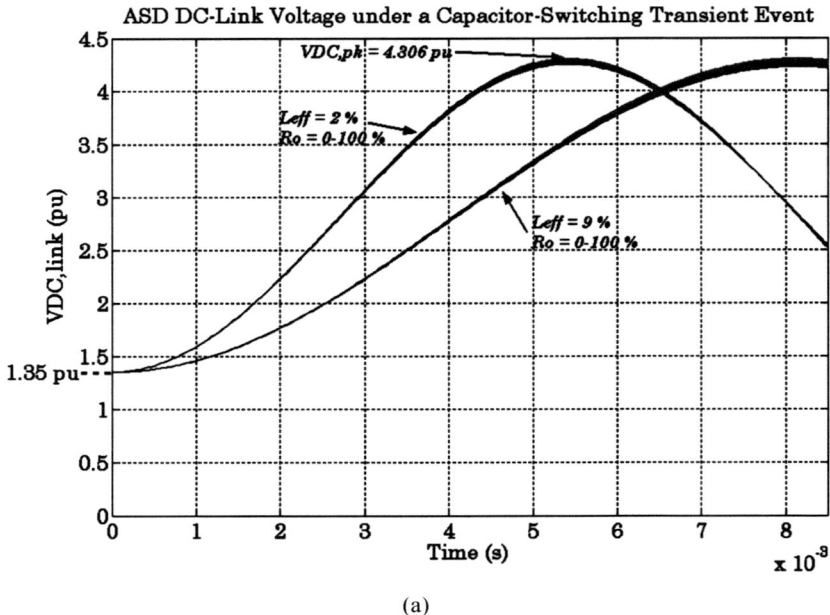

(a)

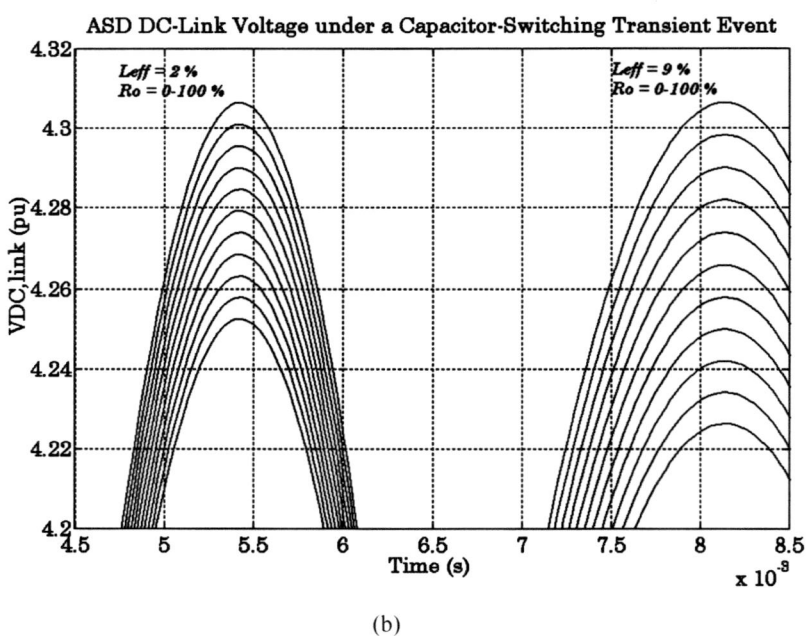

(b)

Figure 5.5. ASD DC-link voltage transient response under a CST event with $L_{eff}=2L_s+L_d$

5.3 Traditional Transient Mitigation Methods on ASDs

5.3.1 Review of Power-system Mitigation Techniques

Several methods of controlling transient overvoltages during a CST have been investigated in the literature [6, 10–12, 14]. Utility mitigation options are suggested to minimize the overvoltage transient during a capacitor bank being energized at the point of application. These options are discussed in the following subsections.

5.3.1.1 Circuit Breakers with Preinsertion Resistors

The effect of a preinserted resistor is to limit the voltage collapse that occurs on the utility when an uncharged capacitor is energized, thereby reducing the voltage collapse. The resistor forms a voltage divider between the resistor, the electric system (source), and the capacitor bank. The overvoltage transient is minimized due to the majority of the voltage drop occurs across the preinsertion resistor. The preinsertion resistor should be selected of an optimum size equal to the characteristic impedance formed by the source inductance and the capacitor bank [6].

A. Advantages

➢ One of the most effective means of reducing the magnitude of surges due to CST.

B. Disadvantages

➢ Synchronization between the resistor and main contacts is required.
➢ If the resistor is not properly sized, the transient can be higher when the resistor is bypassed.
➢ In many applications, maintenance is required when insertion times are exceeded.

5.3.1.2 Circuit Breakers with Preinsertion Inductors

Preinsertion inductors are inserted in the same fashion as preinsertion resistors into the capacitor-switching circuit between the contacts of the switching circuit during the switching operation. The preinserted inductor is bypassed after 7–12 cycles of the closing operation. Preinsertion inductors are primarily used for limiting inrush currents during back-to-back energization, which can also provide overvoltage mitigation as they are effective at limiting the high rate of dv/dt of the initial voltage collapse.

The method of preinsertion inductors offers higher impedance at the switching transient frequency (when an inductor is inserted) and lower impedance at the power-system frequency (during the bypass time). Hence, the preinsertion inductor method has a better impedance characteristic than that of the preinsertion resistor method.

A. Advantages

➤ Adds additional inductance to the source during energization that results in voltage-transient reduction at the utilization-voltage bus.

B. Disadvantages

➤ Optimum values are limited by energy dissipation constraints, peak inrush current, bypass transient magnitude, physical size and weight.

5.3.1.3 Zero-voltage Closing Control

This method employs controlled closing devices, which are typically high-speed vacuum switches or SF_6 circuit breakers. These devices are controlled by sophisticated electronic controls. The zero-voltage closing (ZVC) control is a controlled method to close or energize the bus capacitor bank near voltage zero to minimize voltage and inrush current transients. The basic function of ZVC devices is to energize each phase of the capacitor bank at the point when capacitor and system voltage differential is zero. Timing of ZVC devices is properly adjusted according to the capacitor-bank configuration. That is, grounded banks are closed at each phase zero-crossing point. Ungrounded banks are energized at a proper close timing sequence where the first phase is closed when it crosses zero, the second phase is closed at the time the first and second phases cross each other, and the third phase is closed on its zero crossing.

A. Advantages

➤ More accurate, repeatable timing reference with the use of new microprocessor technology for control supervision of ZVC devices.
➤ Effective closing tolerance of ±1 ms.
➤ One of the most effective methods of limiting transients.
➤ Automatic calibration as one of the most attractive feature that does not need field-timing adjustments.

B. Disadvantages

➤ Restrike may occur on some ZVC devices (*i.e.* SF_6 breakers/switches could restrike due to gas pressure falling below limits).
➤ High cost due to sophisticated electronic control.
➤ Effectiveness of ZVC control is system dependent.

At the industrial customer site, there are some mitigation techniques to reduce the overvoltage transients associated with CST events and avoid nuisance tripping phenomena on ASDs. These customer mitigation techniques include: AC line input or DC-link reactors, low-voltage arresters and harmonic filters.

5.3.1.4 AC Input Line or DC-link Reactors

ASD manufacturers recommend or sometimes include AC line input or DC-link reactors (inductance) as a standard option. Inductive reactance primarily reduces harmonic current levels and solves the ASD nuisance tripping problem. However, in [1] it is shown that AC line input or DC-link reactance, especially beyond 5% reactance, produce an increased voltage drop [6] (larger DC-link voltage variation) under steady-state conditions, which may result in an undervoltage trip for the ASD.

A. Advantages

➤ An optimum inductance of 3–5%, based on the ASD-rating (horsepower, hp), will be sufficient to avoid nuisance tripping and reduce the harmonic current level as well as the diode peak current.
➤ Reactors can be easily integrated into the ASD.

B. Disadvantages

➤ A critical evaluation [1] shows that additional line or DC-link reactance (3–5%) can aggravate the nuisance tripping issue.

As a summary of the various utility mitigation techniques, the impedance created by preinsertion resistor and/or inductor methods, used for mitigating CST, limits the magnitude of the switching transient by limiting the extent of the initial collapse of the bus voltage during a capacitor-bank energization. References [6, 10] recommend highly damped preinsertion inductors, with inherent resistance. From a study [10], it is suggested that a preinsertion inductor is the preferred overvoltage control method over a preinsertion resistor and synchronous closing methods. Also, it is noted that a preinsertion inductor is more economical than the preinsertion resistor.

The application of previous mitigation techniques is largely determined by the actual power-system parameters. Proper analyses and simulations best determine the performance of each method of control, for a specific installation. Also, these methods are more effective when ASDs are equipped with 3–5% of AC or DC-link inductors.

5.4 New Mitigation Methods

5.4.1 Proposed Approaches to Mitigate Nuisance Tripping of CST Events

In this section, new methods to mitigate nuisance tripping of PWM ASDs due to utility capacitor-switching events are presented, analysed and discussed. In the first method, a soft-charge resistor, which is available in most ASD systems, is momentarily introduced in the series path of the power flow to effectively damp the CST event. Another proposed approach is presented, which is an electronic damping variation of the first proposed approach.

5.4.1.1 IGBT and Soft-charge Resistor Approach

Figure 5.6 shows a typical topology for a PWM ASD, which consists of a three-phase diode rectifier, DC-link, and a PWM inverter. A soft-charge resistor ($R_{soft\text{-}charge}$) and a bypass relay module are typically employed in the DC-link power flow path of the ASD equipment to prevent overcharge of the DC-link capacitor during initial startup. The soft-charge resistor is bypassed by a relay contactor in the steady state, after a predetermined delay issued by the ASD control circuit. Hence, the DC-link power flows through the ASD relay module without any DC voltage drop in the main power flow path.

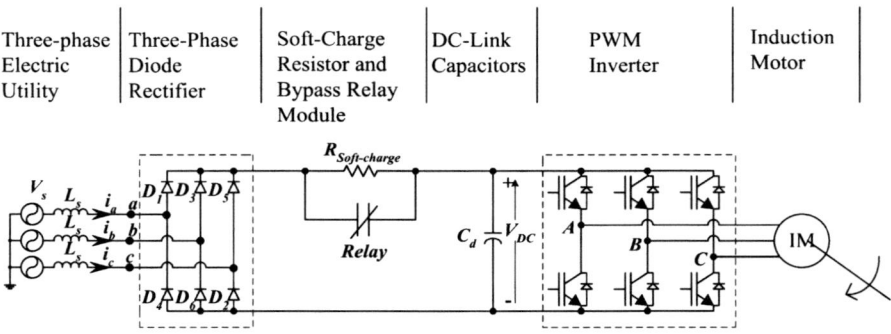

Figure 5.6. Conventional topology for a PWM ASD

Figure 5.7 shows the proposed approach to mitigate nuisance tripping of PWM ASDs due to CSTs events. In this approach, the relay used to bypass the soft-charge resistor, $R_{soft\text{-}charge}$ is replaced by an IGBT. Under normal conditions, the IGBT is enabled and $R_{soft\text{-}charge}$ is effectively short-circuited. However, during a CST event, the IGBT is essentially deactivated and $R_{soft\text{-}charge}$ is introduced in series with the DC-link capacitor C_d such that V_{DC} transient is effectively damped. Furthermore, the value of the effective damping resistance $R_{soft\text{-}charge}$ can be adjusted by suitably modulating the on–off time of the IGBT. This feature facilitates electronic control and suitable damping can be achieved for various types of CST events and varying utility line impedance. Figure 5.8 shows the block diagram for the detection and feedback control scheme to mitigate nuisance tripping of PWM ASDs due to CST events.

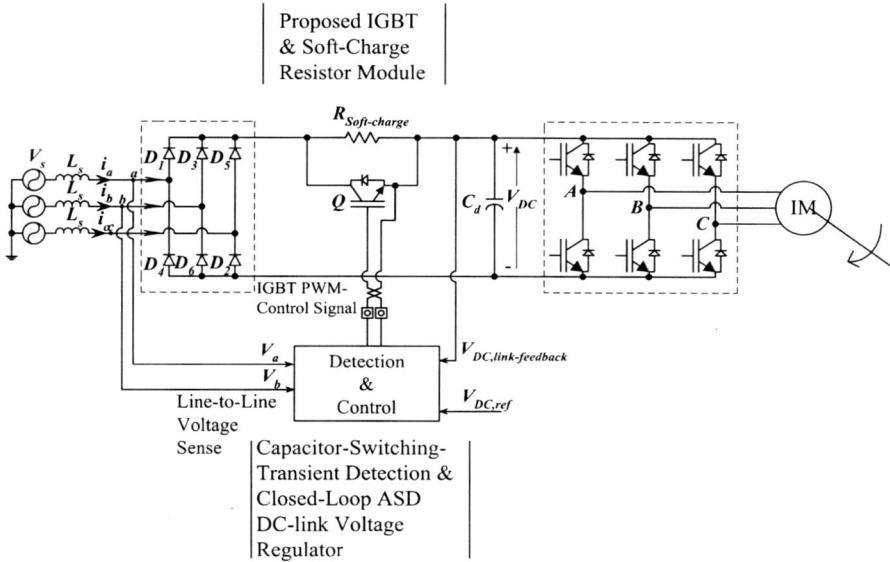

Figure 5.7. Proposed IGBT and soft-charge resistor approach to mitigate nuisance tripping of PWM ASDs due to CSTs events

In Figure 5.8, the detection and control scheme senses the line-to-line voltage, v_{ab}. A differentiator amplifier detects the fast change in the line-to-line voltages due to CST events. The CST-event detection enables a monostable multivibrator, which activates a pulsewidth-detection window given by a preset window time value t_w (1 ms ~ 2 cycles). The pulsewidth-detection window along with the duty cycle of the PWM will activate and deactivate the IGBT according to the logic circuit shown in Figure 5.8. A DC-link-sensing voltage is also implemented to achieve a feedback control loop through a basic error-amplifier scheme. This control scheme will maintain the modulation of the IGBT and $R_{soft\text{-}charge}$ resistor to effectively damp the DC link voltage under a CST event. The effective damping resistance value, R_{damp}, applied to mitigate the overvoltage will be given by the modulating function of the control scheme shown in Figure 5.8. The damping resistance, R_{damp}, will be a direct function of the duty cycle applied to the IGBT and the value of the ASD soft-charge resistance $R_{soft\text{-}charge}$.

116   J. L. Durán-Gómez and P. N. Enjeti

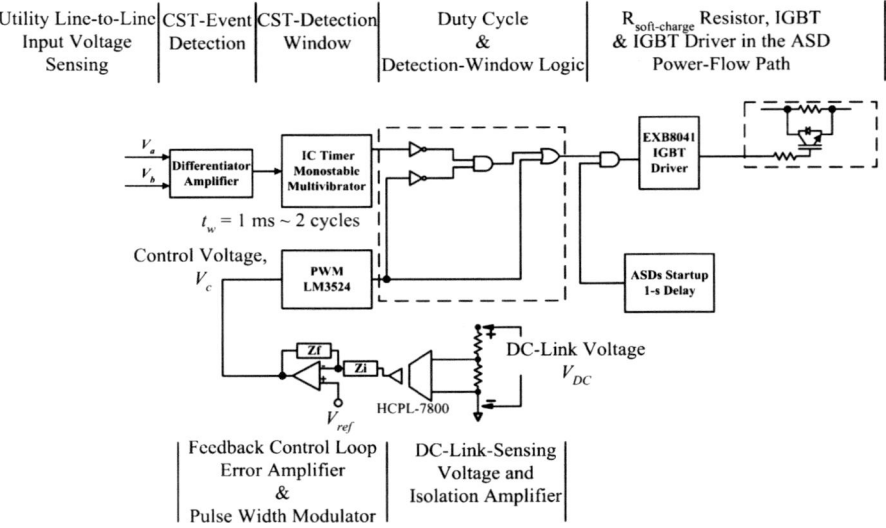

Figure 5.8. Detection and feedback control scheme to mitigate nuisance tripping of PWM ASDs due to CST events

A. Analysis

In this section an indepth analysis is presented to compute the required amount of damping resistance R_{damp} to limit the DC-link voltage overshoot. Figure 5.9 shows the equivalent circuit when a CST event occurs and R_{damp} is introduced in the circuit.

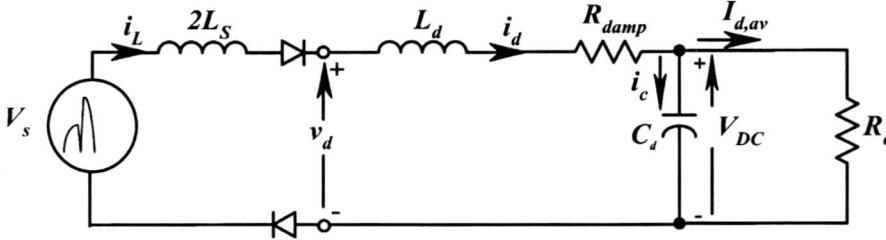

Figure 5.9. Equivalent circuit of an ASD under a CST event with damping resistance, R_{damp}

From Figure 5.9 we have,

$$L_{eff}\frac{d}{dt}i_d(t) = -R_{damp}i_d(t) - v_{DC\_link}(t) - V_{DC}(0^+) + v_s(t), \quad (5.18)$$

$$v_{DC}(t) + V_{DC}(0^+) = R_o i_{d,av}(t), \tag{5.19}$$

$$C_d \frac{d}{dt} v_{DC}(t) = i_d(t) - i_{d,av}(t). \tag{5.20}$$

The solution for the $v_{DC}(t)$ from Equations 5.18–5.20 is,

$$\frac{v_{DC}(t)}{(2\sqrt{2})V_{LL}} = \frac{R_o}{R_{damp} + R_o}\left[1 - \frac{1}{\sqrt{1-\zeta^2}} e^{-\zeta\omega_n t}\sin(\omega_n\sqrt{1-\zeta^2}\,t + \theta)\right] + \frac{1.35}{(2\sqrt{2})} e^{-\zeta\omega_n t}\cos(\omega_n\sqrt{1-\zeta^2}\,t), \tag{5.21}$$

where,

$$\omega_n = \sqrt{\frac{R_{damp} + R_o}{L_{eff} R_o C_d}}, \tag{5.22}$$

$$\zeta = \frac{L_{eff} + R_{damp} R_o C_d}{2\sqrt{(R_{damp} + R_o)(L_{eff} R_o C_d)}}, \tag{5.23}$$

$$\theta = \tan^{-1}\left(\frac{\sqrt{1-\zeta^2}}{\zeta}\right). \tag{5.24}$$

B. Damping Resistance

Equation 5.21 shows the variation of $v_{DC}(t)$ when R_{damp} is introduced into the circuit. By evaluating the Equation 5.21 for,

$$v_{DC}(\infty) = v_{DC}(t), \tag{5.25}$$

$$t \to \infty,$$

and limiting $V_{DC,pk} = v(\infty)$ to $V_{DC,trip} = 130\%$, we have,

$$V_{DC,pk} = 1.3 * 1.35 \text{ pu}. \tag{5.26}$$

Substituting Equation 5.26 into Equation 5.21 and evaluating for $t=\infty$, the required value of R_{damp} can be computed as,

$$R_{damp} = \left[\frac{2\sqrt{2}}{1.35 * V_{dc\_trip}} - 1\right] R_o \text{ pu,}$$

$$R_{damp} = (0.61) R_o \text{ pu.} \quad (5.27)$$

The damping resistance as a function of the duty cycle (D) (Figure 5.8) and soft-charge resistance can be defined as

$$R_{damp} = (1-D) R_{soft-charge}. \quad (5.28)$$

5.4.1.2 A Proposed IGBT Pair and Resistor Damping Approach

A second proposed method to effectively mitigate the CST effect on ASD is illustrated in Figure 5.10. The proposed method shown in Figure 5.10 is a proposed IGBT pair and resistor damping approach installed in the DC-link of the ASD equipment. The approach consists of an IGBT pair, which is shunt-connected as a bidirectional switch to the secondary side of a transformer. A damping resistor is also connected to the secondary winding of the transformer in order to damp the overvoltage DC-link transient generated due to the utility CST event. In the event of a CST disturbance, the CST detection and damping control switches on and off the IGBT pair to effectively introduce the damping resistance and mitigate the overvoltage transient.

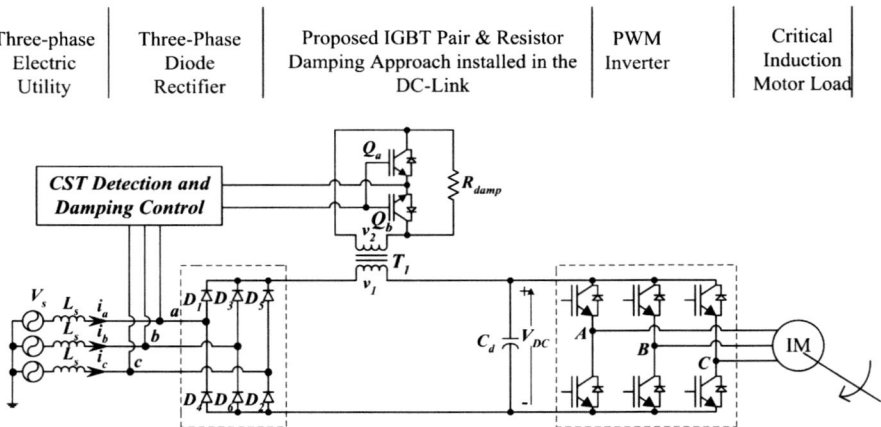

Figure 5.10. Proposed IGBT pair and resistor damping approach installed in the DC-link of the ASD equipment

5.5 Example of Experimental Results

Practical experimental results are discussed in this section based on a design example as well as on extensive simulation results. This will help to illustrate one of the mitigation methods employed to alleviate the nuisance tripping of CST events typically affecting power-distribution systems.

5.5.1 Design Example

In this section, a design example is presented for a 480-V, 60-Hz, 16-kVA commercial ASD equipment. Figure 5.7 shows the hardware implementation of the proposed approach. The following quantities are defined:

Utility line-to-line voltage = $V_{base} = V_{LL}$ = 480 V = 1 pu
Line impedance = $Z_{L,SYS}$ = 3%Z_{base}
ASD power rating = S_{ASD} = 16 kVA = 1 pu
Power factor = pf = 0.7
Power-system frequency = 60 Hz
Damping ratio [16] = λ = 0.2
Integrated ASD $R_{soft-charge}$ = 20 Ω
Output power, $P_{asd} = pf(S_{asd})$ = 0.7 pu
DC-link capacitor, C_d = 1950 μF = 9.45%(Z_{base})
ASD DC-link inductance, L_d = 3%Z_{base}/ω
ASD line input inductance, L_s = 3%Z_{base}/ω
ASD output resistance (30% *full load*), R_o=8.7 pu
Damping resistance, R_{damp} = 0.0 pu
Semiconductor switching-power device[=]Toshiba IGBT MG90V2YS40, I_C=40 A, V_{CE} = 1700 V.

5.5.1.1 Experimental Setup

Figure 5.11 shows the experimental setup for generating a capacitor-switching transient event. A programmable high-power AC power system is employed to supply the AC voltage to the ASD equipment. The programmable AC power source shown in Figure 5.11 can be used for a variety of power test applications. This type of three-phase AC power source has a full power-line disturbance simulation and generation capabilities. Experiments on voltage sags (dips), transients and other power quality disturbances can be simulated and generated by this AC power source. However, for this experimental setup a three-phase power switch was used for generating the CST event as shown in Figure 5.11. The L_s and C_{sw} values are altered to generate a variety of CST events, to which the ASD is subjected in a laboratory setting.

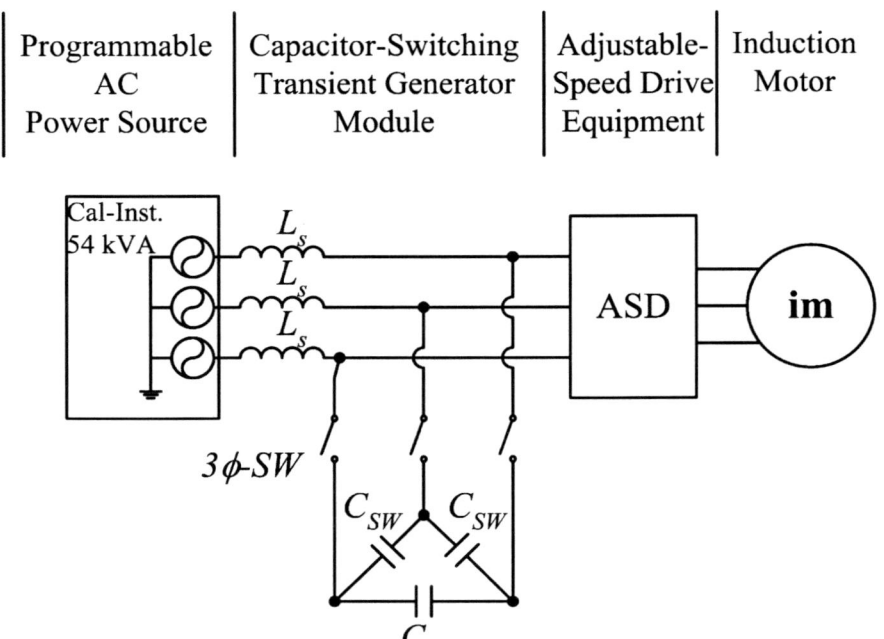

Figure 5.11. Experiment setup for initiating a CST event

Table 5.1 shows the characteristic transient components of the CST events as a function of C_{sw}.

As the C_{sw} value is changed, the CST natural frequency of oscillation f_n is altered. Also, V_{pk} and $V_{pk,crit}$ (as defined in Equations 5.5 and 5.6) are listed.

Table 5.1. Characteristic transient components in a utility CST event affecting at the input of a 480-V, 16-kVA, 60-Hz, ASD equipment

C_{SW} (µF)	Q_C (kVAr)	Z_0 (Ω)	f_n (Hz)	V_{pk} (V)	$V_{pk,crit}$ (kV)
30	7.82	3.0	593	685	1.37
60	15.64	2.1	420	693	1.39
90	23.45	1.7	342	700	1.4
140	36.48	1.4	275	712	1.42

5.5.1.2 ASD DC-link Voltage Peak under a CST Event

The maximum DC-link overvoltage transient in the ASD under a CST event is computed from Equation (5.17). From Figure 5.5a a peak voltage close to 4.3 pu ($V_{DC,pk}$ = 2064 V) can be shown. The ASD DC-link overvoltage reaches its maximum value in less than half a cycle (8.33 ms). A commercial ASD could be tripping just at a small per cent (25%) above the nominal DC-link voltage.

5.5.1.2 Damping Resistance for the Proposed Approach

In order to mitigate the CST disturbance on the ASD DC-link voltage, the proposed approach uses a damping resistance ($R_{soft\text{-}charge}$) as shown in Figure 5.7. Table 5.2 illustrates the required damping resistance values, R_{damp} computed from Equation 5.27 for different ASDs output levels. The value of the damping resistance R_{damp} (pu), shown in Table 5.2, represents the required effective resistance value to effectively damp the overvoltage ASD DC-link voltage caused under a CST event for different ASDs output levels. The electronic damping achieved by the modulation of the semiconductor power device (IGBT) shown in Figure 5.7, is suitably adjusted by the control scheme shown in Figure 5.8 and given by Equation 5.28.

Table 5.2. Damping resistance at different ASDs output levels to mitigate the ASD DC-link overvoltage under a CST event

ASD Load (pu)	0.1	0.2	0.4	0.6	0.8	1.0
R_{damp} (pu)	15.9	7.95	3.98	2.65	1.99	1.6

5.5.2 Simulation Results

As a summary of the simulation results, the proposed approach attempts to mitigate the overvoltage transient in the ASD DC-link voltage. Two 480-V, 16-kVA ASDs were simulated that are connected in a typical power-distribution system and affected by a utility CST event as has been shown in Figure 5.2. The plot at the top of Figure 5.11 shows the CST effect on the utilization bus voltage. This plot illustrates the high overvoltage transient (close to 2 pu) and the natural resonance frequency that are dependent on the L–C parameters of the power system. The middle plot in Figure 5.12 shows the DC-link voltage for ASD-2 that experiences the overvoltage condition as no mitigation technique is applied. The ASD DC-link overvoltage is momentarily increased (> 1.3 pu) to 1.2-kV. This condition could result in an ASD overvoltage trip due to exceeding the overvoltage trip level. The lower plot in Figure 5.12 shows the DC-link voltage on ASD-1 when the proposed mitigation technique is applied to reduce or maintain the voltage within nominal or acceptable ASD limits. A simulation of the CST detection circuit, along with an ASD DC-link feedback voltage are used to mitigate the momentary DC-link overvoltage.

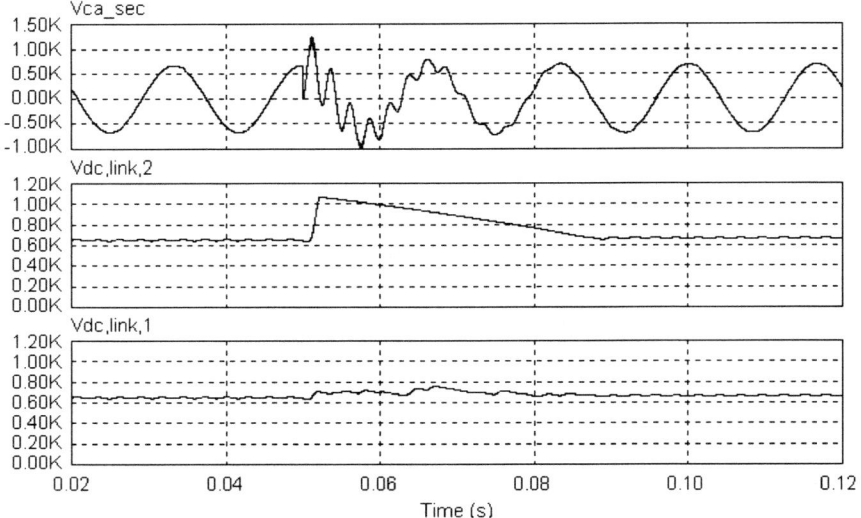

Figure 5.12. Simulation results of the proposed approaches for two 480-V, 16-kVA ASDs connected in a typical power-distribution system

From the simulations results shown in Figure 5.12, the feasibility of the proposed approach to mitigate the ASD DC-link overvoltage can be concluded.

5.5.3 Experimental Results

In this section, experimental results for a simulated utility CST event on a commercial 480-V, 16-kVA, 60-Hz ASD are presented. The test setup consists of an ASD equipment powered from a 480-V, 54-kVA programmable AC power source along with a delta-connected capacitor bank. Figure 5.11 shows the experiment setup.

A damping resistance is computed from Equation 5.27 and explained in Section 5.4.4.1, where R_{damp} = 76.36 Ω. The ASD equipment embodies a soft-charge resistor, which can be used as the damping resistance. However, the proposed approach achieves the damping effect of the CST event by means of an average damping modulation generated through the soft-charge resistor ($R_{soft-charge}$) along with the integrated IGBT device (Figure 5.7).

Figure 5.13a shows the ASD DC-link voltage under a CST event for a 30% ASDs output load without the proposed approach. The DC-link increases up to 708 V without tripping the ASD (top trace of Figure 5.13a). Figure 5.13b shows the experimental results with the proposed approach. The top trace in Figure 5.13b shows the ASD DC-link voltage, which has been effectively damped due to the proposed approach. The middle and bottom traces in Figure 5.13b show the line-to-line voltage and the voltage across the soft-charge resistor, respectively.

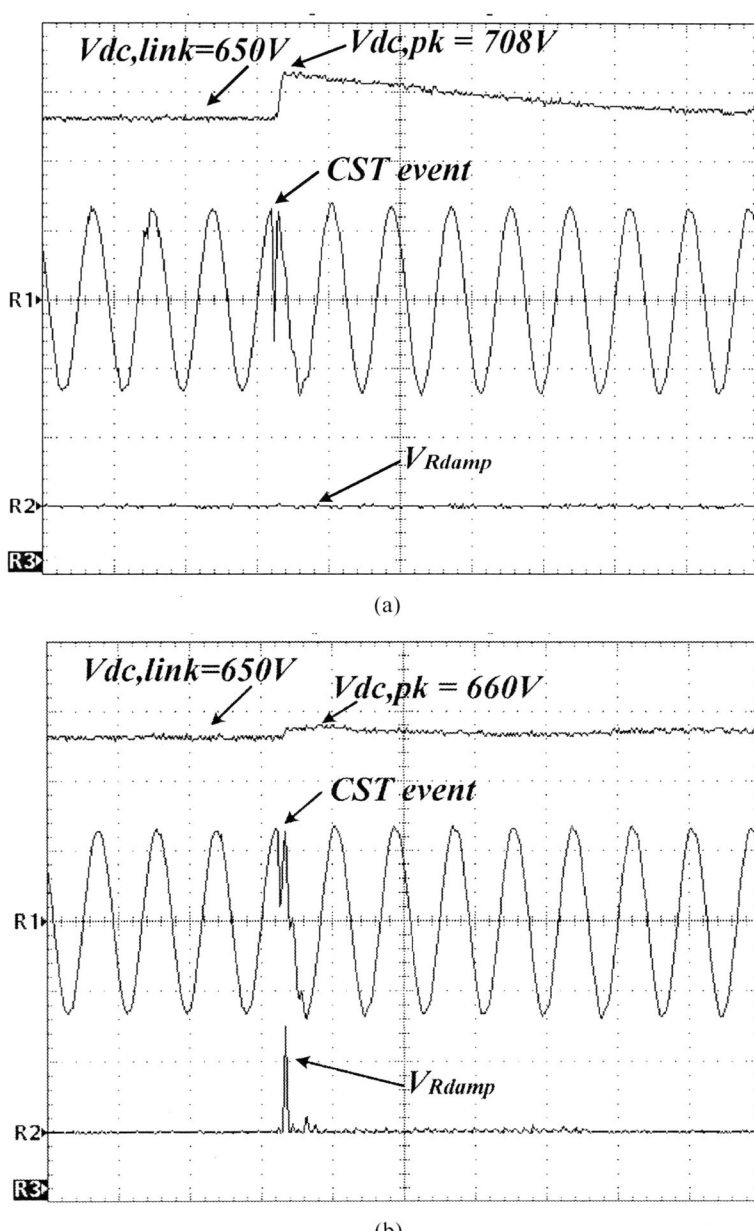

Figure 5.13. Experimental results on a 480-V, 16-kVA, 60-Hz ASD equipment with, C_{SW}=60 μF and L_s=800 μH (3%), **(a)** CST event on ASD without the mitigation approach, **(b)** CST event on ASD with the proposed mitigation approach. **(a)** Trace R1: 500 V/div., trace R2: 200 V/div., trace R3: 100 V/div, time base 20 ms/div. **(b)** Trace R1: 500 V/div., trace R2: 200 V/div., trace R3: 100 V/div, time base 20 ms/div

It is clear, in the bottom trace (Figure 5.13b) that the overvoltage transient generated by the CST event is properly damped across the damping resistance via the pulsewidth-modulated action of the IGBT (Q), during the utility disturbance. The momentary high energy generated by the CST event is dissipated in the damping resistance (Figure 5.13b) during the first half of a cycle at the inception of the CST event. The mitigated action of the proposed approach continues for a 5-cycle interval to reduce and maintain the ASD DC-link voltage within acceptable limits (Figure 5.13b).

Figure 5.14 shows another experimental result, where the voltage peak ($\approx$ 1.6 pu) of the line-to-line voltage at the inception of a CST event (middle trace, R1) is illustrated. However, the proposed approach helps to mitigate the overvoltage DC-link voltage in the ASD equipment. The lower trace in Figure 5.14 shows the damped overvoltage across the damping resistance.

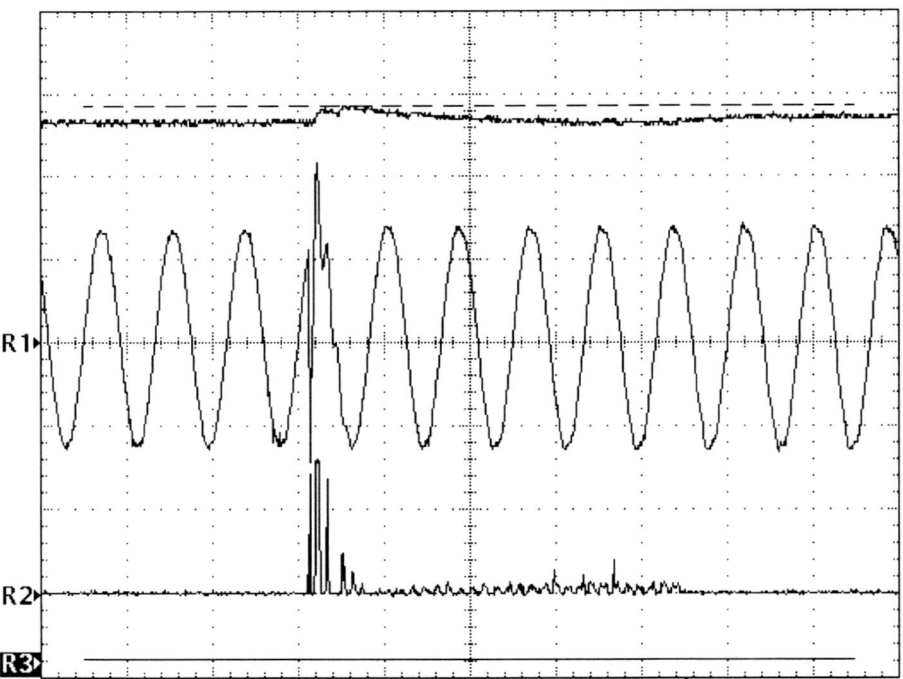

Figure 5.14. Experimental results on a 480-V, 16-kVA, 60-Hz ASD equipment with, C_{SW}=60 µF and L_s=800 µH (3%), CST event on ASD with the proposed mitigation approach. Trace R1: 500 V/div., trace R2: 200 V/div., trace R3: 100 V/div., time base 20 ms/div

Figure 5.15 shows the experimental results on a 480-V, 16-kVA, 60-Hz ASD equipment for CST events. Figures 5.15a and b show the overvoltage transient on the ASD DC-link due to the CST events under two combinations of the *LC* characteristic. The upper traces in Figures 5.14a and b show the ASD DC-link voltage with a voltage peak close to 700 V. No mitigation method was employed in these two experimental results.

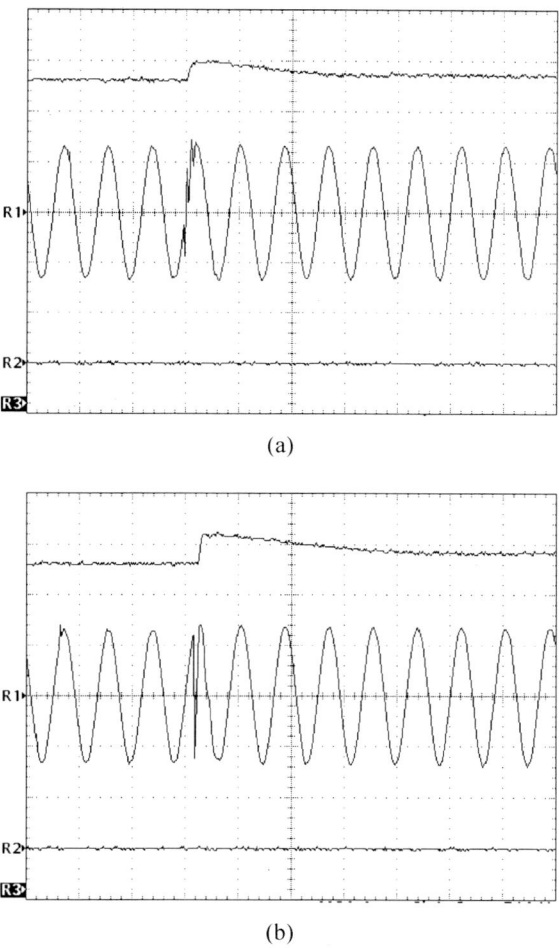

Figure 5.15. Experimental results on a 480-V, 16-kVA, 60-Hz ASD equipment under a CST event without any transient mitigation method: **(a)** C_{SW} =30 µF and L_s=800 µH. Trace R1≻500 V/div., trace R2≻100 V/div., trace R3≻100 V/div., time base 20 ms/div.; **(b)** C_{SW} =90 µF and L_s=800 µH. Trace R1≻ 500V/div., trace R2≻100 V/div., trace R3≻100 V/div., time base 20 ms/div

Figure 5.16 shows the experimental results on a 480-V, 16-kVA, 60-Hz ASD equipment for CST events. In Figures 5.16a and b the effect is shown of the CST event on the ASD DC-link voltage as the proposed approach is employed to reduce the voltage peak. The ASD DC-link voltage is effectively reduced close to the nominal DC-link voltage as depicted in the upper traces shown in Figures 5.16a and b. Hence, the nuisance tripping of ASD equipment can be avoided with the proposed CST mitigating approach.

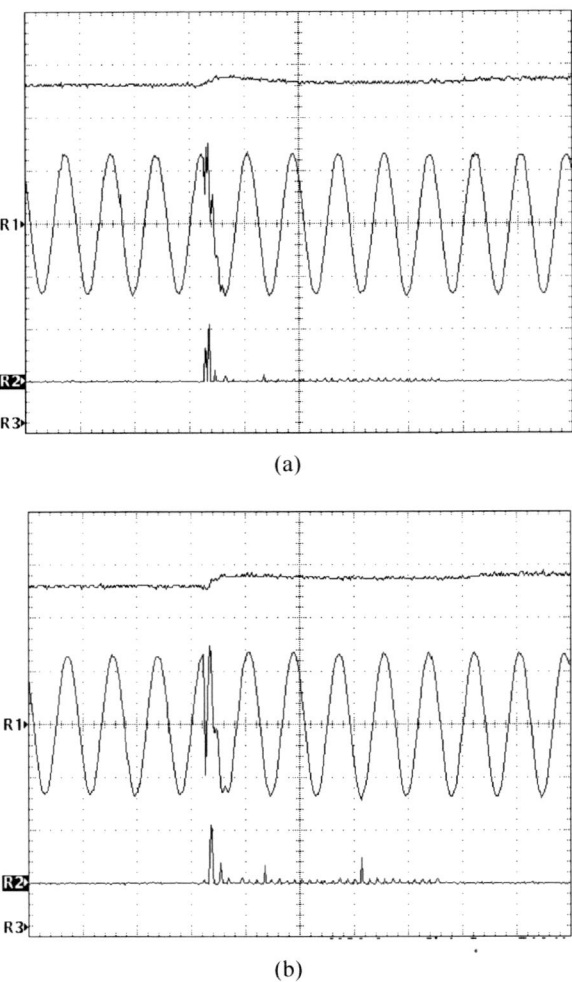

Figure 5.16. Experimental results on a 480-V, 16-kVA, 60-Hz ASD equipment under a CST event with a transient mitigation method: **(a)** $C_{SW} = 30$ μF and $L_s = 800$ μH. Trace R1➤500 V/div., trace R2➤100 V/div., trace R3➤100 V/div., time base 20 ms/div.; **(b)** $C_{SW} = 90$ μF and $L_s = 800$ μH. Trace R1➤500 V/div., trace R2➤100 V/div., trace R3➤100 V/div., time base 20 ms/div

5.6 Conclusions

In this chapter, new approaches to mitigate nuisance tripping of PWM ASDs due to utility CST events have been proposed. Simulations and experimental results have shown the feasibility of the illustrated and proposed new approaches to reduce the overvoltage transient in the ASD DC-link voltage in order to avoid nuisance tripping. The approaches can be easily integrated into ASD equipment as add-on options. The main advantages of the new proposed approaches are:

- electronic damping for CST is achieved by low-cost modifications to ASD hardware;
- They adapt to several utility resonance conditions.

A disadvantage of these methods would be the cost of the new IGBT-based bypass module in the power flow path of the ASD.

References

[1] Durán-Gómez JL, Enjeti PN, Woo BO. Effect of voltage sags on adjustable speed drives: a critical evaluation and an approach to improve performance. IEEE Trans. Ind. Applicat 1999;35:1440–1449.
[2] Grebe TE, Power quality and the utility/customer interface. Conf. Rec. Southcon 1994:372–377.
[3] Durán-Gómez JL, Enjeti PN, von Jouanne A, An approach to achieve ride-through of an adjustable speed drive with flyback converter modules powered by supercapacitors. Conf. Rec. IEEE-IAS Annu. Meeting 1999;3:1623–1629.
[4] McGranaghan MF, Zavadil RM, Hensley G, Singh T, Samotyj M. Impact of utility switched capacitors on customer systems, Part II – Adjustable-speed drive concerns. IEEE Trans. Power Deliv 1991;6:1623–1628.
[5] McGranaghan MF, Zavadil RM, Hensley G, Singh T, Samotyj M, Impact of utility switched capacitors on customer systems, magnification at low voltage capacitors. IEEE Trans. Power Deliv. 1992;7:862–868.
[6] Bellei TA, O'Leary RP, Camm EH. Evaluating capacitor-switching devices for preventing nuisance tripping of adjustable-speed drives due to voltage magnification. IEEE Trans. Power Deliv. 1996;11:1373–1378.
[7] Wagner VE, Strangas E. PWM drive filter inductor influence on transient immunity. IEEE Ind. Applicat. Mag. 1998;4:39–45.
[8] Wagner VE, Staniak JP, Orloff TL. Utility capacitor switching and adjustable-speed drives. IEEE Trans. Ind. Applicat. 1991;27:645–651.
[9] Grebe TE, Tang L. Analysis of harmonic and transient concerns for PWM adjustable-speed drives using the electromagnetic transients program. Conf. Rec. ICHPS V 1992:41–47.
[10] Bhargava B, Khan AH, Imece AF, DiPietro J. Effectiveness of pre-insertion inductors for mitigating remote overvoltages due to shunt capacitor energization. IEEE Trans. Power Deliv. 1993;8:1226–1238.
[11] Skeans DW. Recent development in capacitor switching transient reduction. Conf. Rec. T&D World Expos. Sub. Section 1995:1–13.
[12] Grebe TE, Gunther EW. Application of the EMTP for analysis of utility capacitor switching mitigation techniques. Proc. Inter. Conf. ICHQP 199:583–589.

[13] Alexander RW. Synchronous closing control for shunt capacitors. Conf. Rec. IEEE-PES Winter Meeting 1985: pp. 1–7.
[14] Reid WE. Capacitor application considerations-utility/user interface. Proc. Pulp and Paper Industry Tech. Conf. 1991:70–79.
[15] Liu KC, Chen N, Voltage-peak synchronous closing control for shunt capacitors. Proc. IEE Gener. Transm. Distrib. 1998;145:233–238.
[16] Greenwood A. Electrical Transients in Power Systems, New York: Wiley, 1991.
[17] Mohan N, Undeland TM, Robbins WP. Power Electronics: Converters, Applications and Design. New York: Wiley, 2003.

6

Modern Arrangement for Reduction of Voltage Perturbations

Ryszard Strzelecki, Daniel Wojciechowski and Grzegorz Benysek*

Department of Electrical Engineering,
Gdynia Maritime University,
81–87 Morska Str, Gdynia, Poland.
Emails: rstrzele;dwojc@am.gdynia.pl

Department of Electrical Engineering, Informatics and Telecommunications*
University of Zielona Góra,
50 Podgórna, Zielona Góra, Poland.
Email: G.Benysek@iee.uz.zgora.pl

The contents of this chapter encompass general problems and the most important issues of power-supply-quality improvement in AC systems. In the context of the above, consideration is given to evaluation of bilateral interactions of receivers with an electrical power-distribution system and methods of their reduction. Also are discussed the basis of operation of the most important compensation-filtration devices and their applications that are joined to the system in parallel or in series. The main emphasis is placed on application of modern power-electronic devices, considered a even more in detail in the following chapters.

6.1 Introduction

Rational decision making relating to the power-supply-quality improvement (feeder voltage) can be made only as a result of evaluation of different technoeconomic aspects of possible actions and solutions. This decision process, in simplified form is presented in Fig. 6.1 [1]. In addition, it requires to answer the following questions:

- To what extent a voltage disturbance in the feeder influences the action of power-supply devices (under different loads) of the feeder?
- Which devices must be protected from power-supply disturbances?
- What is the cause of supply disturbances and how do these disturbances propagate?
- What methods and technical solutions that mitigate voltage disturbances are available?

The first and second questions result from different sensitivities of devices to supply disturbances [2]. Many contemporary devices, with respect to an applied technical solution, are resistant to some disturbances. For these devices, lower voltage quality in the feeder is less important. In the electrical power system, first of all the "sensitive points" must be protected where supply disturbances cause large economic loses or may even lead to a failure.

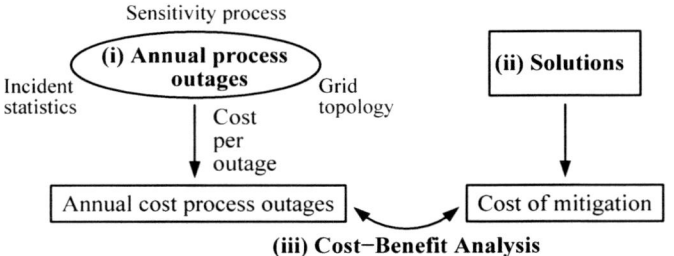

Figure 6.1. The major aspects to finding the best technoeconomic solution

The answers to the third and fourth questions refer to the choice of the points where the devices mitigating supply disturbances are joined and what type of devices were used [3]. The disturbances may be mitigated by both reduction of the cause (source of the disturbances) and reduction of the effects in selected nodes (bus bars) of the electrical power system. In the first case, the quality of the voltage in the whole system is improved. However, it usually requires the application of a compensation-filtration device of relatively high rating power. In the second case only the voltage quality for power receivers that are joined to the selected node is improved. In this case, adequate devices that mitigate voltage disturbances have a lower power rating. This usually involves favorable relation cause – result and mitigation of the disturbances as their distance to the source of the disturbances increases.

6.2 Influence of Load on the Power Quality

The influence of load on voltage disturbances in individual nodes of the electrical power system can be evaluated on the basis of a general 1-phase model with N nodes shown in Fig. 6.2. The values of internode electrical impedance $Z_{i/j}$, between nodes "i" and "j", and node impedance Z_i, between node "i" and the point of reference (potential V_0) result from the system parameters. Same current sources, for example a source I_1 with current value E/Z_1 and internal impedance Z_1, simulates system's supply sources by external sinusoidal voltage with basic pulsation ω_S. Other sources I_i, generated by loads independent from voltage, shape current disturbances with pulsation ω. This system model for pulsation ω can be described by an equation of node potential in a form as follows [4]:

$$[Y] \times [V] = [I]$$
(6.1)

where [V] – vector of nodes voltage, [I] – vector of source currents, [Y] – matrix of self-admittances $Y_{i,i}$ and of mutual admittances $Y_{i,j}$ calculated according to the equations:

$$Y_{i,j} = -1/Z_{i/j}, \quad Y_{i,i} = (1/Z_i) + \sum_{\substack{j=1 \\ j \neq i}}^{N} (1/Z_{i/j}) \quad \text{for} \quad i,j=1, 2,...,N \text{ and } i \neq j. \quad (6.2)$$

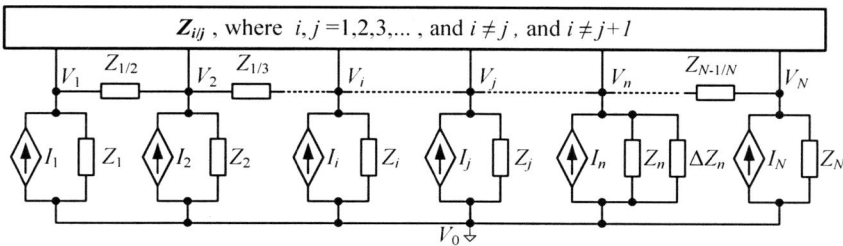

Figure 6.2. General model of electrical power-supply system with N nodes

The solution of Eq. 6.1 is vector [V] that represents the components of voltage with pulsation ω in the nodes of a grid. The solutions formulated in a form:

– general: $[V] = [Z] \times [I]$,

– for voltage in i node: $V_i = Z_{i,i} I_i + \sum_{\substack{j=1 \\ j \neq i}}^{N} Z_{i,j} I_j$, (6.3)

where $[Z]=[Y]^{-1}$ – matrix of self-impedance $Z_{i,i}$ and mutual $Z_{i,j}$, where $i \neq j = 1, 2,...,N$, allows evaluation of the influence of input impedance $Z_{wej(i)} = Z_{i,i}$ of the system (considered on the site of "i" node) and current I_i disturbances on voltage distortion in a node "i".

On the basis of Eq. 6.3 we can also analyse the influence of an additional linear element (for example an LC filter) with impedance ΔZ_n in the system node "n" (Fig. 6.2). The additional impedance ΔZ_n, causes a change in value of all the elements of matrix [Z] according to the dependence [5]:

$$Z'_{i,j} = Z_{i,j} - Z_{n,j} Z_{i,n} / (Z_{n,n} + \Delta Z_n). \quad (6.4)$$

When the value $Z'_{i,j}$ determined from Eq. 6.4, is substituted for the value $Z_{i,j}$ in Eq. 6.3, we obtain a new equation describing voltage in a node "i":

$$V'_i = \sum_{j=1}^{N} Z_{i,j} I_j - \frac{Z_{i,n}}{Z_{n,n} + \Delta Z_n} \sum_{\substack{j=1 \\ j \neq i}}^{N} Z_{n,j} I_j. \quad (6.5)$$

On the basis of Eq. 6.3 and 6.5 we can evaluate the relative change of voltage in the node "*i*", caused by an additional element in the node "*n*" with impedance ΔZ_n. The value is calculated on the basis of the equation:

$$\delta U_{i(n)} = \left|\frac{V_i'}{V_i}\right| = \left|1 - \frac{Z_{i,n}}{Z_{n,n} + \Delta Z_n}\left[\sum_{j=1}^{N} Z_{n,j} I_j \bigg/ \sum_{j=1}^{N} Z_{i,j} I_j\right]\right| = \left|1 - \frac{Z_{i,n}}{Z_{n,n} + \Delta Z_n}\frac{V_n}{V_i}\right|. \tag{6.6}$$

In particular, when *i*=*n* (if we observe voltage changes in the node with an added element), then:

$$\delta U_{n(n)} = |V_n'/V_n| = |\Delta Z_n/(Z_{n,n} + \Delta Z_n)|. \tag{6.7}$$

Eq. 6.7 can also be determined directly from the Norton – Thevenin theorem concerning the change of a system with an equivalent current source [4].

If in the node "*n*", instead of element Z_n, element Z'_n is incorporated then in Eq. 6.4 to 6.7 we should insert the value ΔZ_n, which is calculated on the basis of:

$$\Delta Z_n = -Z_n Z'_n / (-Z_n + Z'_n). \tag{6.8}$$

In a similar way we should take into account the parallel connection of an additional element Z''_n to the element Z_n. In this case, in order to determine ΔZ_n, in Eq. 6.8 we should insert the value:

$$Z'_n = Z_n + Z''_n.$$

The voltage disturbances in nodes involve current changes in particular limbs of the circuit. Even a low-voltage change between two nodes can cause significant current changes in the connecting limb (depending on its impedance). However, the original cause is always current disturbances, generated by loads. Direct evaluation of their influence on currents in the limbs of the system is indicated.

The influence of a current source I_n on the current $I_{(i/j)}$ in the limb between the "*i*" and "*j*" system node (Fig. 6.2) can be evaluated on the basis of the "current disturbances mitigating" coefficient [5]. The modulus of the coefficient, defined as the causality relationship of the current $I_{(i/j)}$ to current I_n (with assumption that currents from other sources $I_k=0$ for $k \neq n$), is calculated for pulsation ω on the basis of the equation:

$$\left|K_{(i/j),n}^I\right| = \left|I_{(i/j)}/I_n\right| = \left|(V_i - V_j)/(I_n Z_{(i/j)})\right| = \left|(Z_{i,n} - Z_{j,n})/Z_{(i/j)}\right|, \tag{6.9}$$

where $I_{S(i/j)}$, $Z_{(i/j)}$ – current and impedance of the limbs between the "*i*" and "*j*" nodes;

$Z_{i,n}$, $Z_{j,n}$ – mutual impedance between the "i" and "j" nodes, and node "n", where the I_n source is added. From Eq. 6.9, after taking into consideration all current sources I_n, $n=1,2,...,N$ (Fig. 6.2) it also results that the changes of current component with pulsation ω in the limb connecting the nodes "i" and "j" do not exceed the value of:

$$\left| I^{max}_{(i/j)} \right| \leq \sum_{n=1}^{N} \left| K^I_{(i/j),n} \right| \left| I_n \right|. \tag{6.10}$$

A more detailed calculation should be made when taking into account the phase shift between currents I_n and the phase characteristics of calculated "current disturbances mitigating" coefficients', which depend on the difference in the impedances $Z_{i,n} - Z_{j,n}$.

6.2.1 Investigation Results

Equation 6.6 and 6.9 allow us to analyse the influence of receivers on the quality of a power supply and find confirmation in the research results on the simple distribution system, which is presented in Fig. 6.3. The system is symmetrically supplied by a 3-phase voltage source U_S with the frequency 50 Hz and an efficient value of line voltage of 400 V. Furthermore, the example shows specific cases of influence of different loads on voltage in the buses 2 and 3, and on the currents in two separated feeders with impedances $Z_{S1} = Z_{S2} = 0.1+j\times\omega\times3.10^{-4}$ Ω.

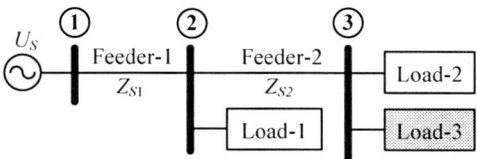

Figure 6.3. Single-line diagram of a simple power-distribution system

The results, which are presented in Fig. 6.4, relate to the direct influence of starting a squirrel-cage induction motor. The distribution system was loaded as follows:

Load-1. Symmetrical R-L impedance: $S=43$ kVA, $\cos\varphi=0.54$
Load-2. Symmetrical ohmic resistance: $P=15$ kW
Load-3. Squirrel-cage induction motor: $S=16$ kVA, $\cos\varphi=0.8$.

During starting the motor in a time of 1.25 s, the effect of voltage reduction on supply buses occurred, which was caused by starting currents. After finishing motor starting, the voltage stabilizes at the level slightly lower than the supply voltage. The voltage drop on the impedances Z_{S1} and Z_{S2} of the feeder does not exceed the value allowed by standards.

The following research related to the evaluation of the unbalanced load influence. For the purpose of the research the following elements were added to the system:

Load-1. Squirrel-cage induction motor (IM1): $S=16$ kVA, $\cos\varphi=0.8$
Load-2. Squirrel-cage induction motor (IM2): $S=16$ kVA, $\cos\varphi=0.8$
Load-3. Nonsymmetrical impedance R-L (1-faza): $S=25$ kVA, $\cos\varphi=0.54$.

The research results in Fig. 6.5, with particular focus on pulsation of the torque of motors with frequency 100 Hz, provide information about unbalanced voltages in the power bus that occurred during the research. Voltage unbalance decreases when the distance to the source of distributions increases (load-3).

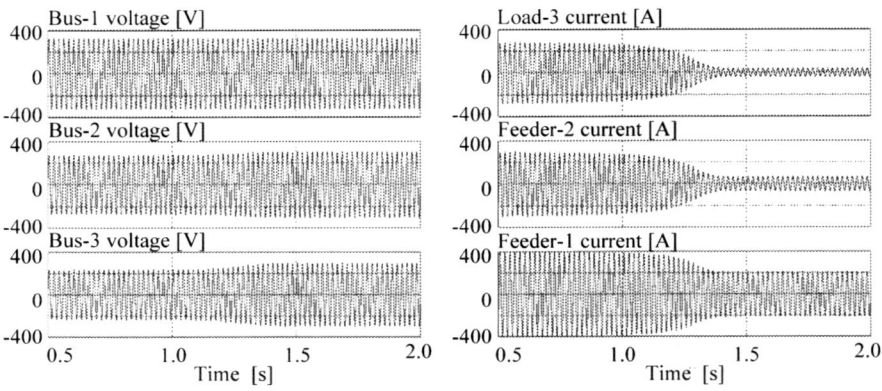

Figure 6.4. Voltage sags caused by starting a squirrel-cage induction motor

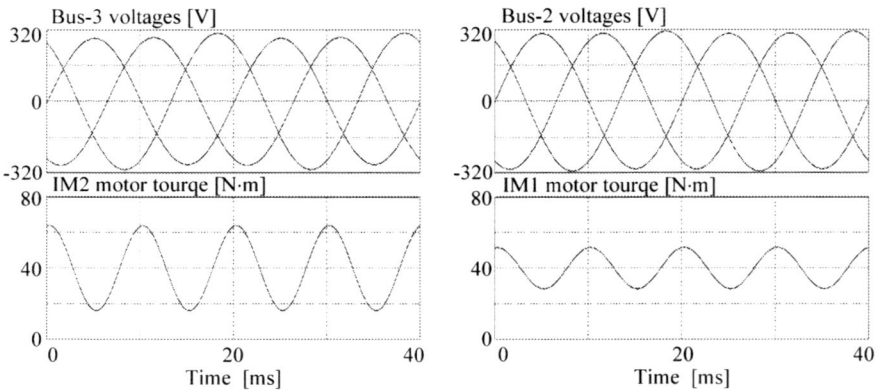

Figure 6.5. Unbalance in the bus voltages caused by unbalanced load

The research also encompassed the evaluation of the influence of the load current DC-offset. The exemplary results obtained are shown in Fig. 6.6. It is noticeable that DC components occurred in the busbar voltages. The results obtained relate to the following load:

Load-1. Symmetrical R-L impedance: $S=43$ kVA, $\cos\varphi=0.54$

Load-2. Symmetrical ohmic resistance: $P=15$ kW
Load-3. Single phase, half-period diode rectifier with R loads: $P=14$ kW.

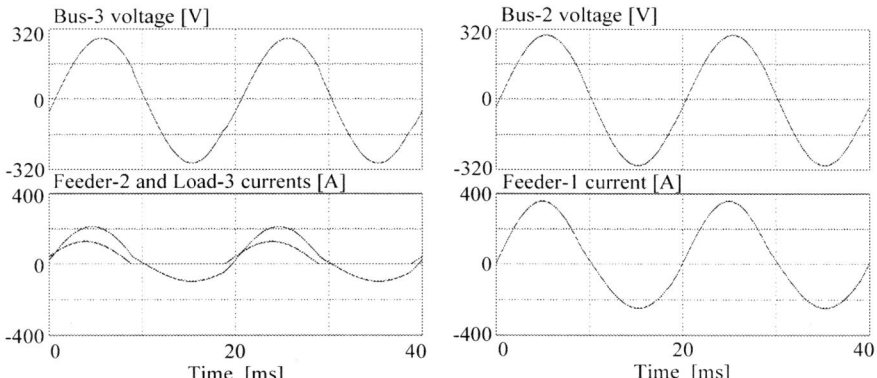

Figure 6.6. Effect of the DC offset in loads

The following research case refers to the effects of the influence of higher harmonics of the load current. The harmonics source served a 3-phase full-bridge rectifier, whereas the distribution system was loaded as follows:

Load-1. Symmetrical R-L impedance: $S=43$ kVA, $\cos\varphi=0.54$
Load-2. Symmetrical ohmic resistance: $P=15$ kW
Load-3. Three-phase, full-bridge diode rectifier with R loads: $P=23$ kW.

As follows from the current courses and voltages presented in Fig. 6.7, the higher harmonics induced by a rectifier into the distribution system (with relatively low power) first of all caused current distortions of a feeder, which is the closest to the source. In a distant feeder, because of its additional load, current distortions are lower. The same relation holds for to low-voltage distortion in busbars, mainly caused by notches.

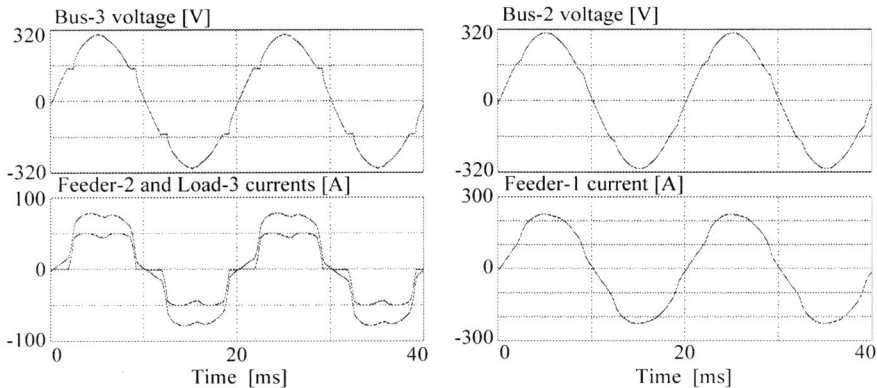

Figure 6.7. Harmonic distortion caused by a 3-phase full-bridge rectifier load

The reason for notches in voltage courses, presented in Fig. 6.7, are commutation processes in the rectifier. The longer these processes, the more noticeable are the notches in the power bus [6]. To bring out the effect, the load of the distribution system (Fig. 6.3) was changed as follows:

Load-1. Symmetrical ohmic resistance: $P=20$ kW
Load-2. Absent
Load-3. Three-phase, full-bridge diode rectifier with R-L loads: $P=29$ kW.

The results are presented in Fig. 6.8. A change in load, together with an additional reactor on the site of the DC rectifier, also caused higher current and voltage distortions than in the previously described case (Fig. 6.7). It should be noted that distortions in a power bus are often a reason for many negative effects: resonance, motors overheating, false reactions of control systems and secure systems.

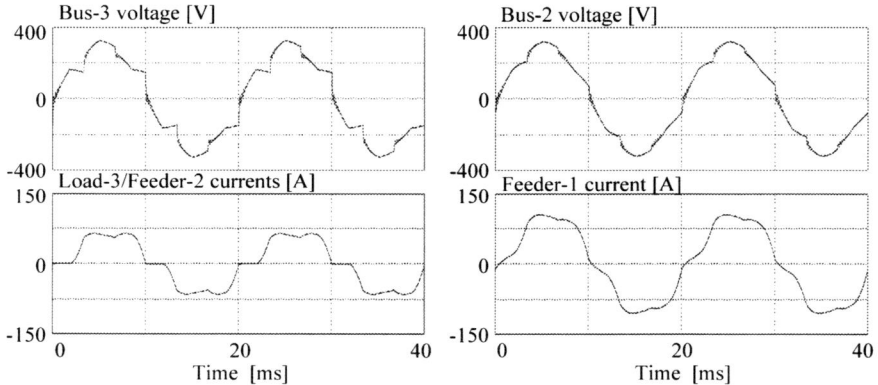

Figure 6.8. Notch in bus voltages

6.2.2 Other Important Influences of Loads

6.2.2.1 Light and Heavy Load Influence

The conducted experiments indicate that the main worry of the customers is too low a voltage in the feeder. In accordance with IEC 60038 standards, the rated voltage in the feeder LV should amount to 230 V ±10%. The requirement of allowable voltage deviation ±10% applies also to the feeders MV (up to 110 kV). The most common reason for a voltage drop is too heavy load on the feeder.

Fig. 6.9 shows a simple 1-phase equivalent diagram of a feeder with a single linear load. The assumed symbols U_S and R_F and X_F represent, respectively supply voltage, resistance and reactance of the feeder. The resistance R_L and the reactance X_L are the elements of the load equivalent diagram. The diagram easily explains the reasons for a voltage drop in a point of common coupling (PCC). Additionally, it illustrates phasor diagrams for three hypothetical loads. The voltage is influenced by a load-power change as well as a load-power factor ($\cos\varphi_L$).

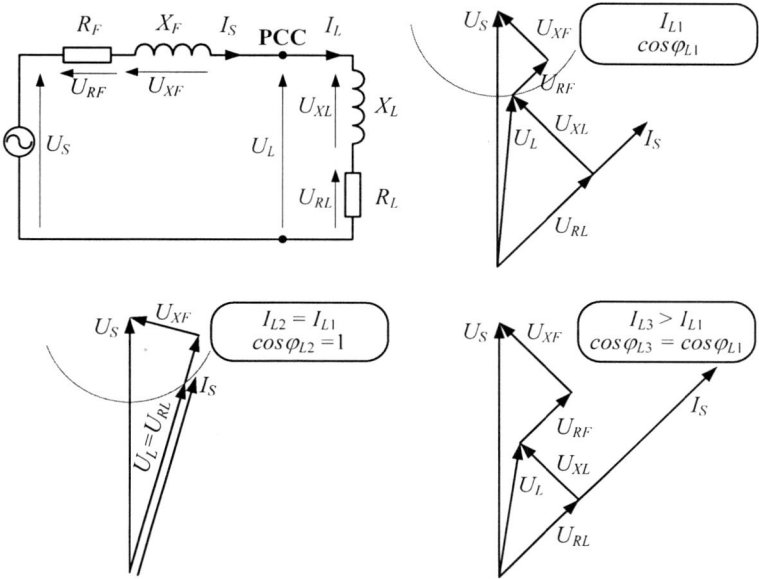

Figure 6.9. Simple equivalent-circuit diagram of a feeder with load

Fig. 6.10 shows a typical voltage profile for a feeder under light and heavy load conditions. The figure shows that under heavy load, the voltage at the end of the line will be less than the allowable minimum voltage. However, under the light load condition, the voltage supplied to each customer will be within the allowable limit. Calculation of the voltage profile and voltage drop is one of the major tasks in distribution-system design. To calculate the voltage drop, the feeder is divided into sections [7,8]. The sections are determined by the loads current $I_1, I_2, \ldots, I_N$.

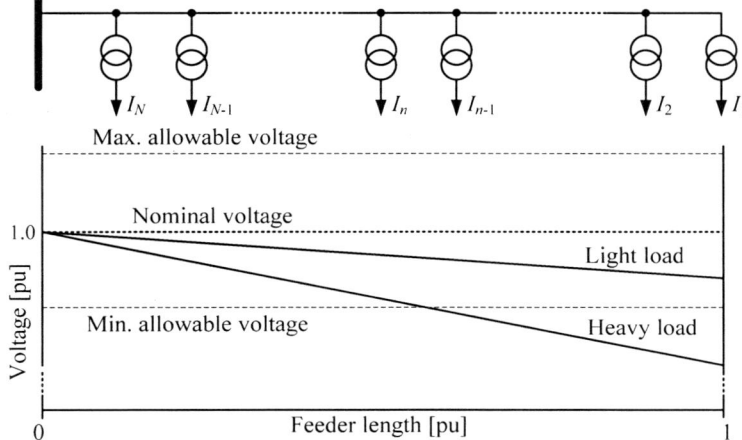

Figure 6.10. Typical voltage profile along a feeder

Feeder parameters are selected in a way that finally guarantees the correct voltage level in the case of the heaviest load. In a dispersed installation it causes the necessity to significantly redimension the feeder (owing to the length of circuits) or to install additional systems of transformer substations. Fig. 6.11 shows an example of the change of voltage profile after installation of an additional transformer Tr. The transformer allows the right voltage level to be maintained near the point of installation. As the distance from the substation increases, in the case of heavy voltage, the voltage decreases faster than in the feeder without the transformer [8]. Fig. 6.12 shows the influence of the circuit length on the extent to which the transformer is used (transformer capacity factor) for various sections of overhead line. The long feeders cause the power grid to be "soft". Improvement of the parameters can be achieved by decreasing the inner impedance of the feeder. However, these actions require relatively large investment expenditures, among the others relating to the application of more expensive apparatus, owing to short-circuit strength. Taking into consideration the shape of the twenty-four-hour load curve, selection of the parameters for the heaviest load causes the supply network to be used in accordance with assumptions for only a few hours a day.

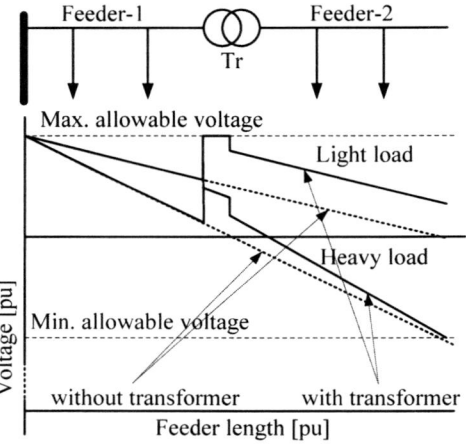

Figure 6.11. Typical voltage profile along a feeder with and without an additional transformer

6.2.2.2 Interharmonics and Subharmonics, Flicker Phenomena
Interharmonics, always present in the power system, have recently become of more importance since the widespread use of power-electronic systems results in an increase of their magnitude. Interharmonics are currents (or voltages) with a frequency that is a non integral multiple of the fundamental supply frequency [9]. The term "subharmonics" does not have any official definition − it is a particular case of interharmonics of a frequency less than the fundamental frequency.

Basic sources of the interharmonics currents include:

- Arcing loads, in particular in arc furnaces

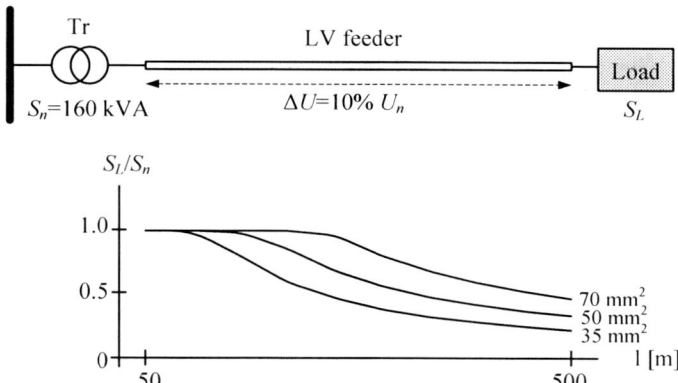

Figure 6.12. Transformer capacity factor as function of the feeder length and section

- High-power variable-load electric drives, in particular in welding machines
- Static converters, in particular in direct and indirect frequency converters

In general, we can distinguish two basic mechanisms to generate current interharmonics [10]. The first mechanism is generation of components, as a result of current rapid changes in equipment and installations. Interharmonic disturbances are generated by loads operating in a transient state, either continuously or temporarily, or, in many more cases, when the amplitude modulation of currents occurs. The second mechanism is the asynchronous switching (*i.e.* not synchronized with the power-system frequency) of semiconductor devices in static converters. In many kinds of equipment both mechanisms take place at the same time. The greatest effect of the presence of currents interharmonics are voltage fluctuations in the rms voltage magnitude (Fig. 6.13), depending on the magnitudes of the current components and the supply-system impedance at that frequency. These voltage fluctuations may result in light flicker (luminous flux fluctuation), possibly annoying for customers, even for the voltage variations of only a few tenths of a percent. Fig. 6.13 illustrates the way in which a small voltage change produces a noticeable effect on the luminous flux of a bulb.

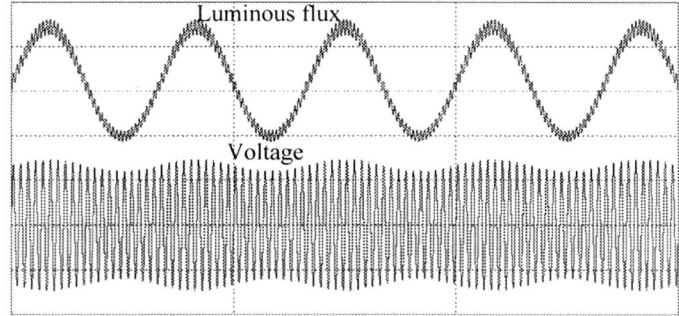

Figure 6.13. Change in luminous flux resulting from a voltage fluctuation

6.3 Reduction of Load Influence on the Voltage Profile

Devices need to lessen load influence or group of loads on a supply grid, especially on line voltage, are generally called compensators. Compensators are mostly shunts to loads, and depending on their position in the electrical system we can distinguish:

- separate compensator – local
- group compensator
- central compensator.

The above methods of compensators arrangements are presented in Fig. 6.14.

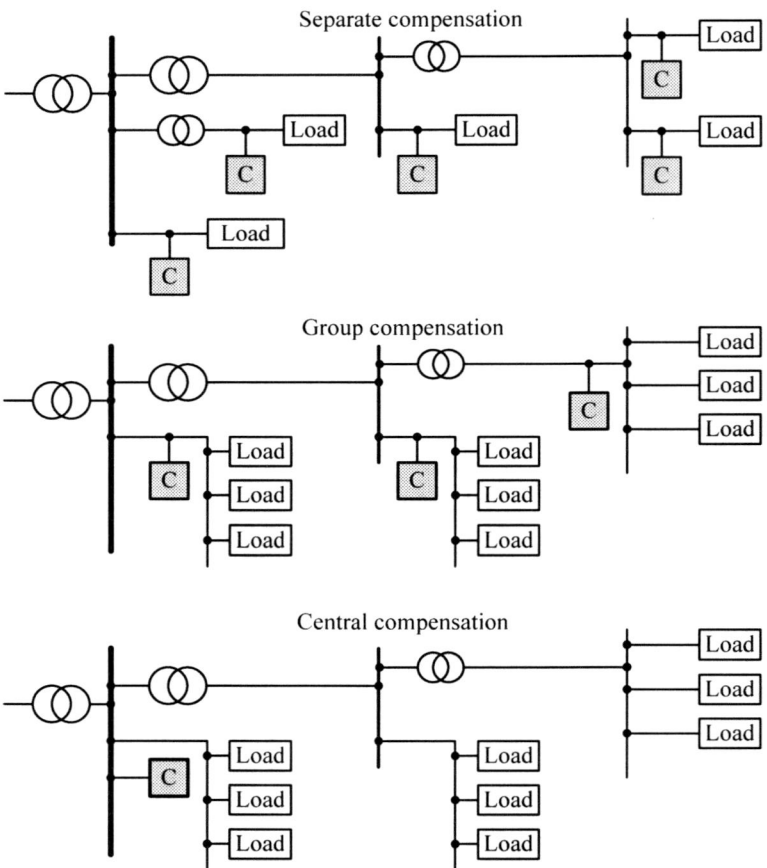

Figure 6.14. Methods of location of compensators

Method 1 – separated compensation – applies to direct cooperation between compensator and load, introducing to the feeder current disturbances. It has been proposed that the load and the compensator together make up a linear load for the feeder of resistance character. Then, only active power is absorbed from the feeder. In addition to the power advantages the method leads to lower investment in installation. The main disadvantage of separated compensation is not making any use of a compensator in the time when the load is disconnected. This method seems to the most useful in the case of compensation of load current harmonics and fast-changing fluctuations of the load reactive power.

Methods 2 and 3 – group and central compensation – allow the use compensators independently from the loads. These methods require additional switchgear and security apparatus, and the compensators should allow control of the compensation current. The range of compensation current changes should be established on the basis of twenty-four-hour analysis of the compensated loads. In general, the group of loads is characterized by frequent changes of the total load. This requires application of compensators controlled automatically. It should be noted that while these are constant development of the power-electronic systems, and the tendency to dispersed generation, the effects of the central compensation decrease. Central compensators are not able to assure effective generation of the system, especially in the case of long distances to different loads. Therefore, the most used is nowadays the group compensation, and for power heavy loads the separated compensation.

6.3.1 Principle Compensation of the Load Influence

We can distinguish two main methods for compensation of load influence on the voltage in power busbars (at the PCC point): parallel (current) and series (voltage). In the case of parallel compensation, the compensator generates currents in anti phase to undesired components of the load current, and the same diminishing them in the current of a feeder. In series compensation, task for a compensator is such a voltage change in load that the current absorbed by the load will not contain any undesired elements. The second method can be applied in practice only for selected loads and relatively seldom.

6.3.1.1 Parallel Compensation

The principle of parallel compensation and phasor diagrams, which illustrate the principle are presented in the Fig. 6.15. The compensator is represented by the current source I_C. The other symbols are the same as in Fig. 6.9. The presented diagrams show three different cases of load compensation. In the first case the load is not compensated. The voltage in the PCC node is lowered with respect to the supply source voltage with value $\Delta U_{S1} > 0$. After adding a compensator (case 2) the voltage drop U_F in the feeder decreases and the voltage in the node PCC ($0 < \Delta U_{S2} < \Delta U_{S1}$) increases. Case 3 concerns the situation of overcompensation. A light load and considerable compensation current caused voltage increase in the PCC node over supply voltage ($\Delta U_{S3} < 0$). At the same time it should be stressed that similar phasor diagrams can also concern individual harmonic components in the case of nonlinear load. The representation provides the possibility to present such a load with a linear circuit and noncontrollable monoharmonic current sources or voltage, modelling distortions caused by the load.

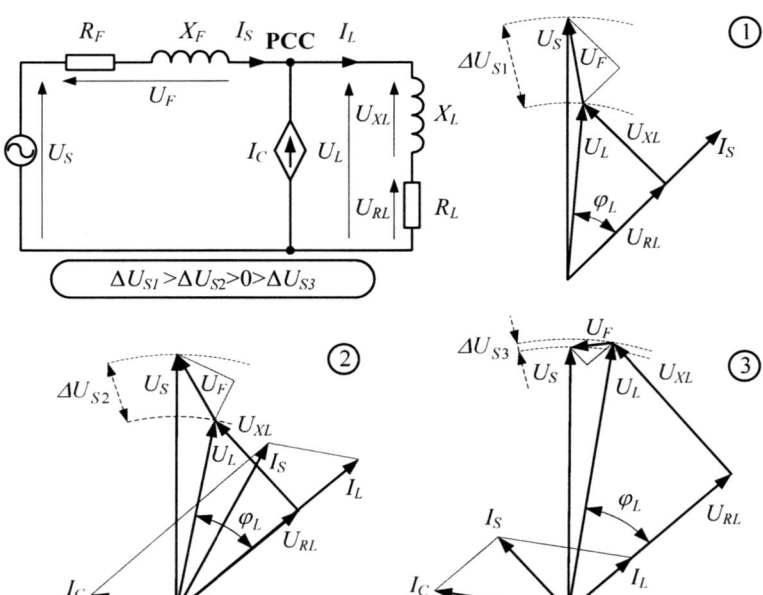

Figure 6.15. Illustration of the parallel compensation of load influence

In general the current I_C of a parallel compensator, minimizing the effect of load influence on the voltage in the feeder, can be determined as a result of minimization of the functional:

$$\left\| \bar{U}_S - \bar{U}_L \right\| \to \min \quad \text{or} \quad \left\| (\bar{I}_L - \bar{I}_C) \right\| = \left\| \bar{I}_S \right\| \to \min, \tag{6.11}$$

where $\bar{U}_S$ and $\bar{U}_L$ – vectors of components of voltage sources and load, $\bar{I}_L$ and $\bar{I}_C$ – vectors of components of load current and compensator, $\bar{I}_S$ – vector of components of the feeder current, $\left\| \ldots \right\|$ – symbol of the norm selected with respect to the compensation priority (rms, THD, etc.). Minimization must take into consideration the following limitation:

$$Re(\bar{U}_L \bar{I}_C^*) = P_C = 0, \tag{6.12}$$

where P_C – average active power absorbed/developed by the compensator, $\bar{I}_C^*$ – vector of the current components, conjugated with the vector $\bar{I}_C$, $Re(\ldots)$ – the real part. From limitation (Eq. 6.12) it follows that in the interval of active power averaging, any compensator can be treated as a reactance element. If the condition 6.12 is not met, the compensator should be additionally supplied or loaded. In practice, aside from the limitation we should also take into consideration the value of the compensation current, limited for the given compensator. Also, additional conditions can be taken into consideration.

6.3.1.2 Series Compensation

Series compensation of the load influence on the voltage in supply rails can be applied only when voltage changes in load, which occur in this case, are not harmful and not too significant. The possibility should be considered to apply the compensation only when the load is characterized with very low impedance Z_L for the compensated current components. The case is described by the equivalent diagrams presented in Fig. 6.16 (Tevenin – Norton theorem). The lower the impedance Z_L, the lower the equivalent voltage U_D caused by compensated component of the load current – current I_D. To compensate the component only a low voltage of the series compensator is necessary that is generated in antiphase to the voltage U_D. Figure 6.17 illustrates the principle of series compensation with the example of a linear load.

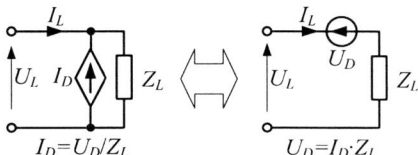

Figure 6.16. Equivalent circuits of the load

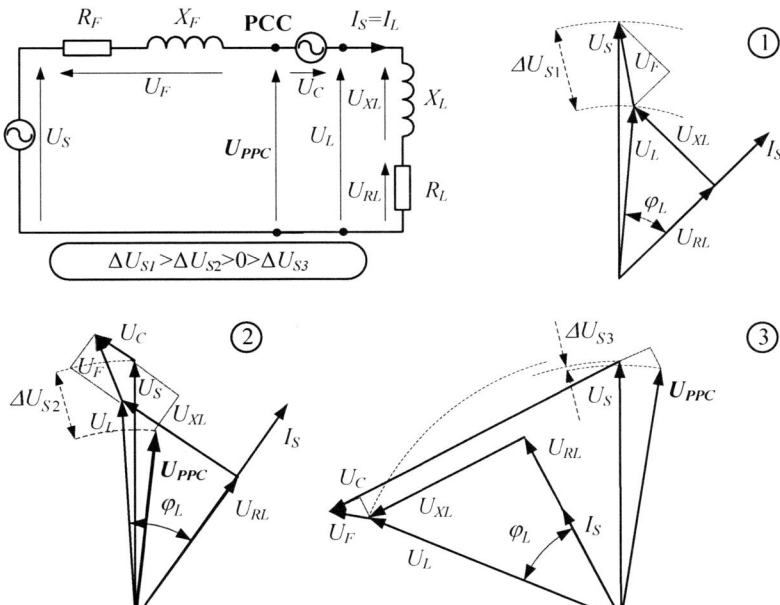

Figure 6.17. Illustration of the series compensation of load influence

The compensator in Fig. 6.17 is represented by the voltage source U_C. The presented phasor diagrams show three cases. The first case is the same as the one presented in Fig. 6.14. After the compensator has been added (the second case) the voltage in the feeder U_F decreases, voltage U_L in the load increases only a little, and voltage U_{PCC} in the node PCC ($0 < \Delta U_{S2} < \Delta U_{S1}$) increases even less. In case 3, aberrations of the voltage levels U_L and U_{PCC} from the supply voltage level U_S are very small, however, the voltage U_C is very high. It involves a relatively high load impedance for the compensated component (explained above) and the limitation:

$$Re(\overline{U}_C \overline{I}_S^*) = P_C = 0, \tag{6.13}$$

where $\overline{I}_S^*$ – vector, conjugated with the vector of the current components $\overline{I}_S$, $\overline{U}_C$ – vector of compensation-voltage components.

The compensators without an additional power source, connected in series between the PCC node and the load (Fig. 6.17), are suggested in practice as series active filters of harmonics of a diode rectifier with an output capacity filter [11 – 13]. In this case, the cause of voltage distortion V_{PCC} is compensated. Such an application is discussed further in Sec. 6.3.2. The series compensators, installed before the PCC node, compensate only the effect of the current load, *i.e.* the appropriate voltage drop in the feeder. In this context, they should be rated among voltage-distortion compensators, discussed in Sec. 6.4.

6.3.2 Review of Selection Problems

6.3.2.1 Load-unbalance Reduction

Reduction of supply-voltage unbalance, produced by unbalanced load currents, can be obtained by reduction of the supply system impedance. However it involves large investments, often economically unfounded. Another traditional way is a suitable arrangement of unbalanced loads in the feeder. However it is sometime impossible and ineffective. An adequate and very simple example is shown in Fig. 6.18. Although the 3-phase supply source (U_A, U_B, U_C) is balanced and symmetrically loaded, the voltage at the end of the feeder (U_a, U_b, U_c) is unbalanced. It is caused by different voltage drops in the sections of the conductor of a cable in the feeder. To eliminate the influence of unbalanced loads, especially with high power, first of all balancing compensators are applied.

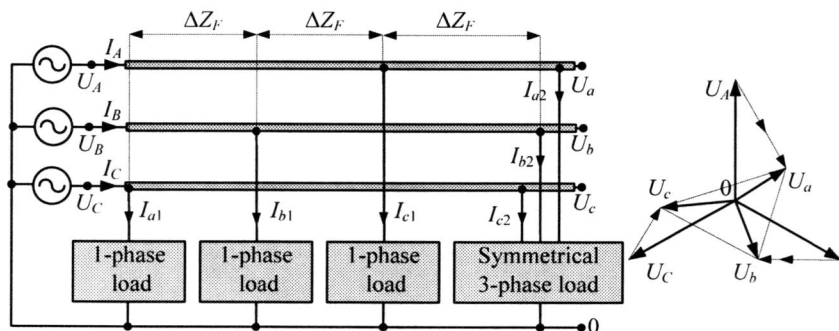

Figure 6.18. The example of ineffective symmetrization through loads arrangement

As results from electrical circuits theory, any unbalanced load of the 3-phase feeder without a neutral cable can be exchanged for an equivalent balanced load and connected to it in series a 2-phase load. This partition of the unbalanced load is presented in Fig. 6.19, where Y – admittance. It facilitates analysis of the voltage influence on the value of a negative-sequence load current. The analysis also determines the way to supply two 1-phase loads with approximately equal apparent power. The load with lower phase angle (larger $\cos\varphi_L$) should be supplied with a phase-lag voltage with respect to voltage, supplying the second 1-phase load. The above conclusion also allows the right way to supply an unbalanced 3-phase load to be determined, minimizing the negative sequence of absorbed current. The above-described principle of unbalanced load can be realized by application of a so-called Scott transformer, presented in Fig. 6.20a.

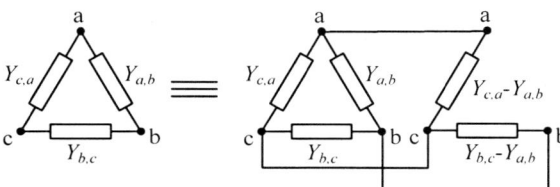

Figure 6.19. Division of nonsymmetric load into a symmetric section and 2-phase section

The Scott transformer is especially fitted for supplying two 1-phase loads with equal apparent power, *i.e.* heating coils of induction furnace, two sectionalization supply lines of electric traction [14,15]. By changing the tap we can match the phase shift $\varphi_{A,B}$ between voltages supplying two loads within the following interval:

$$\pi/3 \leq \varphi_{A,B} = \pi - arctg\left[\sqrt{3}\,(z_C + z_B)/(z_C - z_B)\right] \leq 2\pi/3,$$

where z_B and z_C – number of turns of winding (Fig. 6.20a). Because the Scott transformer is affixed between the feeder and the load, according to the assumed classification (Sec. 6.3.1) it is rated among the series compensators of load influence on the voltage in the power rails.

Balance of the load, which can not be divided into the balanced and 2-phase parts, must be carried out with a shunt arrangement, usually fulfilling the purpose of balancing and reactive power compensator. Its parameters should be selected in a way that assures that, seen from the feeder's side and together with the load, it fulfills the general condition for circuits balance:

$$Y_{a,b} + aY_{b,c} + a^2 Y_{c,a} = 0, \tag{6.14}$$

where $a=exp(j2\pi/3)$. The balancing device must also fulfill Eq. 6.12. Fulfilling the conditions (Eqs. 6.12 and 6.14) is possible thanks to Steinmetz's principle [15,16]. According to this rule, presented in Fig. 6.20b, every 1-phase load of resistance character (or any other load after substitution susceptance compensation) can be balanced by means of wattles elements LC of such values that the values of phase currents rms satisfy the equation:

$$I_{a,b}/\sqrt{3} = I_{b,c} = I_{c,a}.\tag{6.15}$$

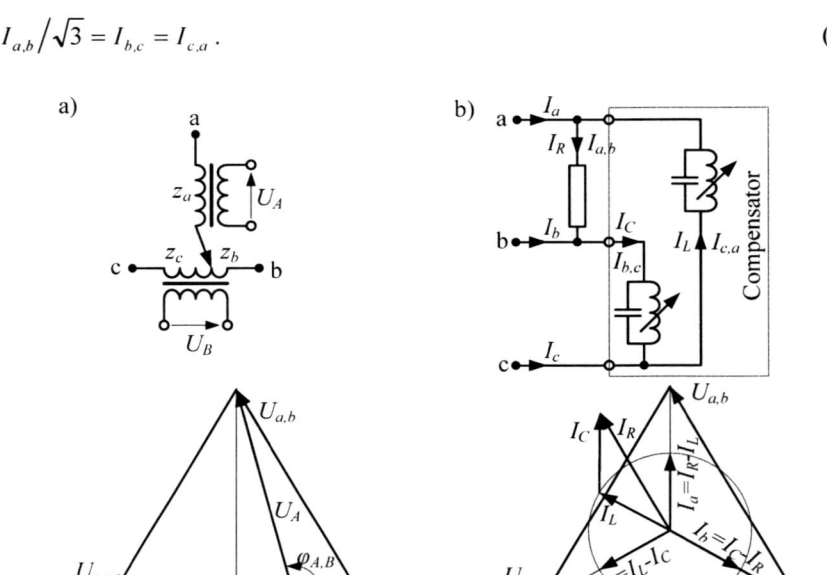

Figure 6.20. Scott transformer (a) and Steinmetz compensator (b)

The balanced methods presented above (Fig. 6.20) concern only linear loads. In the case of a variable load parameter, a change in the parameters of compensator elements is required. For this purpose in a Scott transformer, tab thyristor switches are applied. In the Steinmetz compensator, instead of single LC elements, FC/TCR systems (fixed capacitor/thyristor controlled reactor), TSC (thyristor switched capacitor) or TCS/TCR (thyristor switched capacitor/thyristor controlled reactor) are applied. Despite range the complex construction, these solutions are not efficient enough for rapidly changing loads, with the frequency 10–15 Hz [15]. This disadvantage does not concern compensators with arrangements of the type D-STATCOM (distribution static synchronous compensator) and D-SSSC (distribution static synchronous series compensator) [17,18]. Their exemplary application and application of Scott's transformer and Steinmetz's compensator to supply 1-phase rail traction from the 3-phase grid is presented in Fig. 6.21. D-STATCOM systems are discussed in more detail in Chap. 7, and D-SSSC systems in Chap. 8.

6.3.2.2 Load Reactive Power Reduction
Reactive power compensation is used to unload the feeder from the passages of reactive currents and the same reduction of voltage drops [19,20]. The influence of

compensation on the voltage profile in the feeder is presented in the Fig. 6.22, where $I_{L(Re)}$ and $I_{L(Im)}$ – active and reactive components of the load current, I_C – compensator current (reactive), I_F – feeder current.

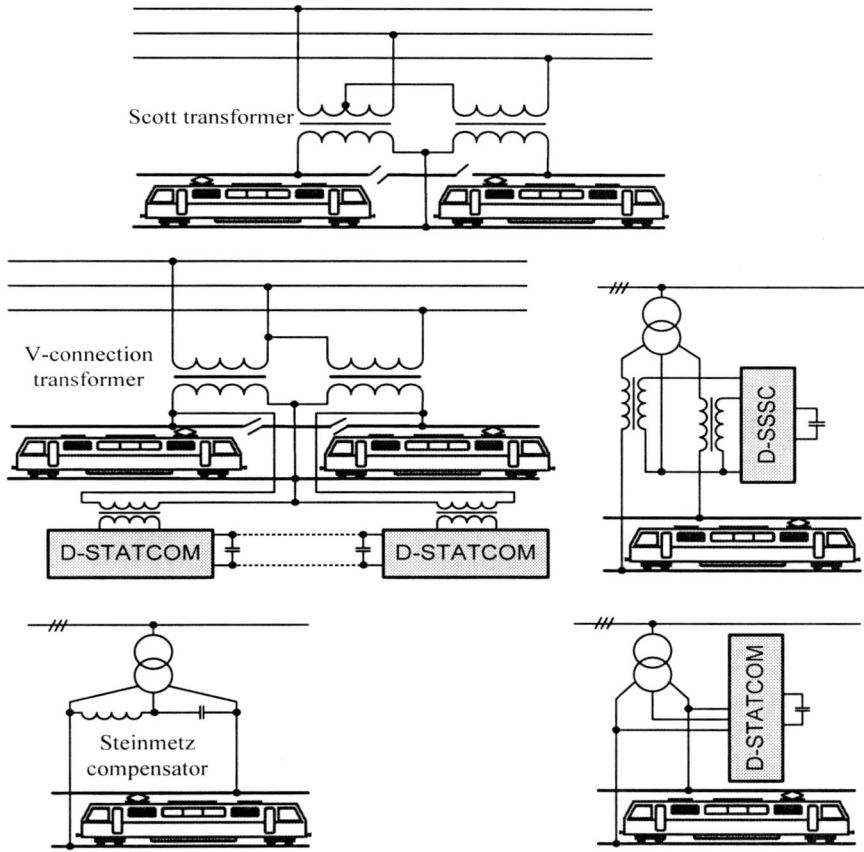

Figure 6.21. AC railway connections to a 3-phase grid using different arrangements

The traditional method of reactive power compensation is application of synchronous machines and capacitor banks. Nowadays, application of synchronous machines to reactive power compensation is justified only in the cases where these machines were installed for a different major purpose, and their rated power significantly exceeds the load's demand. More commonly used are the capacitor banks. However, when designing them we should remember occasionally occurring series and parallel resonance in the feeder. Ignorance of this has been many times the cause of damage and failure in supply systems. It would not have happened if beforehand a simple and effective way to counteract resonances had been applied, which is adding damping reactors in series with a capacitor bank. Such a series circuit should be of capacitive character in the fundamental harmonic frequency, however, in higher harmonic frequencies it should be of inductive character. It follows that in the case when in the supply system the lowest higher harmonic is 3rd order, reactor reactance damping in the fundamental frequency should be higher than $0.12X_C$, where X_C – capacitor reactance in fundamental frequency. In the case when only 5th or higher harmonics occur, reactor reactance can be lower, however, it should be higher than $0.04X_C$. Reactor reactance must not be unlimitedly high because the reactor reduces the efficiency of capacitor banks as a reactive power compensator, moreover, it causes an increase of voltage in capacitors.

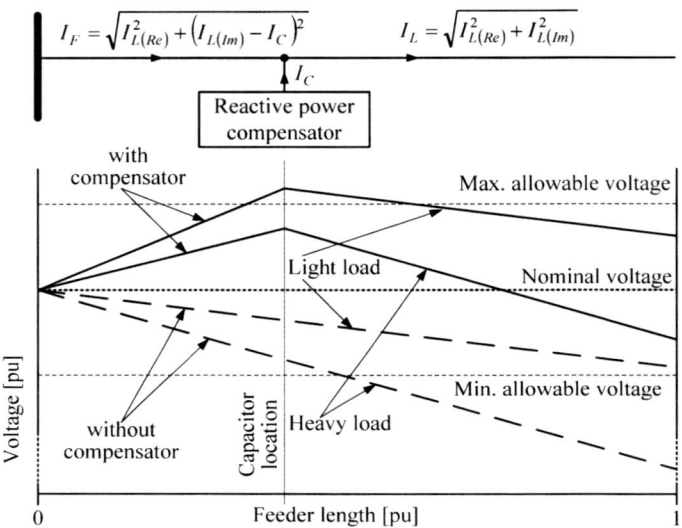

Figure 6.22. Effect of the reactive power reduction on the voltage profile

Undeniable advantages of compensating capacitor-like devices:
- Possibility to install at any point in the feeder;
- Ease of construction and development;
- Not large active power losses (in the case of undistorted currents and voltages)
- Easy installation and low investment and utilization costs.

In contrast to these advantages, their significant disadvantages should be presented, such as:
- Low operation rate and low mechanical resistance of contactor and stepwise character of reactive power control, which makes their application difficult or even can diminish compensation effects in the case of quick changes of load reactive power over a broad range.
- Occurrence of current surge, dependent on the moment of switching capacitor banks to the feeder and sensitivity of capacitors to higher-harmonic current overloads.

These disadvantages can be partially eliminated by application of stepping reactive power compensators with thyristor switches, automatically switching the correct sections of the capacitor banks. Also applied are controllable compensators with smooth reactive power changes by means of inductive current controllers, adjoined in parallel [3,15,21 – 23]. The example of solution of a capacitor-bank thyristor switch arranged in a triangle and the example of a controllable reactive power compensator are presented in Fig. 6.23.

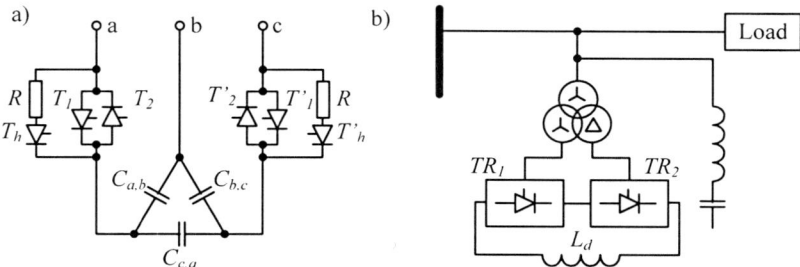

Figure 6.23. Examples of thyristor switch (**a**) and controllable compensator (**b**)

In the switched system, presented in Fig. 6.23a, connection of the capacitor banks takes place in two stages. In the first stage, the capacitor $C_{a,b}$ is switched, and switching follows preliminary capacitor charging to the voltage equal to the amplitude of the grid. For that purpose resistor R and thyristor T_h are selected. When the voltage in the capacitor reaches its maximal value it is followed by connecting the main thyristors T_1 and T_2. In the second stage of the process, the two left capacitors $C_{b,c}$ and $C_{c,a}$ are switched on. This stage is similar to the first stage with the only difference being: the switching moment of the main thyristors T'_1 and T'_2, and the voltage to which the capacitors $C_{b,c}$ and $C_{c,a}$ are preliminarily charged.

In the controllable capacitor system, presented in Fig. 6.23b, to the capacity reactive current of capacitor banks is added an inductive reactive current absorbed by 12-pulse thyristor rectifiers (TR_1 and TR_2), loaded by a smoothing reactor L_d. The reactive current value is regulated by a value change of the output voltage rectifier. The voltage is very low and voltage regulation is carried out through the change of the thyristor firing angle around the value $\pi/2$. The capacitor, with regards to symmetric typology and symmetric control, generates equal reactive power in each phase. Thereby, it can not compensate the negative component of the reactive power, in the case of unbalanced load compensation. Other solutions for controllable compensators, allowing component compensation opposite to the load current, in D-STATCOM systems [17,24,25], are also presented in Chap. 7.

6.3.2.3 Reduction of Load High Harmonics

There are many known methods that result in reduction of undesired harmonics generated by loads. The traditional reduction means are resonant LC filters [3,4,9,19,26]. These filters consist usually of branches of capacitors and reactors connected in series, illustrated in Fig. 6.24. The number of branches depends on the number of filtered harmonics (in Fig. 6.24 for 3-, 5-, 7-, 11- and 13-harmonics). The capacity and inductance in each branch is matched on the basis of the voltage resonance condition for the selected harmonic frequency. Because the impedances of series *LC* circuits during resonance are low, therefore, the filter shunts adequate load current harmonics. In this way resonant filters decrease their influence on the voltage in the supply bus.

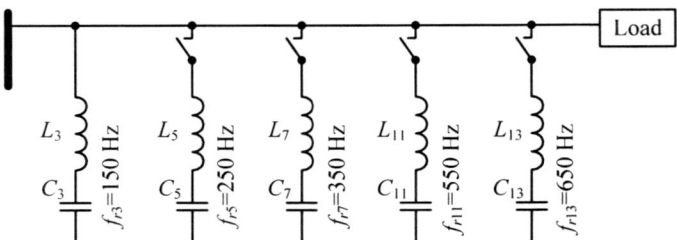

Figure 6.24. Example of connection of typical shunt *LC* filters

LC filters of higher harmonics are also often used in simultaneous compensation of reactive power [20,23,26]. In these cases, the filter is divided into small groups, suitably switched depending on the compensated reactive power. It could be done, for example, as is presented in Fig. 6.24, although this is not the most advisable way, it is very cost effective. When the demand for capacity reactive power decreases, the branches of the filter are disconnected one after another, starting with the branch tuned in to the highest harmonic frequency. This usually does not lead to a significant increase of the voltage distortion in the supply bus, because reduction of the reactive power is mostly connected to reduction of the load current, and also the absolute value of harmonic components.

Figure 6.25 illustrates the filtration qualities of a typical three-branch resonance *LC* filter, with branches tuned to the frequency of 5- and 7-harmonic of high quality factor and a branch tuned to 11-harmonic of low quality factor. As we can see from the filtration characteristic $I_F(\omega)/I_L(\omega)$, filter *LC* additionally causes strengthening of some specific frequencies. This is the result of series resonance between the feeder inductance and filter elements. Parallel resonances occur at as many frequencies as there are branches tuned to various harmonics, of which the filter is constructed, and always below the frequency of the series resonance of this branch. For that reason application of *LC* filters tuned only to higher harmonics is unacceptable when lower harmonics or interharmonics can occur in the feeder. Installation of the *LC* filter should be preceded by analysis of the frequency feeder characteristics and the load. The influence of ageing of filter elements should be also evaluated [9,5].

Some utilities use sound frequencies to control street lighting, night storage heating, and other systems for demand-side management of the load in their system. Care must be taken not to short out these signals and make them ineffective. The closer the signal frequency is to the resonant frequency of an acceptor circuit, the lower the impedance of that circuit is at that signal frequency. When the installation is fed from a dedicated transformer the associated inductance may well be high enough to ensure that there is no effect on the signaling frequencies. Otherwise it may be necessary to install a parallel *LC* rejection filter – tuned to the signaling frequencies as shown in Fig. 6.26 (with a utility that uses 183.3 Hz signals, 13/3 of the fundamental frequency).

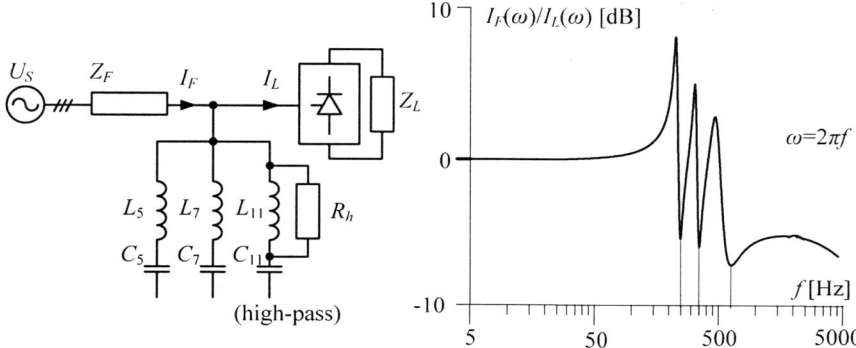

Figure 6.25. Filtering performance of the shunt LC filters

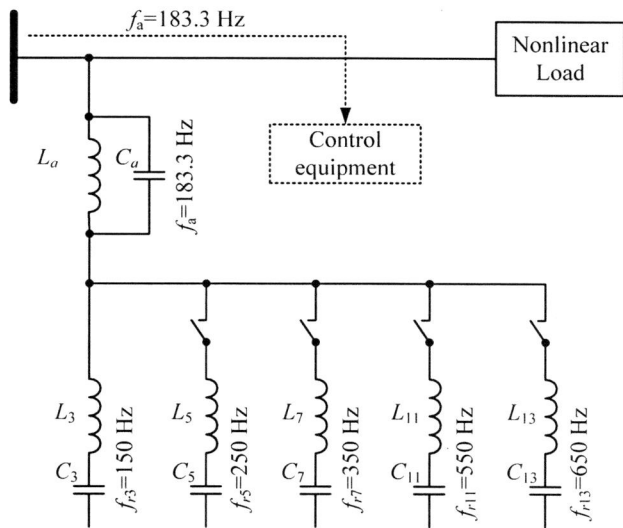

Figure 6.26. Shunt *LC* filter with rejection circuit against losing sound-frequency signals

Rejection filters are used in series connection with the load [3,12]. In these cases filtration almost does not depend on the feeder impedance. However, series connection is possible and reasonable only when the load impedance for higher harmonics is low and lower than the impedance for the fundamental frequency. This load is typically a diode rectifier with capacity filter, deforming feeder current (without filter) at the level THD(*I*) 60–90%. Furthermore, application of the shunt *LC* filter is not necessarily a fully effective solution. More effective seems to be a rejection filter. The typical application of *LC* rejection filters in series connection to the load, and exemplary current and voltage waveforms are presented in Fig. 6.27.

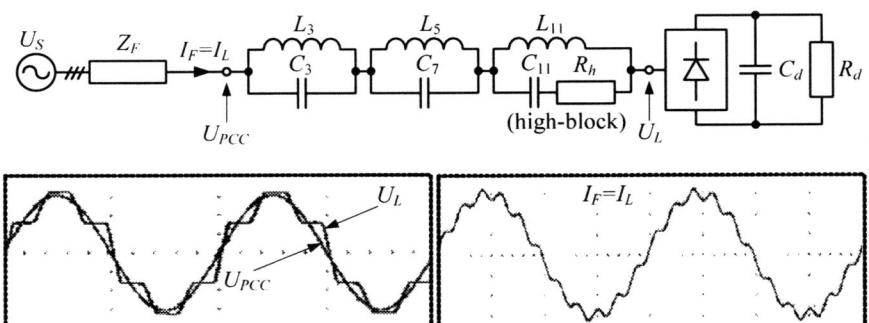

Figure 6.27. Typical application of the series *LC* filters

Aside from presented the above *LC* shunt filter systems (Figs. 6.24–6.26) and series systems (Fig. 6.27), in the case of specific requirements concerning filter effectiveness, we can apply the shunt-series or series-shunt connections. Worthy of attention and very effective is also a filter, proposed under the name "Lineator™". The connection system of this filter, mainly used for adjustable-speed drives, and results of the sample test for the typical frequency converter supplying a motor with 110 kW power, are presented in Figs. 6.28 and 6.29.

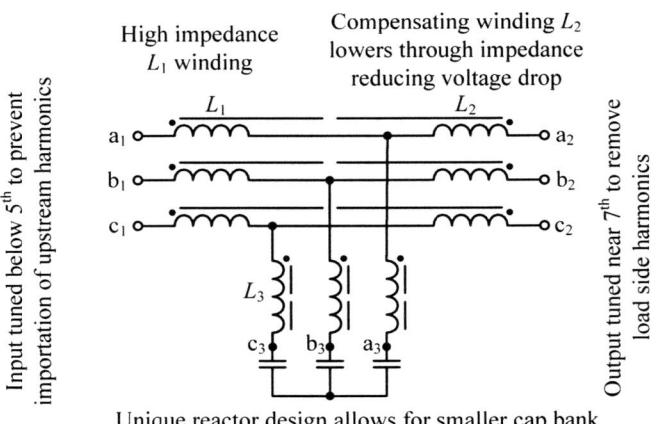

Figure 6.28. Lineator™ internal connection diagram (with permission of Mirus Internat. Inc.)

In the 1990s recent years, including and even a few years before, competition for reactive filters became power–electronic systems with semiconductor devices with turnoff capability, called active power filters (APF) [12,23,27,28]. Their undoubted advantages are:

- smaller overall dimensions and weight, with smaller number of smaller main reactive elements;
- low dependency on the parameter change of the feeder and the load;
- elimination of occurrence of undesired resonances and strengthening of some specific frequencies;
- ability for filtration of broad spectrum frequencies and free shaping of the spectrum while taking into consideration the division/separation into different components, resulting from the power theory.

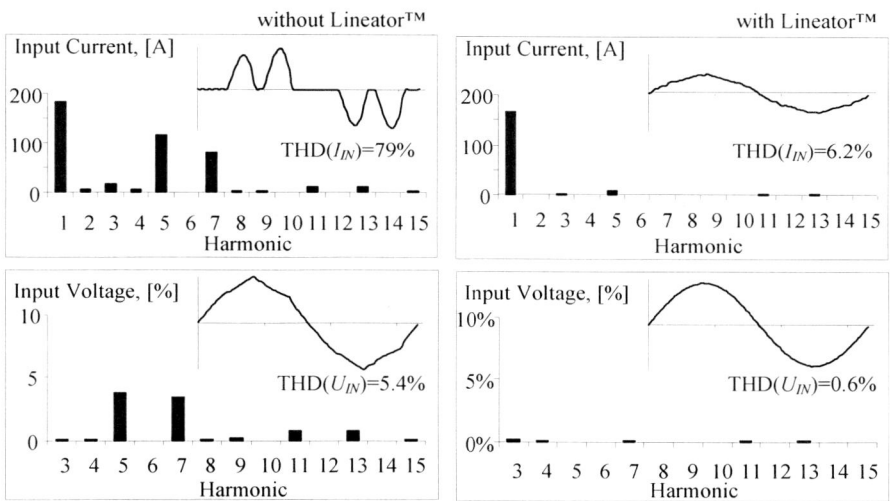

Figure 6.29. Lineator™ performance testing results (with permission of Mirus Internat. Inc.)

The final advantage is their unique ability for filtration compensation of the interharmonic components, and subharmonics, dc offset and active power surges. The systems used for this purpose must be equipped with energy storage of increased capacitance. APF systems, their abilities, control and applications are broadly discussed in Chap. 9. At this point it is worth paying attention to the disadvantages of shunt APF applications to reduction of the higher harmonics generated by diode rectifiers with a capacity filter. Since the rated power of these filters depends on the parameter THD(I_L), i.e. in a given case it is situated within limits 60% to 90%. Moreover, a shunt APF must be designed for peak current, depending on the crest factor of the load current. Application of series APF systems is more rational.

In Figs. 6.30 and 6.31 are presented the examples of application of series APF systems to reduce higher harmonics, generated by diode rectifiers with a capacity filter [11,29]. The advantage of the solutions is the relatively small rated power of the D-SSSC systems, about 30% of the load-rated power. In both cases a series

APF serves as the voltage source, feedback controlled, in proportion to the actual value of total current higher harmonics in the feeder (Fig. 6.30) or in a neutral conductor (Fig. 6.31). The factor of proportionality defines the value of the equivalent resistance, induced into the feeder or neutral conductor for higher harmonics.

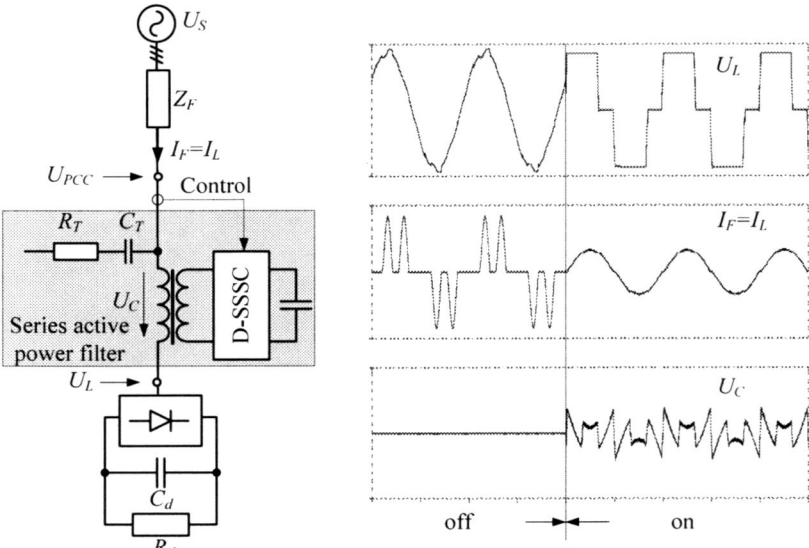

Figure 6.30. Series APF for reduction of load-current high-harmonic and test results

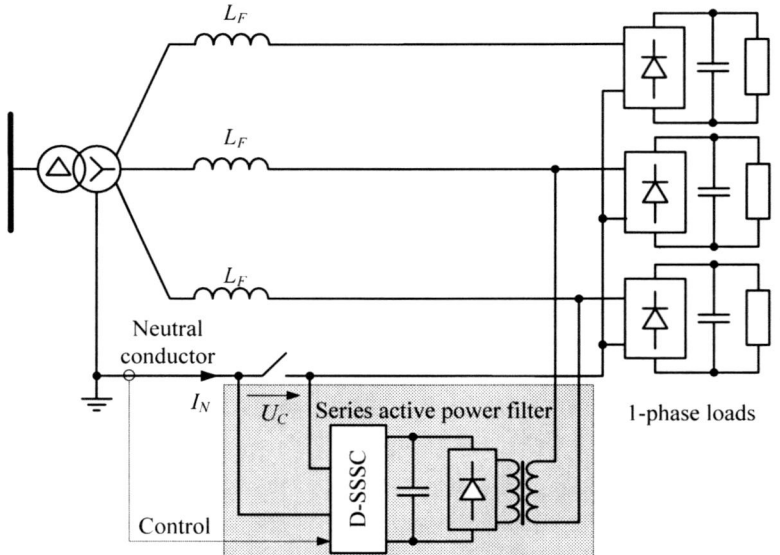

Figure 6.31. Series active power filter for reduction of neutral current high-harmonic

6.4 Mitigation of the Voltage Disturbance

6.4.1 Basic Concept and Methodological Questions

Current developments in the mitigation procedures of the voltage disturbances can only be understood after the basic concept has been made clear [32]. Fig. 6.32 illustrates the basic concept of voltage quality, and highlights the most significant relations between compatibility levels, voltage characteristics and planning levels.

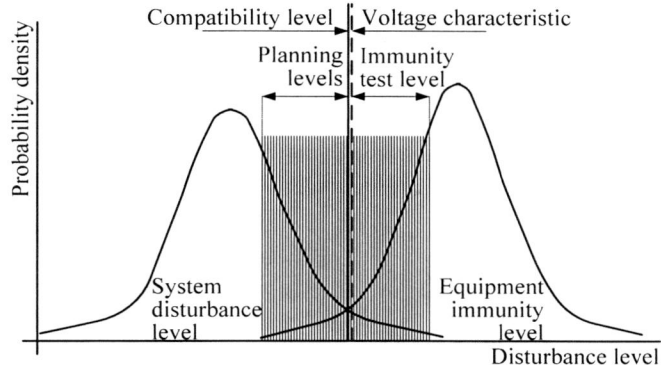

Figure 6.32. Basic voltage-quality concepts covering the whole system

Compatibility levels are reference values used for coordination of emission and immunity of equipment making up or being supplied by a network. Taking into account the time and space distribution, the levels correspond to a 95% probability, where the limit values are not exceed in the whole power system. The level of tolerance is built in with consideration given to the impossibility to control every part of the network at any moment. An assessment of real-level interference, if compared with the compatibility levels would have to be done for the whole network – which is too difficult and almost pointless. Therefore, no evaluation method is defined in relation to the compatibility levels. Additionally, we should stress that these are references rather than any operational values. *Voltage characteristics* are the characteristics at the customer supply points in LV and MV utility networks under regular operating conditions (standard EN 50160). They represent virtually guaranteed limits covering all points of the network based on a solely time-related statistic. *Planning levels* are used for planning when evaluating the impact of all combined customer loads on the network. The planning levels are specified (by the utility) for every network voltage level and can be considered with reference to the quality targets. The planning levels are usually equal to or lower than the compatibility levels. The method used to estimate the real levels (to be compared with the planning levels) uses solely time-related statistics, as well as the method for the voltage characteristic. Therefore, the planning levels rather than the voltage characteristics should be considered for the future network changes or disturbing loads connection. Considering the whole system (Fig. 6.32), interference is bound to arise at certain times and the distributed levels of disturbance and immunity overlap. We have already noted that voltage characteristics can be either equal to or greater than the compatibility levels, whereas the planning levels are usually equal to or lower than the levels of compatibility. The levels of immunity tests are the outcome of an agreement between users and manufactures that is adequate to the mitigation methods.

Mitigation methods due to voltage disturbance can be subdivided into four categories as shown in Fig. 6.33. Modifications in the process or equipment itself ("a" and "b" in Fig. 6.33) are considered to be the cheapest to implement, however they are not always possible. These modifications include both the process adaptation without additional hardware, as well as protection equipment installation that unfortunately protects only part of the process. Modifying the grid ("d" in Fig. 6.33), even though a very interesting option, is not always possible and turns out to be very expensive. The only methods that could be commonly applied are mitigation arrangements installed between the sensitive process (or equipment) and the AC grid ("c" in Fig. 6.33), The arrangements are discussed in the following parts of this section.

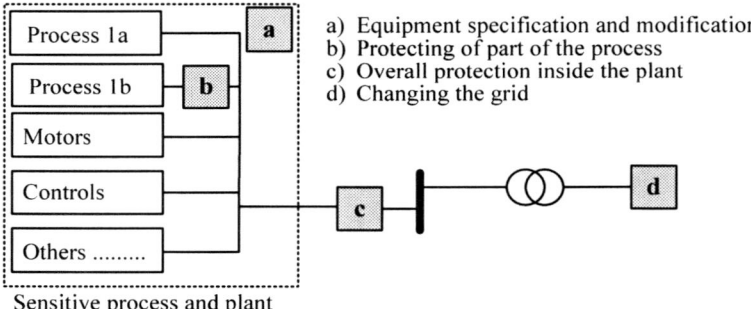

Figure 6.33. Possible mitigation methods

A mitigation arrangement between the AC grid and the process (or equipment) can be installed in all plant facilities; it requires only a little information about the electricity grid and the vulnerable process. As interest in PQ phenomena has increased, the number of different mitigation appliances has also risen. Since many of these appliances contain brand-specific and often exotic names, the comparison is not very easy. As the section intends to briefly describe these mitigation arrangements, they are ranked into the following categories: isolation and protection, compensation, and immunization against all disturbance.

6.4.2 Modern Protection and Reconfiguration Devices

Typically, industrial and commercial arrangements for protection end reconfiguration use electromechanical switches. Electromechanical switches commonly need from 1 up to 10 seconds for effective performance, which highly depends on the strength of the current passage in these switches. Many of the sensitive loads (customers), such as voltage sags and swells, are susceptible to distribution-system disturbance. For these kind of loads the application of only electromechanical protection or configuration devices is not sufficient. In many cases it would be enough to reduce switch activity to only a few milliseconds. To make the problem clearer we should consider a simple example that illustrates the influence of the short-circuit on the power system.

Short-circuits occurring in the electricity grid, which are caused by for example digging, are the most common cause of voltage sags. Fig. 6.34, which presents a single-line diagram of a MV subgrid, aims to explain the occurrence of a voltage sag due to short-circuit at MV. We can easily calculate the voltage drop at the MV busbar and each of the parallel MV feeders in Fig. 6.33 due to a symmetric 3-phase circuit by using a simple voltage-divider rule [4, 31]:

$$U_{sag} = U_N \cdot Z_S/(Z_S + Z_{Fl}), \qquad (6.16)$$

where Z_S – source impedance of the transformer, Z_{Fl} – fault impedance between busbar and fault, U_N – pre-event voltage (normal), U_{sag} – voltage drop during the sag. The voltage level remains until the faulty feeder is isolated from the remaining grid by the protection device. The time between fault initiation and isolation is the sag duration, determined by the response time of the protection device and the demand for selectivity in the grid, which is typically for fast electromechanical switches between 100 ms and 1.5 s. The application of modern semiconductive switches in protection devices allows this time to be reduced to milliseconds [3,31 – 33]. During this time the short-circuit current will not manage to increase up to the value that would cause a voltage drop at the MV busbar, which is noticed by most sensitive customers in parallel MV feeders.

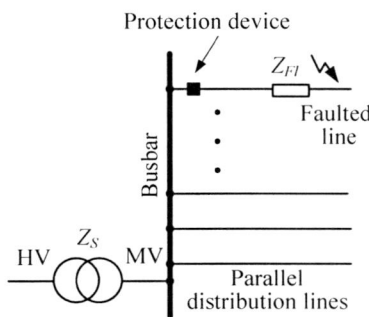

Figure 6.34. Voltage sag due to a symmetrical 3-phase short-circuit in a MV radial grid

Fig. 6.35 presents schematic diagrams of basic types of semiconductive protection and reconfiguring devices, into which the following types are included:
- Static current limiter (SCL) – presented in Fig. 6.35a – limits a fault current by quickly inserting a series inductance in the fault path

- Static current breaker (SCB) – presented in Fig. 6.35b – breaks a faulted circuit much faster than a mechanical circuit breaker
- Static transfer switch (STS) – presented in Fig. 6.35c – connected in the bus-tie position where a sensitive load is supplied by two feeders. It protects the load by quickly transferring it from the faulty feeder to the healthy feeder.

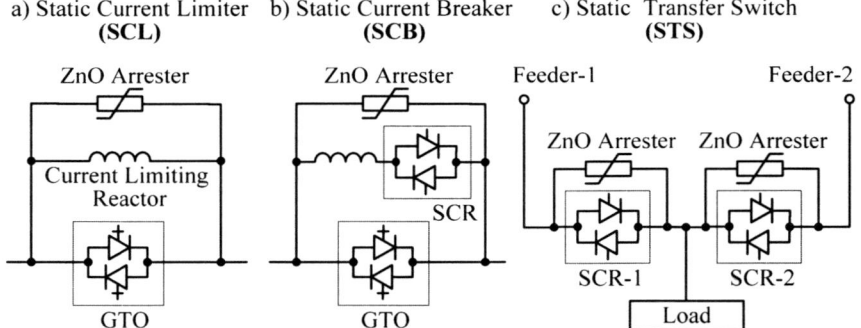

Figure 6.35. Types of semiconductor protection and reconfiguring devices

6.4.2.1 Static Current Limiter
An SCL (Fig. 6.35a) is a parallel connection of an antiparallel gate-turnoff thyristor (GTO) switch with snubbers and current limiting, and zinc oxide (ZnO) arrester. A GTO can be switched off at any time by applying a negative pulse. Therefore, it can interrupt a current instantaneously. Under regular (unfaulted) operating conditions the GTOs are gated for full condition. If a fault occurs, the GTOs are turned off as soon as the fault is detected. The GTO response time is closed within a few microseconds. Once the GTOs are turned off the fault current is delivered to the snubbers capacitor, which limits the rate of rise in voltage across the GTOs. The voltage across the antiparallel GTO switch rises until it reaches the clamping level established by the ZnO arrester. The same voltage also appears across the current limiting reactor. When the voltage reaches its clamping level, the current across the reactor rises linearly. The linear rise will continue until it is equal to the instantaneous level of current passage in the line. Thus, the current is limited by the total effective series impedance, *i.e.* by a combination of the limiting reactor and the faulted feeder impedance.

6.4.2.2 Static Current Breaker
A SCB (Fig. 6.35b) has very similar topology to the SCL typology. The only difference is that the limiting reactor is connected in series with an antiparallel thyristor (SCR). The GTOs are the regular current-carrying elements and if the fault is detected, the GTOs go through a number of subcycle auto reclose operations. For a persistent fault, the GTOs are turned off, whereas the SCR is turned on and the fault current starts flowing though the current-limiting inductor. The current is eventually cut off by blocking the SCR.

An alternative SCB typology, shown in Fig. 6.36a, is also applied. When applying this SCB solution, the current in the unfaulted (normal) state flows thought the vacancy conventional breaker (VCB). When a fault is detected, the GTOs are turned on simultaneously and opening signals are given to the VCB. The fault current starts flowing through the GTO switch and when completely commutated, it is interrupted by turning the GTO switch off.

We should consider the SCB (or SCL) installation possibilities with the example of an electricity subgrid, presented in Fig. 6.36b. If 1 is an SCB and 4 is a conventional breaker, then for a fault downstream from 4, breaker 1 will operate before breaker 4. In this way both faulty and healthy feeder that are supplied by transformer Tr_1 will be disconnected. A potential installation point of the SCB is also the bus-tie location 3 and thereby does not require coordination with any other protection device. For example, for a fault in the transformer Tr_1 side of the subgrid, the SCB will open the bus-tie and thereby prevent transformer Tr_2 from feeding the fault. The SCB can also be connected at location 7 (Fig. 6.36b.). In this case, a fault in the load can be quickly isolated by the SCB without affecting the other protective devices.

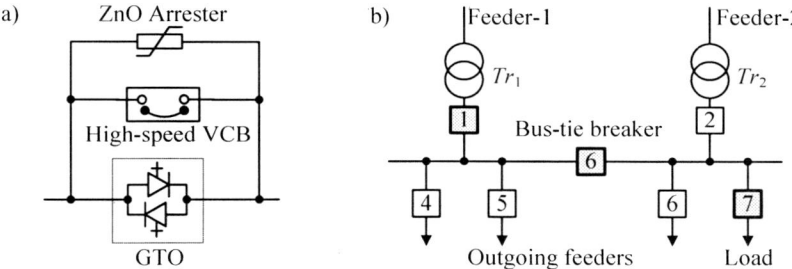

Figure 6.36. Alternative SCB topology (**a**) and example of the SCB installation (**b**)

The best position to place a current limiter (SCB or SCL) is the output of the main incoming transformers, *i.e.* locations 1 and 2 (Fig. 6.36b). Any fault in the network will be limited before it can cause any coordination problem. In this case the current tap setting of the downstream overcurrent relays can be set at lower values. A limiter at the bus-tie location 3 is the most effective as it has lower losses under normal operating conditions. Since the current flowing through this position for a fault at any part of this circuit is maximum, the rating device at this location must be very high.

6.4.2.3 Static Transfer Switch
A STS (Fig. 6.35c) is used to protect a sensitive load from sag/swell and other fault effects in the preferred feeder. Although the load is fed from two alternative and independent feeders it is usually supplied by the preferred feeder, on the example of feeder-1, and the load current flows through the antiparallel thyristors SCR-1. If a voltage sag or interruption is detected in this feeder, the antiparallel thyristors SCR-2 is turned on. Once the load current starts flowing through the thyristors

SCR-2 the thyristors SCR-1 is turned off. This switching action between thyristors SCR-1 and SCR-2 can cause a fault current, supplied by an alternative feeder (feeder-2) before the SCR-1 is cut off. To prevent flow of the fault current, adequate procedures to synchronize thyristor switches can be applied, this, however, slows down the STS reaction and its operation approximately from ~ 2 up to 20 ms or even a little longer.

6.4.3 Compensation Devices

Compensating voltage-disturbance devices can be generally divided similarly to the compensators of load influence, into shunt and series devices (Sec. 6.3.1). However their control is different and therefore their characteristics. Series compensators effectively mitigate voltage disturbances in the PCC node, in the cases when their cause is load connected to the node. The application of series compensators will be discussed in Chap. 8. If the voltage disturbances "come from outside" (from the supply source), then the efficiency of series compensators is usually poor, and the poorer, "more rigid" the feeder (*i.e.* the lower is the impedance of the feeder). Let us consider, as an example, the linear system presented in Fig. 6.37, where ΔU_S – disturbances of supply voltage U_S, I_C – compensating current, ΔU_L – voltage disturbances in load Z_L, Z_F – feeder impedance. For this system it is easy to determine that:

$$\Delta U_L = \Delta U_S \cdot \frac{Z_L}{Z_L + Z_F} - I_C \frac{Z_L \cdot Z_F}{Z_L + Z_F} \quad \text{and} \quad \Delta U_L = 0 \Leftrightarrow I_C = \frac{\Delta U_S}{Z_F}. \quad (6.17)$$

Additionally, we assume that the short-circuit source current $I_{SC} = U_S/Z_F$ is 10 times higher than the load current (*i.e.* $I_{SC} / I_L = 10$ or $Z_L/Z_F = 10$) and that $\Delta U_S = 0.1 U_S$. In this case, as can be concluded from Eq. 6.17, full compensation of small voltage disturbances in the PCC node requires application of a shunt compensator with rated power at least equal to the load rated power. Compensation of voltage disturbances "injected" on the supply source side is much more effective when applying further discussed series compensators [30,34 – 36].

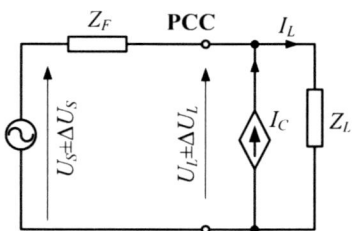

Figure 6.37. Illustration of the shunt compensation of the voltage-source disturbance

The fundamental principle of voltage series compensators operation, illustrated in Fig. 6.38, is quite simple. By inserting a voltage U_C of the required magnitude and frequency, synchronized with the distribution feeder voltage U_S, the series compensator can restore the load-side voltage $U_{PCC} = U_L$ to the desired amplitude and waveform even when the source voltage is unbalance or distorted. Series compensators and their application are discussed in detail in Chap. 8. This section presents only their general characteristics.

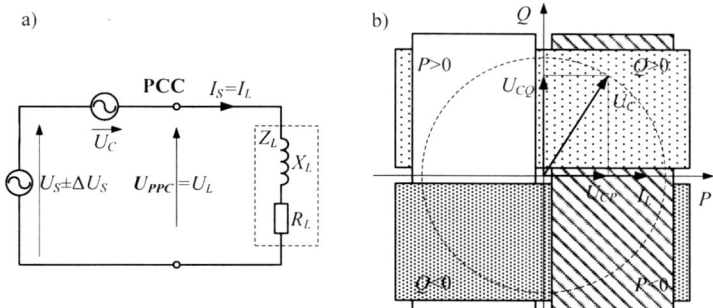

Figure 6.38. Illustration of the principle of series compensation

As shown in Fig. 6.38b, the series compensator can generate or absorb real P and reactive Q power. About this, what power absorbs or generates compensator decide voltage components U_C of fundamental frequency: active U_{CP} and reactive U_{CQ}. This component of injected voltage U_C depends on the compensator control algorithm, directly related to its realized compensation function. It should be stressed that the real power exchanged at the series compensators output AC terminals must be provided by the compensators input terminal by an additional energy source (external energy source or storage systems). The reactive power exchanged between series compensators and the distribution system is internally generated by compensators without passive reactive components. When a series compensator does not absorb or generate active power, the voltage injection U_C must be in quadrature with the line current equal to the load current $I_S = I_L$, i.e. $U_{CQ} = U_C$ and $U_{CP} = 0$ (Fig. 6.37b). This requirement is obviously related to limited compensation ability of the series compensator without an additional energy source.

Let us consider a simple example of application of a series compensator without an additional energy source, to regulate (stabilize) voltage $U_L = U_{PCC}$ in PCC node in a symmetric linear circuit presented in Fig. 6.38a, of the source sinusoidal voltage U_S. In this case, because voltage injection U_C is in quadrature with the line current, performance of the compensator is equivalent to connection to the feeder of a series reactance with the value $X = U_C/I_S$. The reactance is of capacitive character and the voltage U_C, taking into account a direction arrow, leads phase current I_S. For the load R-L, it allows voltage $U_{PCC} = U_L$ to increase above the source voltage U_S, depending on the U_C voltage value and load phase angle φ_L. Changing the character of the equivalent reactance from capacitive to inductive we can also decrease the U_{PCC} voltage below source voltage U_S. The possible range of the voltage U_{PCC} changes up (+) and down (−), and is determined by the dependency:

$$\Delta U_{S(\pm)} = |U_{PCC}| - |U_S| = \pm U_C \sin\varphi_L + \sqrt{U_S^2 - U_C^2 \cdot (1 - \sin^2\varphi_L)} - U_S, \qquad (6.18)$$

where increasing voltage is possible only when $0 \leq U_C \leq 2 \cdot U_S \sin\varphi_L$. The constrain of voltage injection U_C does not occur in the case of U_{PCC} voltage decrease.

As results from Eq. 6.18 the efficiency of application of a series compensator, without an additional energy source, regulation of voltage $U_{PCC}=U_L$ is strongly limited by the load phase angle φ_L. The narrower the angle the narrower the range of regulation up (+) as well as down (−). The case is illustrated in the phasor diagrams in Fig. 6.39 and the regulation characteristics of the compensator in equivalent reactance function X are presented in Fig. 6.40. In the specific case of a resistance load ($\cos\varphi_L=1$) regulation of the voltage U_{PCC} is possible only below the source voltage U_S.

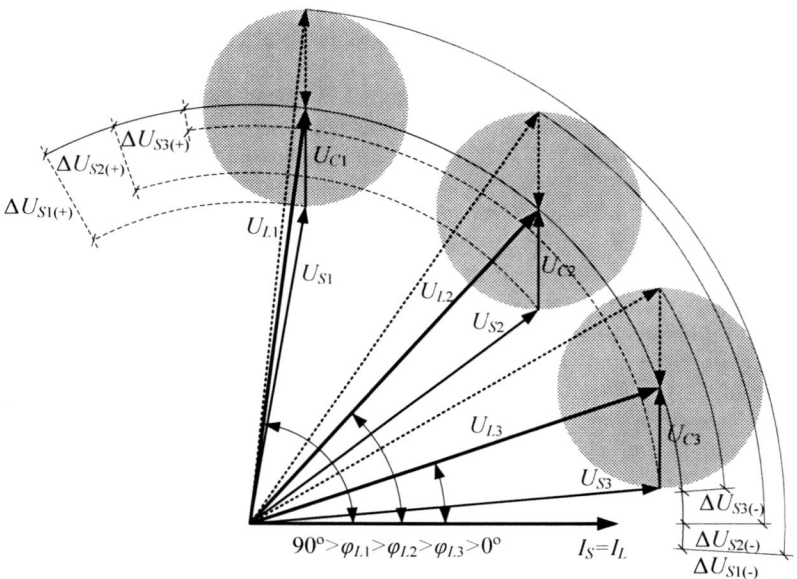

Figure 6.39. Illustration of the series compensator limit without additional energy sources

To illustrate the complete performance of the series compensator without additional energy sources in the grid, Fig. 6.41 presents instantaneous current i_S and instantaneous voltages u_S and u_C in the 3-phase symmetric system, of which the equivalent diagram has been presented before in the Fig. 6.38a. Courses were obtained from application of the controller shown in Fig. 6.40. When equivalent reactance X is of inductive character, transient processes during the regulation are smoother than in the case when reactance X is of capacitive character. The change in character of the equivalent reactance from inductive into capacitive of the value $X = -2$ allows somewhat higher voltage $U_{PCC}=U_L=|U_S - U_C|$ (or load current i_L) than the supply voltage U_S to be achieved. Further increase of the equivalent capacitive reactance, as expected, at first caused voltage U_{PCC} to increase up to the maximal value $U_S/\cos\varphi_L$, and then its decrease (the cases $X = -10$ and $X = -26$) [35].

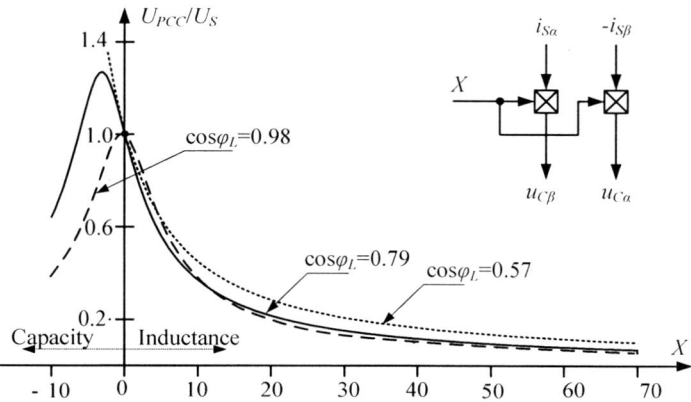

Figure 6.40. Control characteristics of the series compensator without additional source

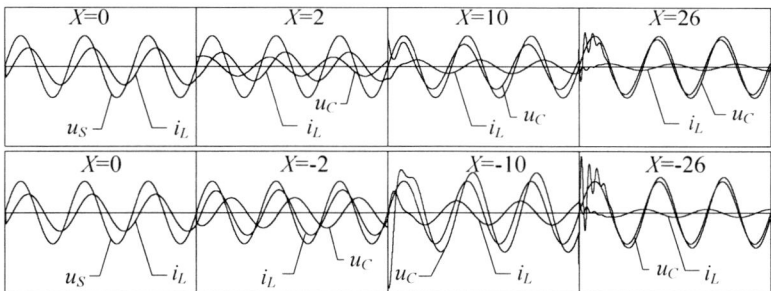

Figure 6.41. Influence of the equivalent reactance X on the current and voltages – examples

Because of the limitations associated with unfeasibility to generate or absorb real power, presented above series compensators without an additional energy source are most effective when applied to compensate higher harmonics or an imbalance voltage in the PCC node of the feeder. In these cases, as results from the power theory [37], the compensator must be equipped only with small energy storage (for example a capacitor), of capacity depending on pulsation of instantaneous active power. These applications are discussed in Chap. 8. The same chapter discusses in detail series compensators with an additional energy source, of which fundamental agreements are shown in Fig. 6.42.

Systems of series compensators with an additional energy source (also presented in Fig. 6.42) are predisposed especially for voltage-sags compensation. An additional energy source frees the range of voltage U_{PCC} in the node PCC from the load and allows the quality factors of this voltage to be kept at the demanded level, independently of the kind of U_S supply-voltage disturbances. The allowable duration and depth of voltage disturbances U_S, which when compensated require active power P, obviously depends on the energy capacity of the additional energy source. With regards to their features of U_{PCC} voltage parameters restoration back to the demand parameters, systems of series compensators with an additional energy source are called dynamic voltage restorers (DVR) [1,30,36,38].

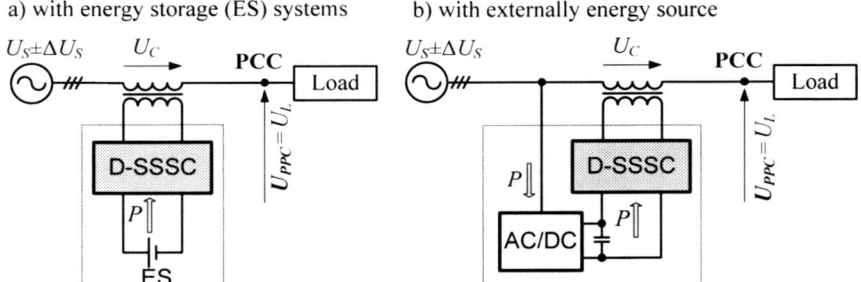

Figure 6.42. Basic agreements of the series compensators with additional energy source

6.4.4 Immunisation – Standby Power-supply Devices

Many sensitive loads, including IT equipments and those connected with safety, defense, health and also devices connected with important technological processes, must be protected against catastrophic failure in the electrical grid, *i.e.* total lack of voltage. The most dangerous are primary failures, *i.e.* failures in parts of the electrical grid occurring relatively close to the load and those occurring directly in the feeder supplying the load. In these cases, compensation devices are not sufficient mitigation arrangements and we must apply devices that protect sensitive loads from catastrophic failure – duplicated feeders from the grid or standby power supply devices. The first solution where are reconfiguration devices were used, *i.e.* STS (Sect. 6.4.2.3), is only effective if the two connections are electrically independent, *i.e.* a predictable single failure will not cause both network connections to fail at the same time. This depends on the network structure, and, often, this requirement cannot be met without the use of very long (and expensive) lines. More often, especially for the loads of power up to several MW, the second solution is definitely more efficient – standby power-supply devices. In this case, various engine generating sets (EGS) and, static uninterruptible power supply (UPS) systems are used [27]. Appropriately designed EGSs can meet most requirements for reserve power sources as well as continuous power supply. On the other hand, the main and crucial disadvantages of a ESGs, especially high-power units, are noise (the average noise level is from 70–95 dB), they are large and heavy, and they require large fuel storage, air intake and exhaust systems. Consequently, these generators are usually installed in separate buildings, relatively distant from occupied buildings. A delay in their starting time is not always acceptable.

Static UPS systems are now commonly used as standby power supplies for critical loads where the transfer time must be very short or zero. Modern 3-phase static UPS systems, with power-electronic switches with turnoff capability, are readily available in ratings from 10 kVA up to 4000 kVA. As well as providing a standby supply in the event of an outage. UPSs are also used to locally improve power quality. The efficiency of UPS devices is very high, with energy losses ranging from 3% to 10%. The basic classification of UPS systems, in accordance with norms IEC 62040-3 or EN 50091-3, specifies three main classes:

- UPS-VFD (output voltage and frequency dependent from mains supply).
- UPS-VFI (output voltage and frequency independent from mains supply).
- UPS-VI (output voltage independent from mains supply).

Basic UPS structures of the individual class are presented in Fig. 6.43

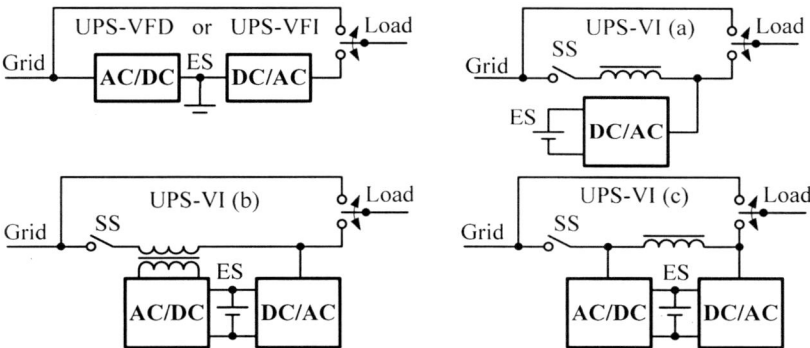

Figure 6.43. Block diagrams of a typical structure of the standard UPS

UPS-VFD, also called standby UPS, line-preferred UPS or offline UPS, has two operating modes. Normally, the power for the load is provided from the main input, optionally through a filter to remove transients or to provide a measure of voltage regulation. The rectifier AC/DC provides the charging current for the energy storage (ES) element – a battery. In "stored energy" mode, the load power is provided from the ES through the inverter DC/AC. The change from "normal" to "stored energy" mode occurs when the main supply voltage is out of tolerance via a bypass line. The UPS-VFD arrangement is the simplest, most compact and the least expensive topology, however, it has some serious disadvantages. The main disadvantages are lack of a real isolation of the load from the AC line, no output voltage or frequency regulation, long switching time, and poor performance with nonlinear loads. Sometimes, to improve UPS-VFD characteristics (performance) three-winding or ferroresonant transformers are applied, this, however, increases costs, overall dimensions and decreases the efficiency.

In contrast to the UPS-VDF, the UPS-VFI provides power for the load in the case of power outage, overvoltage, and undervoltage situations in the AC line, provides power for the load in "normal" mode. In this mode of the UPS-VFI, the load is supplied via a rectifier (AC/DC converter) – a charger with a energy-storage element, most commonly an accumulator battery bank – inverter (DC/AC converter). Thereby for UPS-VFI we also use names such as: online UPS, double-conversion UPS and inverter-preferred UPS. The batteries are connected to the DC link and are charged continuously. They are rated in order to supply power during the backup time, when the AC line is not available. The DC/AC converter (inverter) is rated at 100% of the load power since it must supply the load during the "normal" mode of operation as well as during the backup time. In "stored energy" mode the inverter supplies the load with energy from the batteries. As far as the load is concerned, nothing has changed – the power is supplied through the inverter, but now the source of energy for the inverter is different. There is absolutely zero transfer time and therefore this structure is ideal for very sensitive loads and this is the main advantage of UPS-VFI systems. Other advantages are: very wide tolerance to the input voltage variation, very precise regulation of the output voltage, possibility to regulate or change the output frequency, simple application on the high-frequency transformer isolation. The bypass with a static switch (Sect. 6.4.2.3) provides redundancy of the power source in the case of UPS failure or overloading. We should note that, if the facility is to be used, the frequency of output must be synchronized to that of the main supply. This can be achieved by a locked-phase control loop.

Structures of the systems UPS-VI(a), UPS-VI(b) and UPS-VI(c) shown in Fig. 6.43, are commonly also called line-interactive UPS systems. The easiest arrangement UPS-VI(a), consists of a series inductor, a DC/AC converter, and a battery set (ES system). The DC/AC converter is bidirectional, *i.e.* acts as a rectifier to charge the battery when mains power is available, but acts as an inverter to produce standby power from the battery when mains power is inaccessible. The UPS-VI(a) has three modes of operation. In "normal" mode the load is supplied with conditioned power via the static switch (SS). The DC/AC converter operates to provide output-voltage conditioning and to charge the secondary battery. The output frequency is equal to the main supply frequency. In "stored energy" mode the load is supplied with energy from the battery via the DC/AC converters. The static switch is open to prevent power being fed back onto the main supply. This type of UPS may also have a "bypass" mode in which it allows the load to be connected directly to the main supply in the event of a UPS failure or for maintenance purposes.

The main advantages of the UPS-VI(a) systems are a simple design, and as a result, high reliability and lower cost when compared to UPS-VFI systems. They also have a good, internal capacity to improve the power factor of the load, especially harmonic suppression for the input current [27]. Its main disadvantage is the lack of effective isolation of the load from the main supply. Additionally, frequency control is not possible. Furthermore, the output-voltage conditioning in "normal" mode is not good because the inverter is not connected in series with the load. In this respect more complex structures line-interactive UPS, presented in Fig. 6.43: UPS-VI(b) and UPS-VI(c) are more efficient. Their characteristics are discussed in detail in the literature. In practice, only UPS-VI(b) structures have been applied, more commonly known under the name of delta-conversion UPS.

Note that a similar typology to delta-conversion UPS applies to systems like unified power quality conditioners (UPQC) [39], discussed in Chap. 9.

6.5 Usability of the Modern Power Electronics

Applying power electronics (PE) in various electrical power engineering devices, *i.e.* to improve power quality, started in the 1960s. At the beginning systems with conventional thyristors SCR were applied, used mainly in controlled capacitor compensators (see Fig. 6.23 and Chap. 7) and various kinds of switchgear (for example STS). New solutions and many more application opportunities for power engineering devices in power quality appeared when the high-power semiconductor devices with turnoff capability were introduced [40].

In 1982 the companies Hitachi, Mitsubishi and Toshiba elaborated thyristor GTO (gate-turn-off) 2500 V and 1000 A. A year later also improved were the transistors IGBT (insulated gate bipolar transistor). Since then transistors IGBT and thyristors GTO and their modifications with gate/control-circuits – thyristors GCT/IGCT (gate control thyristor/ integrated gate control thyristor), and also ETO (emitter turnoff) thyristors, were developed [41]. Contemporary research aims especially at an increase of allowable voltages, currents, switch frequencies and decreased loses. Great attention is given to thyristors ETO, because their switching ontime may be much shorter than the time of GTO or IGCT. Laboratory research also concerns power semiconductor devices, on a different basis from silicon, with higher breakdown voltage and higher allowable temperature [42]. However, it should be noted that parameter of these days offered high power, turnoff semiconductor devices are in high-level. To support the above statements, in Table 6.1 we present repeatable upper values of currents and voltages of commercial semiconductor devices (data at the end of 2005). The highest currents for upper voltages of about 6 kV, are for thyristor GTO and some IGCT. However, their maximal frequency in the switch is not too high (400–800 Hz). In this aspect transistors IGBT are better, produced even for somewhat higher

Table 6.1. Parameters range of the modern high-power turnoff semiconductor devices

Firm Devices	POWEREX	ABB	WESTCODE	MITSUBISHI
GTO	FG6000AU-120D (6000 V, 4000 A)	----------	G3000TC600 (6000 V, 3000 A)	FG6000AU-120D (6000 V, 6000 A)
GCT & IGCT	FGC800B-130DS (6500 V, 800 A) FGC6000AX-120DS (6000 V, 6000 A)	----------	----------	GCU15CA-130 (6500 V, 1500 A) GCU35AC-120 (6000 V, 3500 A)
IGBT	CM600HG-130H (6500 V, 600 A)	5SNX 20H2500 (2500 V, 2000 A)	T0900TA52E (5200 V, 1800 A)	CM900HB-90H (4500 V, 1200 A)

voltages but lower currents. The majority of power semiconductor devices are offered as complete construction modules together with heatsink agreement. Examples of these modules are shown in Fig. 6.44. This basis, together with modern control systems and control methods, modern passive elements and construction technologies allows modern power electronics converters of very high rated power and with various applications to be built [43,44 – 46].

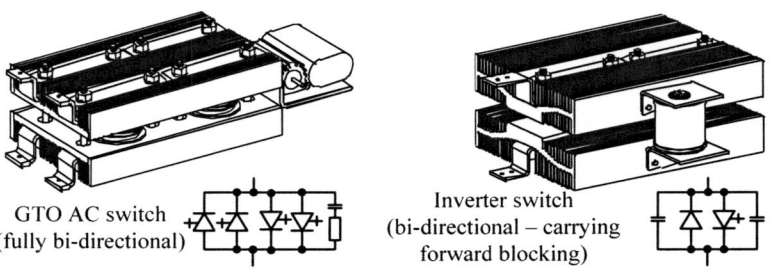

Figure 6.44. Example of a construction of the high power electronics modules

6.5.1 General Model of Power-electronics Converters

PE converters, independent of their function, and construction details and application can be divided into two general classes:

- Direct converters – where the main reactive elements are connected only to input or output terminals of the converter and can be considered as part of the source or the load. A rectifier and voltage inverters with an external LC filter are an example of the direct converters.
- Indirect converters – including main reactive elements inside their structure. There are usually very few elements. Therefore, indirect converters are mostly analysed as connections of direct converters with reactive elements included among them.

The general model of direct converters in the form of a switch matrix with "N" inputs and "m" outputs and examples of realization of the switchers, determining specific characteristics of application systems, is presented in Fig. 6.45. Inputs and outputs are of course changeable, however, they must still have different character, *i.e.* if there is voltage input (voltage source or capacitor) then the output must be current (current source or reactor), and *vice versa*. The presented model is described by the following equations:

$$\begin{bmatrix} u_{a0} \\ u_{b0} \\ u_{c0} \\ \vdots \\ u_{m0} \end{bmatrix} = \underbrace{\begin{bmatrix} S_{Aa} & S_{Ba} & S_{Ca} & \cdots & S_{Na} \\ S_{Ab} & S_{Bb} & S_{Cb} & \cdots & S_{Nb} \\ S_{Ac} & S_{Bc} & S_{Cc} & \cdots & S_{Nc} \\ \vdots & \vdots & \vdots & & \vdots \\ S_{Am} & S_{Bm} & S_{Cm} & \cdots & S_{Nm} \end{bmatrix}}_{\|M\|} \times \begin{bmatrix} u_{A0} \\ u_{B0} \\ u_{C0} \\ \vdots \\ u_{N0} \end{bmatrix}, \qquad (6.19a)$$

$$\begin{bmatrix} i_A \\ i_B \\ i_C \\ \vdots \\ i_N \end{bmatrix} = \underbrace{\begin{bmatrix} S_{Aa} & S_{Ab} & S_{Ac} & \cdots & S_{Am} \\ S_{Ba} & S_{Bb} & S_{Bc} & \cdots & S_{Bm} \\ S_{Ca} & S_{Cb} & S_{Cc} & \cdots & S_{Cm} \\ \vdots & \vdots & \vdots & \vdots & \vdots \\ S_{Na} & S_{Nb} & S_{Nc} & \cdots & S_{Nm} \end{bmatrix}}_{|M|^T} \times \begin{bmatrix} i_a \\ i_b \\ i_c \\ \vdots \\ i_m \end{bmatrix} \qquad (6.19b)$$

where $|M|$ – connection matrix, $|M|^T$ – transpose of a matrix $|M|$, S_{ij} – state of the switch S_{ij}, where if the switch is "on" then $S_{ij}=1$, and if the switch is "off" then $S_{ij}=0$, $i=A,B,...,N$ and $j=a,b,...m$.

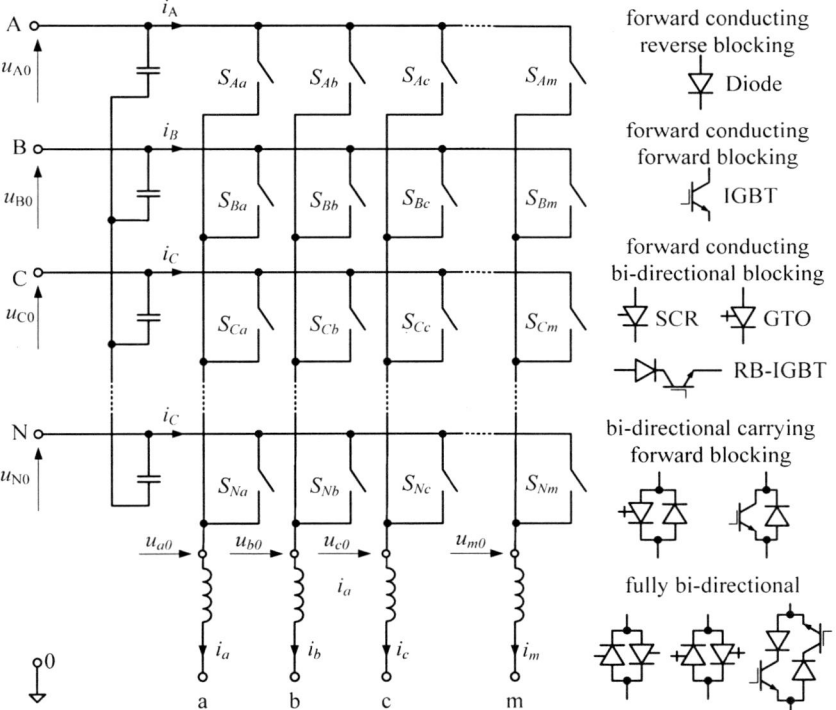

Figure 6.45. The general model of the direct converters and examples of switch realisation

Because switch states S_{ij} can take only values 0 or 1, and depending on time, the natural method to shape output voltage $[u_{a0}, u_{b0}, u_{c0},..., u_{m0}]$ and input currents $[i_A, i_B, i_C,..., i_N]$ in direct converters is pulse modulation [47]. The applied modulation algorithm must include practical limitations [48]. In particular, states of all switches at any given moment can not result in short-circuit or overvoltage. For example, in the presented direct converters model (Fig. 6.45) with input voltage and output current the states S_{ij} of all switches must meet the following requirements:

a) $\sum_{i=A}^{N} S_{ia} = \sum_{i=A}^{N} S_{ib} = \sum_{i=A}^{N} S_{ic} = \cdots = \sum_{i=A}^{N} S_{im} = 1$, b) $\sum_{i=A}^{N}\sum_{j=a}^{m} S_{ji} = m$. (6.20)

A first condition (Eq. 6.20a) means that into one output only one switch can be connected – otherwise an input short-circuit occurs. However, meeting the second condition (Eq. 6.20b) ensures the direction of the output current flow – the number of additional switches must always be equal to the number of outputs. On this basis, the direct converters analysis is made together with synthesis of their algorithms. The analysis of general characteristics of direct converters, can also be carried out in a simplified way. It is assumed that the relative time to connect a switch S_{ij} is equal to the instantaneous value of the modulating continuous function $m_{ji}(t)$, such as $0 \leq m_{ji}(t) \leq 1$ and satisfying Eq. 6.20, i.e.:

$$\sum_{i=A}^{N} m_{ia}(t) = \sum_{i=A}^{N} m_{ib}(t) = \sum_{i=A}^{N} m_{ic}(t) = \cdots = \sum_{i=A}^{N} m_{im}(t) = 1, \quad \sum_{i=A}^{N}\sum_{j=a}^{m} m_{ji}(t) = m, \quad (6.21)$$

In this case, instead of the switch matrix (Fig. 6.45) we obtain a continuous model of direct converters, where each switch S_{ij} is exchanged by an ideal transformer with a transformation ratio equal to the modelling function $m_{ji}(t)$. An example of such a model for 3×3 direct converters (3 inputs, 3 outputs) with fully bidirectional turnoff switches is presented in Fig. 6.46. The same Fig. shows the fundamental method to determine the state of the matrix switches on the basis of modelling functions. The method is discussed with the example of the simplest pulsewidth modulation (PWM) and in relation to the output "a" of a direct converter.

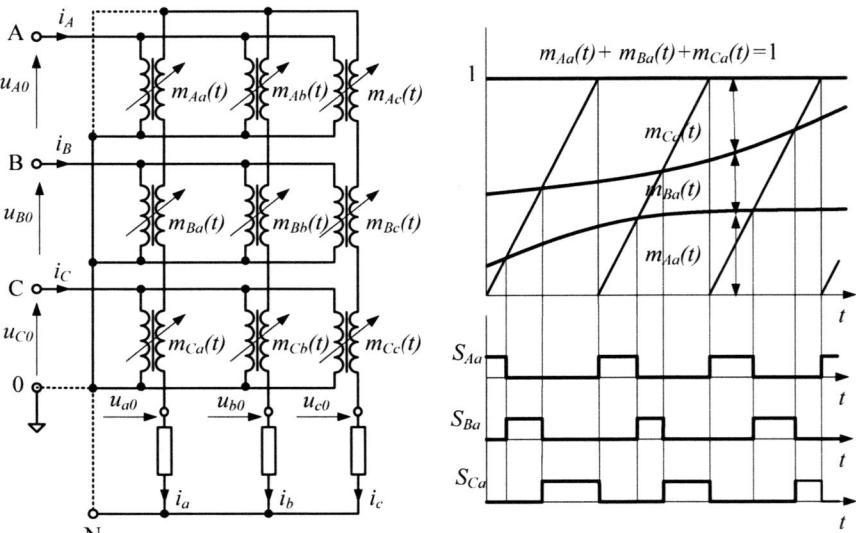

Figure 6.46. The continuous model of the 3×3 direct converters

6.5.2 Basic 3-phase Power-electronics Converters with AC Output

In the following section we will briefly present selected systems of modern PE converters, which either are already applied in PQ improvement agreements, or, according to the authors can also be applied in the near future. More detailed discussion of these and many other systems, omitted from this section can be found in the literature concerning the topic [44,46,49 – 53].

6.5.2.1 Three-phase Voltage-source Inverter

In Fig. 6.47 is presented the basic scheme of a 3-phase voltage source inverter (VSI) supplied by the voltage U_{DC} and an adequate continuous model. When taking into account this model, the general characteristics of 3-phase VSI can be determined from the following equations:

$$\begin{bmatrix} u_{a0} \\ u_{b0} \\ u_{c0} \end{bmatrix} = \begin{bmatrix} m_{Aa}(t) & m_{Ba}(t) \\ m_{Ab}(t) & m_{Bb}(t) \\ m_{Ac}(t) & m_{Bc}(t) \end{bmatrix} \times \begin{bmatrix} u_{A0} \\ u_{B0} \end{bmatrix}, \quad \begin{bmatrix} i_A \\ i_B \end{bmatrix} = \begin{bmatrix} m_{Aa}(t) & m_{Ab}(t) & m_{Ac}(t) \\ m_{Ba}(t) & m_{Bb}(t) & m_{Bc}(t) \end{bmatrix} \times \begin{bmatrix} i_a \\ i_b \\ i_c \end{bmatrix} \quad (6.22)$$

where $u_{A0}=-u_{B0}=U_{DC}/2$. Only sinusoidal modeling functions are further considered:

$$\begin{bmatrix} m_{Aa}(t) \\ m_{Ab}(t) \\ m_{Ac}(t) \end{bmatrix} = \frac{1}{2}\begin{bmatrix} 1 + A\sin(\omega t + 0) \\ 1 + A\sin(\omega t - 2\pi/3) \\ 1 + A\sin(\omega t - 4\pi/3) \end{bmatrix}, \quad \begin{bmatrix} m_{Ba}(t) \\ m_{Bb}(t) \\ m_{Bc}(t) \end{bmatrix} = \frac{1}{2}\begin{bmatrix} 1 - A\sin(\omega t + 0) \\ 1 - A\sin(\omega t - 2\pi/3) \\ 1 - A\sin(\omega t - 4\pi/3) \end{bmatrix}, \quad (6.23)$$

where $\omega=2\pi f$, f – output fundamental frequency, A – modulation factor ($0 \leq A \leq 1$). From Eqs. 6.22 and 6.23 it results that for such modelling functions, the amplitude of the sinusoidal output voltage can not exceed $U_{DC}/2$. It should be stressed that the value is not any boundary value. In the case of vector modulation the amplitude of the sinusoidal voltage can be increased by about 16%. However, if output-voltage overmodulation is allowable (it is even advisable), then in boundary cases

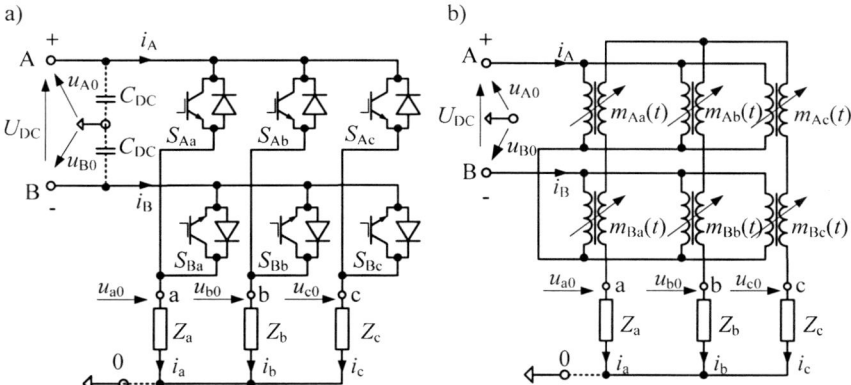

Figure 6.47. Basic scheme of the 3-phase VSI (**a**) and its continuous model (**b**)

the amplitude of a component of fundamental frequency may even reach the value $2U_{DC}/\pi$ [46, 48].

On the basis of Eqs. 6.22 and 6.23 it is easy to show that the input currents of a 3-phase VSI with sinusoidal output voltage are:

$$i_A = i_{DC} + (1/2) \cdot i_0 \quad , \quad i_B = -i_{DC} + (1/2) \cdot i_0 , \tag{6.24}$$

where

$$i_{DC} = (A/2) \cdot [i_a \cdot sin(\omega t) + i_b \cdot sin(\omega t - 2\pi/3) + i_c \cdot sin(\omega t - 4\pi/3)] \quad , \quad i_0 = i_a + i_b + i_c .$$

The above relations indicate the unique characteristic of a 3-phase VSI, which is its ability to generate reactive currents, theoretically without application of any input energy storage, for example as capacitors C_{DC}. This characteristic, resulting from the lack of reactive power in DC circuits, is used in D-STATCOM systems [24,25,28,44,54] (see also Chap. 7). Application in practical systems of relatively small capacitors C_{DC}, of capacity many times smaller than in traditional reactive power compensators, first of all is connected with the limited frequency of the switches state changes. However, if the system D-STATCOM as also used to compensate load high-harmonics and/or inter/subharmonics, and/or unbalance, then capacitors C_{DC} at a VSI input (Fig. 6.47a) must be of the same capacity. The capacity can be estimated on the basis of instantaneous pulsation of the active power $p_L(t) = p(t) = u_{a0} \cdot i_a + u_{b0} \cdot i_b + u_{c0} \cdot i_c$ (connected with compensated components of the load current), by using the equations [23]:

$$C_{DC} > 4 \cdot P_{max} / [\omega_p \cdot (U_{DC(max)} - U_{DC(min)}) \cdot U_{DC(0)}] , \tag{6.25}$$

where

$$P_{max} = max_{t \to \infty} \left[p_L(t) - \lim_{\Delta T \to \infty} \frac{1}{\Delta T} \int_{t}^{t+\Delta T} p_L(\tau) d\tau \right] \tag{6.26}$$

is the instantaneous pulsation amplitude of the active power in the established state, ω_p is the angular frequency of the instantaneous power pulsation p_L, and $U_{DC(0)}$, $U_{DC(min)}$, and $U_{DC(max)}$ – set value of minimal and maximal voltage in capacitor C_{DC}. Equations 6.25 and 6.26 do not depend on the applied control method and the algorithm PWM VSIs.

6.5.2.2 AC Controllers – Regulators

Application of AC controllers to improve PQ (power quality) in an electrical distribution grid encompasses also static switch agreement in transformer tap changing, capacitor switching in static reactive power compensation (see Sect. 6.2.2.2) and various reconfiguration devices (Sect. 6.4.2), as well as regulator agreements for continuous control of the rms value of voltage or current [35,50,52]. Among these last agreements reactive current regulators deserve special attention, used in regulated reactive power compensators and unbalance load (see also the Steinmetz compensator – Fig. 6.20).

The offered AC regulators are usually realized on the basis of SCR thyristors. These systems are mostly presented in the literature [41,44,50], which also indicates their unfavourable effect on a feeder, as a result of the generation of higher harmonics, especially in resistance loads. For an inductive load, which is applied to thyristor reactive current regulators, higher harmonics are much smaller. In support of this statement, Table 6.2 sets together schemes and characteristics of the basic system (№1) and selected modifications (№3 and №3) of thyristor reactive current regulators. Modification №2 is possible only in a 3-phase grid. The most favourable, with regards to current distortions, is modification №3 – for a magnetic coupling coefficient $k_m=0.5$, with total harmonic distortion of current THD(I)< 3%.

Table 6.2. Examples of the thyristor reactive current regulators

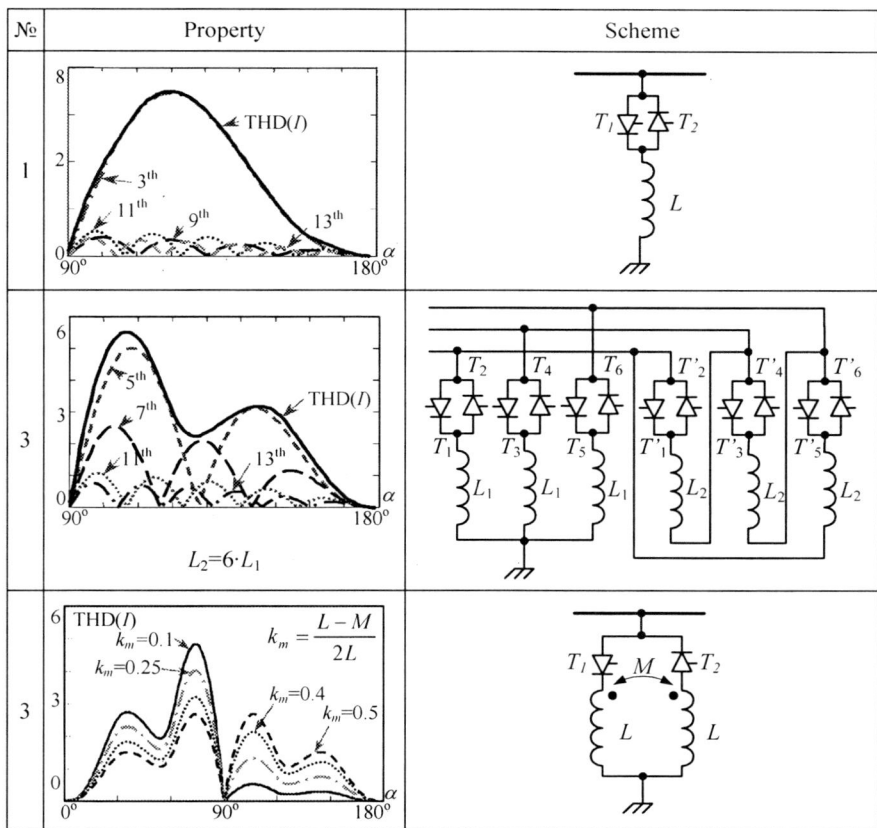

The second significant disadvantage of thyristor AC regulators, mentioned in Sect. 6.3.2.1, is the relatively slow reaction to the control signal, and as a result the

insufficient efficiency in their application to arrangement for compensation of disturbances in a feeder of frequency over 10–15 Hz. Furthermore, the thyristor systems must be synchronized with the supply voltage, which can cause control problems in the case of significant distortions of this voltage. The listed disadvantages are not present in AC regulators on the basis of turnoff semiconductor devices of pulsation frequency f_i many times exceeding the fundamental frequency of the feeder voltage, called also AC choppers [35]. Such regulators do not deform output voltage and input current with regards to PQ standards, additionally they do not require control by using PWM methods, used in applications with inverters.

In Fig. 6.48 is presented an example of 3-phase transistor inductive current regulators and an adequate continuous model. Performance of the regulator, the essence of which lies in cyclical and synchronic change of the switches state S_{Aa}, S_{Ba}, S_{Ca}, and S_{0a}, S_{0b}, S_{0c} (cumulative), is illustrated by the voltage and current forms shown in Fig. 6.49. The higher the pulse frequency f_i the thinner is their frequency spectrum, which is illustrated in Fig. 6.50, and that significantly simplifies filtration. If the relation of the connection time T_{on} of the switches S_{Aa}, S_{Ba}, S_{Ca} to the repeating period $T_i = 1/f_i$ is constant, then modelling functions are also constant and equal to $m_{Aa}(t)=m_{Bb}(t)=m_{Cc}(t)=\delta$, where $\delta = T_{on}/T_i$ – the so-called duty cycle. In this case, the continuous model of a regulator (Fig. 6.48b) is described by the equations :

$$[u_{a0}, u_{b0}, u_{c0}] = \delta \cdot [u_{A0}, u_{B0}, u_{C0}], \quad [i_A, i_B, i_C] = (1/\delta) \cdot [i_a, i_b, i_c], \quad (6.27)$$

analogously to the equation of an ideal autotransformer with a regulated transformation ratio $0 \leq \delta \leq 1$. In conclusion, the change of condition duty cycle δ of the regulator changes the substitute inductance seen from input terminals in accordance with dependences:

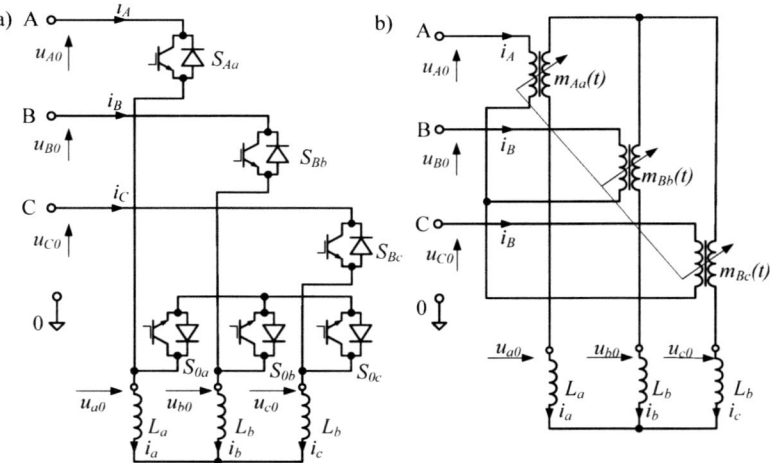

Figure 6.48. Transistorised reactive current regulator (**a**) and its continuous model (**b**)

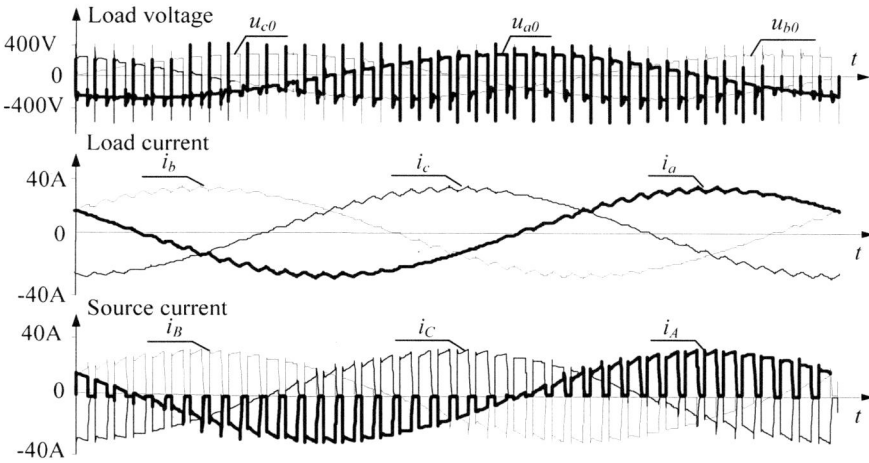

Figure 6.49. Voltage and current waveforms for transistorized reactive current regulators

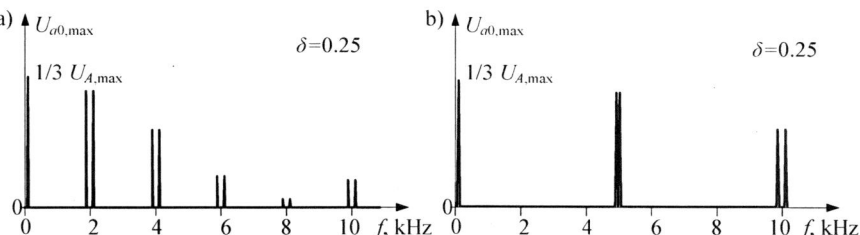

Figure 6.50. Output voltage frequency spectrum for $f_i = 2$ kHz (**a**) and $f_i = 5$ kHz (**b**)

$$L_A = (1/\delta^2)L_a \ , \ L_B = (1/\delta^2)L_b \ , \ L_C = (1/\delta^2)L_c \ . \tag{6.28}$$

In addition to the system shown in Fig. 6.48, there are many other known solutions of AC choppers, each of them serves the autotransformer function of smooth regulation of transformation ratio with ability to its very quick changes. Ability for these changes is assumed in basics of matrix converters operation, usually built as 3×3 direct converters (Fig. 6.45) with turnoff fully bidirectional switches [48,49]. As an example the continuous model (Fig. 6.46), which is a combination of 3 models of AC choppers (Fig. 6.48b) of modulating functions, serves in the simplest case, sinusoidal and phase shift about 120°.

Matrix converters, besides their complicated construction, are nowadays in the stage of intensive prototype research [53]. This results, among others from such characteristics as: smooth bidirectional electrical energy flow; lack of a circuit that mediates voltage/direct current; possibility of change of function, (*i.e.* use as converters AC/DC, DC/DC, DC/AC); regulated input power factor and sinusoidal input current and output voltage. The two last characteristics are illustrated by the oscillographs presented in Fig. 6.51. We believe that power-electronic semiconductor devices technology development, will lead in the near future to many interesting applications of the matrix converters, also in the field of power electrical engineering.

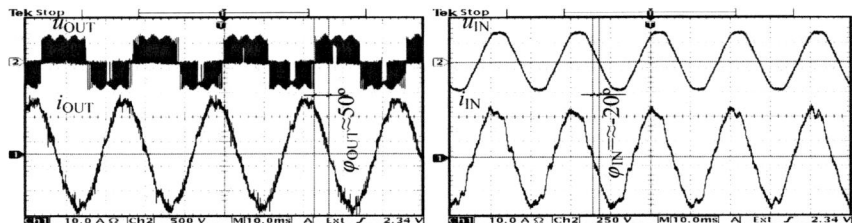

Figure 6.51. Typical output and input waveforms for AC/AC 3×3 switch matrix

6.5.3 Multilevel Voltage Inverters as Arrangement to MV Grid

In the systems VSI with PWM of strong power and medium voltage, a typical 3-phase bridge (Fig. 6.47) is not the most appropriate solution. Not too high a frequency of switches in semiconductor devices of high power (requires a compromise among output-voltage quality and regulation dynamics with application of an output filter), higher voltage hazard and interferences (without any special countermeasures) and sometimes an insufficient value of the peak voltage in semiconductor devices (for a peak voltage of 6 kV, the recommended voltage is about 3.5 kV) would cause in the 1990s a second interest in multilevel inverters [51,54]. Older solutions would then have limited applications [55,56].

As the main advantages of modern multilevel VSI we can count [57,58]:

- increased range of output-voltage amplitude changes;
- greater accuracy in modelling output voltage and current;
- ability to decrease transformation ratio and even elimination of the output transformer for medium voltage;
- more easy adaptation to low-voltage energy storage;
- decreased voltage hazard and current elements (dependent on applied typology);
- decreased level of common-mode disturbances.

Their basic difference as regards typical 2-levels VSI and the general principle of forming output multilevel voltage is illustrated in Fig. 6.52.

All known topologies of multilevel VSI in the literature, without galvanic storage isolation (sources) of energy sources in DC circuits, can be summarized and synthesized on the basis of fundamental modes VSI and the multilayer topology presented for the case of 3 layers in Fig. 6.53. The manner of their synthesis is to properly exclude individual semiconductive switches. However the majority of multilevel VSIs obtained in this way did not find any applications,

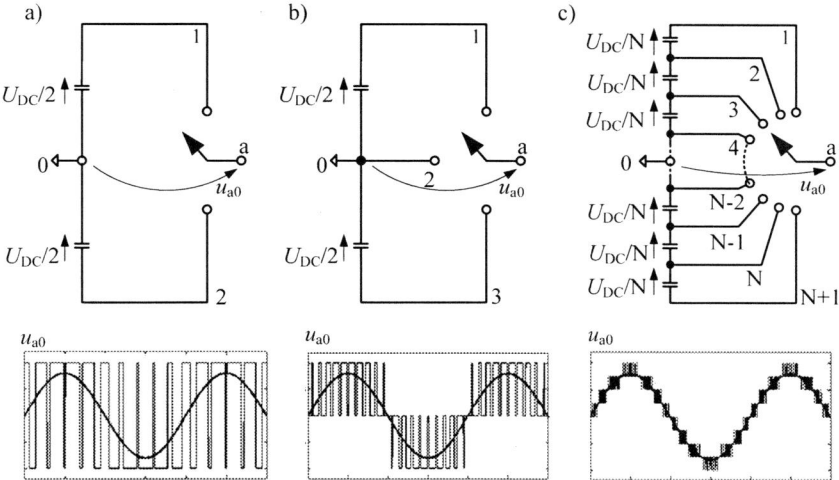

Figure 6.52. General principle of voltage forming in 2 (a), 3 (b) and N+1 (c) levels VSI

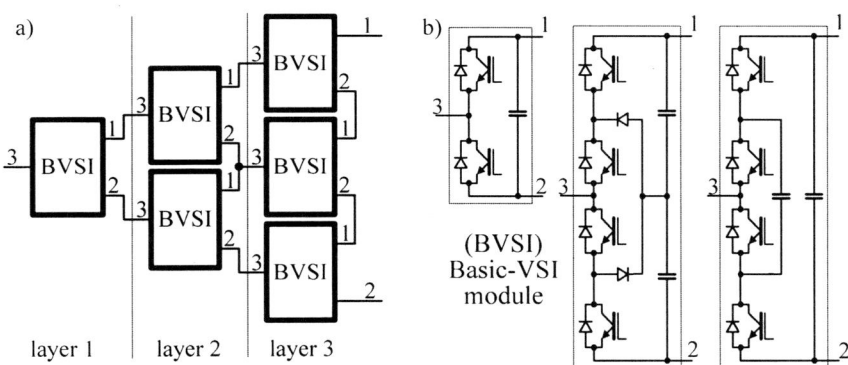

Figure 6.53. Structure of the 3-layer VSI (**a**) and example of the basic-VSI module (**b**)

either because of the complexity of this typology or because of greater losses. In practice, two special cases of multilayer topology realized are: multilevel VSI of DC type (diode clamped), to which we also include VSI of NPC type (neutral point clamped), and multilevel VSI of FC type (flying capacitor) [51,57]. Furthermore, also applied are multilevel VSI of CHB type (cascaded H-bridge), requiring in contrast to the first two, isolated storage systems (energy sources) [13]. The main advantage of CHB-VSI is the possibility of easy development and independent stabilization of the voltage in the DC circuit for each mode H-bridge. However, none of the three selected typologies of multilevel VSI, presented as one branch in Fig. 6.54 has so far gained a leading position.

An interesting solution of multilevel VSI are also hybrid switches typical for 3-phase VSI (2-level) with additional modes H-bridge in every output [59]. An example of this solution, together with characteristic oscillograms of output line-to-line voltage and phase voltage in individual modes, is presented in Fig. 6.55. In this way we can relatively easily improve the operating systems of typical industrial VSI, increasing both the quality and amplitude of the output voltage.

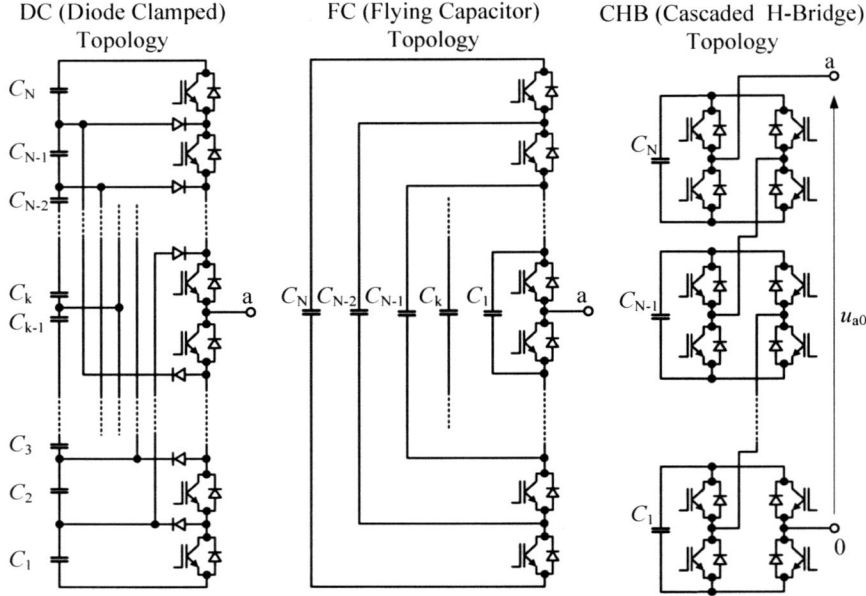

Figure 6.54. Generalized typologies of the most frequently applied multilevel VSI

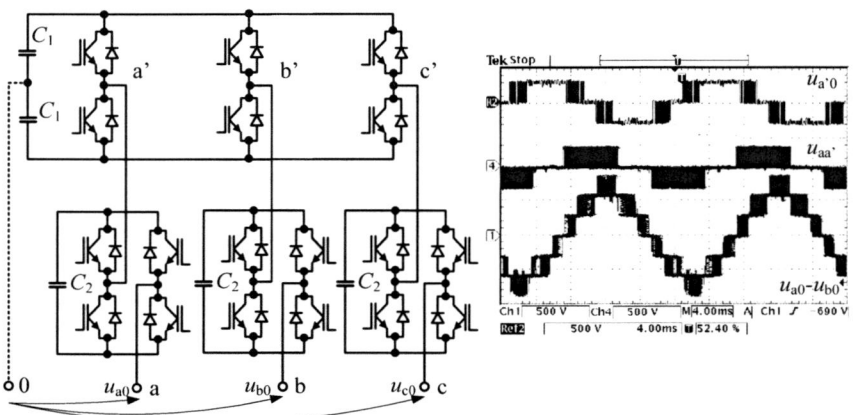

Figure 6.55. Hybrid connection of the 3-phase, 2-level VSI and three H-bridge module

The discussed multilevel inverters obviously do not run the offer of all important solutions that were applied recent years, and first of all they concern only voltage systems – VSI. The issue of the multilevel current source inverter

(CSI), because of its duality when compared to VSI systems was not undertaken, despite the increasing interest given to these systems and achieved results [60]. This was partially caused by lack of significant applications of multilevel CSI.

6.6 Summary

This chapter discussed general issues concerning supply quality in AC systems, and especially the methods and selected countermeasures for power-quality improvement. In Sect. 6.2 the main attention is paid to the cause–effect relation. Simple examples show the characteristic influence of different typical loads on the supply disturbances in the load-connection point and on their propagation distribution systems.

In the majority of the chapter, including Sects. 6.3 and 6.4, we tried to present material in a way that allows us evaluate and compare the potentials of various modern mitigation agreements, without getting into the details of their structure and control. The information content in these sections can be considered as the answers for questions referring to the choice of the connection point and the general type of devices mitigating supply disturbances. Actual solutions, discussed in the following chapters, were presented rather in a pictorial way. Aspect of realization of controlled voltage and current sources as well as regulated reactance were also presented in Sect. 6.5.

Section 6.5 characterizes the potential of modern high-power electronics converters with turnoff semiconductor devices for electrical power engineering, and also voltage-disturbance-mitigation agreements. Therefore, it focuses mainly on operation of the most important and basic power electronic converters with AC output on the basis of a general model of the direct converters, and especially an adequate continuous model. As a supplement, we briefly presented multilevel voltage inverters, which are usually used in applications of high power and medium voltage.

The authors hope that the whole chapter was inspiring for the reader and will result in deeper study of the dedicated literature.

References

[1] Didden M. Techno-economic analysis of methods to reduce damage due to voltage dips. Ph.D. Thesis, Leuven/Belgium: Catholic University of Leuven, 2003.
[2] Green T.C., Arámburo H. Future technologies for a sustainable electricity system: The role of power electronics in future power systems. Cambridge, UK, Cambridge University Press, 2005.
[3] Ghosh A., Ledwich G. Power quality enhancement using custom power devices. Boston: Kluwer Academic Publishers, 2002.

[4] Dorf R.C. et al. The Electrical Engineering Handbook 2nd edn, New York: CRC Press in cooperation with IEEE Press, 1997.
[5] Strzelecki R. Analysis of resonant phenomena in industrial power networks. Archives of Electrical Engineering, XLVI (1), 1997; 103–122.
[6] Kloss A. Stromrichter-Netzrückwirkungen in theorie und praxis: EMC der Leistungselectronik. Aarau/Schweiz: AT Verlag, 1981 (in German)
[7] Pansini AJ. Power Transmission and distribution. Liburn, USA: The Fairmont Press, 1991.
[8] Schlabbach J., Blume D. Voltage quality in the electrical power systems. London: IEE Press, 2001.
[9] Acha A., Madrigal E. Power System Harmonics: Computer Modeling and Analysis. Chichester: John Wiley, 2001.
[10] Arrillaga J., Watson N.R., Chen S. Power Quality Assessment. New York: John Wiley, 2000.
[11] Akagi H. New trends in active filters for power conditioning, IEEE Trans. Ind. Appl. Vol. 32, No. 6, 1996; 1312–1322.
[12] Bhattacharya S., Frank T. M., Divan D., Banerjee B. Active filter system implementation. IEEE Trans. Ind. Appl. Vol.4, No.5, 1998; 47–63.
[13] Wang J., Peng FZ. Unified Power Flow Controller Using the Cascade Multilevel Inverter. IEEE Trans. on Power Electronics, Vol. 19, No. 4 , 2004; 1077–1884.
[14] Chen TH. Comparison of Scott and Le Blanc transformers for supplying unbalanced electric railway demands. Electric Power System Res., Vol. 28, No. 3, 1995, 235–240.
[15] Hanzelka Z. Efficiency of static compensation applied to reduction of the variable loads influence on supply network. Cracow, Poland: Mining and Metallurgy Academy (AGH) Publishing House, 1994 (in Polish).
[16] Sainz L., Pedra J. Caro M. Steinmetz circuit influence on the electric system harmonic response. IEEE Trans. Power Delivery, Vol. 20, No. 2, 2005; 1143–1149.
[17] Schauder C.: STATCOM for Compensation of Large Electric Arc Furnace Installations: Proceedings of the IEEE PES Summer Power Meeting, Edmonton, Alberta, July 1999, pp. 1109–1112.
[18] Yoshioka Y., Konishi S., Eguchi N., Hino K. Self-commutated static flicker compensator for arc furnaces, IEEE/APEC Conf., San Jose, USA, 1996; 891–897.
[19] Dugan R., McGranaghan M., Beaty H. Electrical Power Systems Quality. Knoxville, USA: McGraw-Hill, 1996.
[20] Miller T.J.E. Reactive Power Control in Electric Systems. Chichester, UK: John Wiley Interscience, 1982.
[21] Gyugyi L., Otto RA., Putman TH. Principles and applications of static, thyristor controlled shunt compensators. IEEE Trans. Power App. Syst., Vol. 97, 1978: 1935–1945.
[22] Kincic S., Wan, XT., McGillis DT., Chandra A., Boon-Teck Ooi, Galiana, FD., Joos G. Voltage support by distributed static var systems (SVC). IEEE Trans. Power Delivery, Vol. 20, No. 2, 2005; 1541–1549.
[23] Strzelecki R., Supronowicz H. Power factor correction in AC supply systems and improving methods. Warsaw, Poland: Warsaw University of Technology Publishing House, 2000 (in Polish).
[24] Akagi H., Kanazawa Y., Nabae A. Instantaneus reactive power compensators comprising switching devices without energy storage components. IEEE Trans. Ind. Appl., Vol. 20, No. 3, 1984; 625–630.
[25] Sumi Y., Harumoto Y., Hasegawa T., Yano M., Ikeda K., Matsuura T. New static VAR control using force-commutated inverters. IEEE Trans. Power App. Syst., Vol. 100, 1981; 4216–4224.

[26] Zhezelenko I.V. High harmonic in feeders of the industrial plant. Moscow, Russia: Energoatomizdat, 2004 (in Russian).
[27] Emadi A., Nasiri A., Bekiarov S.B. Uninruptible power supplies and active filters. Boca Raton, USA, CRC Press, 2005.
[28] Gyugyi, L., Strycula EC. Active AC power filters. IEEE/IAS Annual Meeting, Orlando, USA, 1976: 529–535.
[29] Wang Z., Wang Q., Yao W. Liu J. A series active power filter adopting hybrid control approach. IEEE Trans. Power Electron., Vol. 16, No. 3, 2001; 301–310.
[30] Bollen MHJ. Understanding Power Quality Problems: Voltage Sags and Interruption. Piscataway, NJ, USA: IEEE Press, 1999.
[31] Tosado F., Quaia S. Reducing voltage sags trough fault current limitation. IEEE Trans. Power Delivery, Vol. 16, No. 1, 2001; 12–17.
[32] Meyer C., Schröder S., De Doncker RW. Solid-state circuit breakers and current limiters for medium-voltage systems having distributed power systems. IEEE Trans. Power Electron., Vol. 19, No. 5, 2004; 1333–1340.
[33] Meyer C., De Doncker RW. Solid-state circuit breaker based on active thyristor topologies. IEEE Trans. Power Electron., Vol. 21, No. 2, 2006; 450–458.
[34] Chun L., Qirong J., Jianxin X. Investigation of voltage regulation stability of static synchronous compensator in power system", IEEE/PES Winter Meeting, Singapore, Vol. 4, 2000; 2642–2647.
[35] Fedyczak Z., Strzelecki R. Power electronics agreement for AC power control. Torun, Poland: Adam Marszalek Publishing House, 1997 (in Polish).
[36] Fornari F., Procopio R., Bollen MHJ., SSC compensation of unbalanced voltage sags. IEEE Trans. Power Delivery, Vol. 20, No. 3, 2005; 2030–2037.
[37] Depenbrock M., Marshall DA., Van Wyk JD., Formulating requirements for a universally applicable power theory as control algorithm in power compensators. Europan Trans. Elec. Power Eng., Vol. 4, No. 6, 1994; 445–455.
[38] Nilsen JD. Blaabjerg F.: A detailed comparision of system topologies for dynamic voltage restorers. IEEE Trans. Industrial Appl., Vol. 41, No. 5, 2005; 1272–1280.
[39] Aredes M., Heumann K., Watanabe E. H. A universal active power line conditioner, IEEE Trans. Power Delivery, Vol. 13, No. 2, 1998: 545–551.
[40] Benda V., Gowar J., Grant DA. Power semiconductor devices – theory and applications. New York, John Wiley, 1999.
[41] Zhang B., Huang AQ., Liu Y., Atcitty S. Performance of the new generation emitter turn-off (ETO) thyristor. IEEE/IAS Conf., Pittsburgh, USA, 2002; 559–563.
[42] Sugawara Y., Takayama D., Asano K., Agarwal A., Ryu S., Palmour J., Ogata S. 12.7kV ultra high voltage SiC commutated gate turn-off thyristor: SICGT. Int. Symp. Power Semicon. Devices & ICs, Kitakyushu, Japan, 2004; 365–368,
[43] Lee T-YT. Design Optimization of an Integrated Liquid-Cooled IGBT Power Module Using CFD Technique. IEEE Tran. Components and Packaging Technologies, Vol. 23, No. 1, 2000; 50–60.
[44] Rashid MH. (Editor) Power Electronics Handbook. San Diego, USA: Academic Press, 2001.
[45] Stemmler H. State of the Art and Future Trends in High Power Electronics. IPEC Conf., Tokyo, 2000, 2000; 4–13.
[46] Skvarina T.L. The Power Electronics Handbook. Boca Raton, USA CRC Press. 2002.
[47] Holtz J. Pulse-width modulation for electronic power conversion. Proc. of the IEEE, Vol. 82, No. 8, 1994; 1194–1214.
[48] Kaźmierkowski MP., Krishnan R., Blaabjerg F. Control in Power Converters. Selected Problems. San Diego, USA, Academic Press, 2002.
[49] Gyugyi L., Pelly BR., Static frequency changers. New York, John Wiley. 1976.

[50] Shepherd W. Thyristor control in AC circuits. Bradford, UK: Bradford University Press, 1976.
[51] Soto D., Green TC. A comparison of high-power converter topologie for the implementation of FACTS controllers. IEEE Tran. Ind. Electron., Vol. 49, No. 5, 2002; 1072–1080.
[52] Trzyndalowski AM. Introduction to Modern Power Electronics. New York, John Wiley, 1998.
[53] Wheeler P., Rodriguez J., Clare JC, Empringham L., Weinstein A. Matrix converter: a technology revived. IEEE Trans. Ind. Electron., Vol. 49, No. 2, 2002; 276–288.
[54] Lai JS., Peng FZ. Multilevel converters - a new breed of power converters. IEEE Tran. Ind. Appl. , Vol. 32, No. 3, 1996; 509–517.
[55] Kobzev AV. Multizone impulse modulation. Novosibirsk, Russia: Nauka, 1979 (in Russian).
[56] Tonkal VE., Mielniciuk LP., Novosielcev AV., Dychenko JI. Power electronics converters with modulation and high frequency link. Kiev, Ukraine, Naukova dumka, 1981 (in Russian).
[57] Rodriguez J., Lai JS., Peng FZ. Multilevel inverters: a survey of topologies, controls, and applications. IEEE Trans. Industrial Electronics, Vol. 49, No. 4, 2002; 724–738.
[58] Rodriquez J., Pontt J. Lezana P., Kouro S. Tutorial on Multilevel Converters. International Conference on Power Electronics and Intelligent Control for Energy Conversation. Warsaw, Poland, 2005.
[59] Strzelecki R., Jarnut M., Kot E,. Kempski A., Benysek G. Multilevel voltage source power quality conditioner. IEEE/PESC Conf., Acapulco, Mexico, 2003; 1043–1048.
[60] Xu Z. Advanced semiconductor device and topology for high power current source converter. Ph.D Dissertation, Virginia Polytechnic Institute, Blacksburg, USA, 2003.

7

Static Shunt PE Voltage-quality Controllers

Ryszard Strzelecki, Daniel Wojciechowski and Grzegorz Benysek*

Department of Electrical Engineering,
Gdynia Maritime University,
81-87 Morska Str, Gdynia, Poland.
Emails: rstrzele;dwojc@am.gdynia.pl

Department of Electrical Engineering, Informatics and Telecommunications*
University of Zielona Góra,
50 Podgórna, Zielona Góra, Poland.
Email: G.Benysek@iee.uz.zgora.pl

This chapter focuses on power electronics (PE)-based solutions used in distribution networks for reduction of supply-voltage-quality deterioration, from long interruptions to the harmonics. Various PE devices and their combinations are addressed for particular or a range of voltage disturbances. A particular solution is applied taking into account demands for voltage quality and network configuration. In general, there are three types of static shunt PE voltage-quality controllers: The distribution static var compensator (D-SVC), distribution static synchronous compensator (D-STATCOM) and various hybrid arrangements. From the viewpoint of device topology D-SVC and D-STATCOM are identical with SVC and STATCOM, respectively. The difference between them results from place of installation (distribution network or transmission system) and, in consequence, the rated power and control methodology. The SVC and STATCOM are flexible AC transmission systems (FACTS) devices and are used to enhance controllability and increase the power transfer capability in transmission systems [1], whereas D-SVC and D-STATCOM are custom power-compensating shunt controllers used in distribution systems [2]. Their main purpose is voltage regulation, load compensation, but also voltage profile improvement. When operating with an energy-storage system D-STATCOM is capable of compensating active power fluctuations in the feeder.

7.1 Fundamentals of Shunt Compensation

From the "black box" viewpoint, any static shunt PE voltage-quality controller can be modelled with an ideal voltage source connected to the distribution network in point of common coupling (PCC). The principle of voltage regulation using a D-SVC controller is depicted in Figure 7.1. This example is for a two-source simplified system, and can be also used for a radial system with load instead of the second voltage source. Without compensation (Figure 7.1a), the voltage V_m at PCC results from amplitudes of V_{S1} and V_{S2} and angle δ for two-source system, and from V_{S1}, I_S and X for a radial system. With shunt compensation (Figure 7.1b) the voltage at PCC can be regulated by generating or absorbing only the reactive power. The principle of voltage regulation is shown in Figure 7.2. Without compensation, the terminal voltage varies depending on load current I_S, feeder reactance X, and bus voltage. The shunt controller usually has to regulate a terminal voltage at a level that depends on a compensating current, and not at a constant, nominal level. This regulation slope provides a better-defined operating point, especially in the case of low feeder impedance. Additionally, it enforces better load sharing among other controllers connected to the distribution system. In Figure 7.2 there are marked operating points for equivalent susceptance B_1 and B_2.

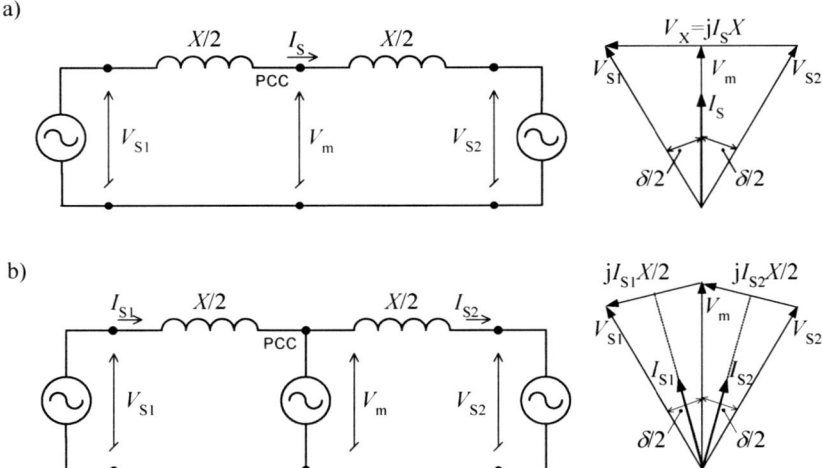

Figure 7.1. Simplified two-source system with (**a**) and without (**b**) shunt compensator

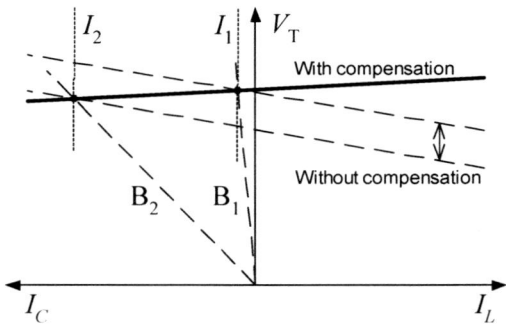

Figure 7.2. The principle of terminal-voltage regulation

The method of voltage regulation by generating or absorbing reactive power can be also used for voltage-profile improvement with use of several controllers that divide a line into N equal segments (Figure 7.3). For this case, active P_S and reactive Q_S powers of sources, as well as the reactive power of a single shunt controller Q_m are equal to, respectively:

$$P_S = NP_{max} \sin(\delta/N), \quad Q_S = NP_{max}(1-\cos(\delta/N)),$$
$$Q_m = 2NP_{max}(1-\cos(\delta/N)),$$
(7.1)

where P_{max} is the maximum active power of the line without compensation. The power components defined by Equation 7.1 are depicted in Figure 7.4.

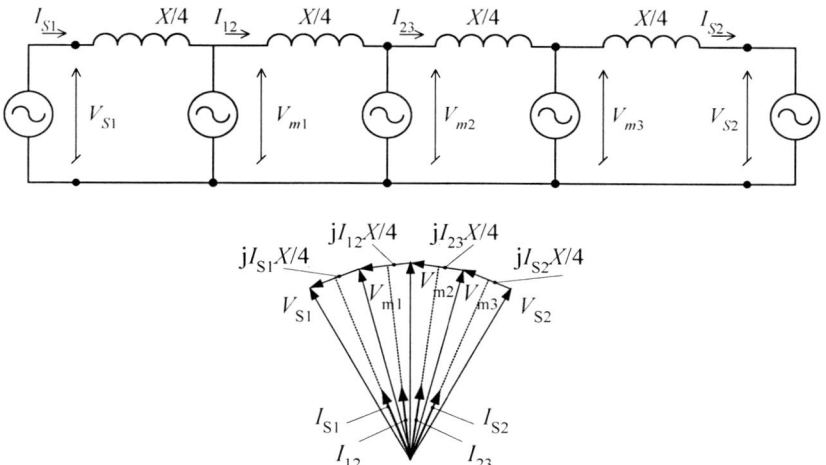

Figure 7.3. The principle of voltage-profile improvement

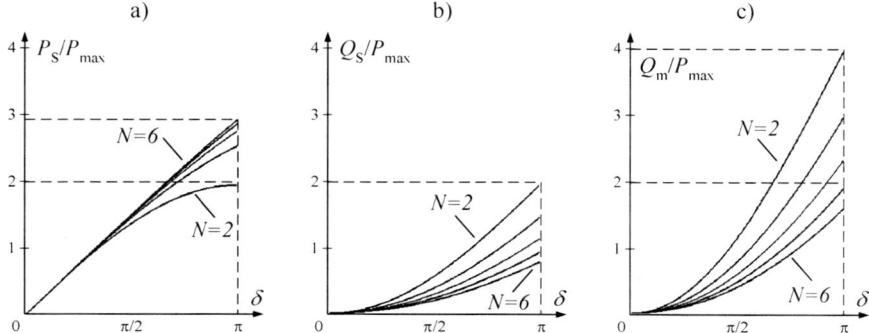

Figure 7.4. Power components with shunt controllers dividing a line into N equal segments

It can be derived from Equation 7.1 that for $N \to \infty$ the power components approach: $P_S \to P_{max}\delta$, $Q_S \to 0$, $Q_m \to 0$, but the total power of shunt controllers is:

$$Q_{mN} = 2N(N-1)P_{max}(1-\cos(\delta/N)), \tag{7.2}$$

and for $N \to \infty$ it approaches $Q_{n\infty} \to P_{max}\delta^2$. Based on Equation 7.1 and 7.2, and considering the costs of investment the optimal number of line segments N created by shunt controllers improving the voltage profile is usually 2 or 3.

7.2 Distribution Static Var Compensator (D-SVC)

The static var compensator (SVC) has been widely used by utilities since the mid-1970s. SVC is based on conventional capacitors and inductors combined with fast semiconductor switches without turn off capability (*i.e.* SCR thyristors). It is capable of providing voltage support at its terminals by controlling the amount of reactive power injected into or absorbed from the power system [3]. In general, when the system voltage is low, the SVC generates reactive power (SVC capacitive), and when the system voltage is high, it absorbs reactive power (SVC inductive). The variation of reactive power is performed by switching three-phase capacitor banks and inductor banks connected on the secondary side of a coupling transformer. The purpose, topologies and control principle of D-SVC is similar to an SVC controller. The possible topologies of D-SVC described below are thyristor-controlled reactor (D-TCR), thyristor-switched reactor (D-TSR), thyristor-switched capacitor (D-TSC) and combinations of these controllers.

7.2.1 Simple D-SVC Controllers

In Figure 7.5 are presented basic circuits of D-TCR/D-TSR and D-TSC, and their $V - I$ characteristics. The topology of a D-TCR shunt controller is the same as D-TSR (Figure 7.5a). It consists of a reactor connected in series with a bidirectional thyristor switch. In both D-TCR and D-TSR thyristors are switched on synchronously to the voltage, but with a delay angle in the range $0 \leq \alpha \leq \pi$ in D-TCR and without delay $\alpha=0$ in D-TSR. In consequence, in D-TCR it is possible to control the equivalent shunt reactance (or susceptance B_L) connected to the network, from 0 to the reactance of inductor L. Due to the phase control of D-TCR thyristors, this type of shunt controller injects into the network higher-order harmonics of current. It is possible to reduce these harmonics by employing multipulse configurations or by providing filtering (passive or active). This drawback is not present in D-TSR, but the reactance (or susceptance B_C) can not be varied continuously. It is only possible to connect or disconnect fixed reactance. The $V - I$ characteristic shows that maximum compensating current for D-TCR and compensating current for D-TSR are proportional to voltage, which degrades their transient behaviour under system-voltage disturbances. The next factor limiting transient performance of this type of controller is a relatively slow response time, which is equal to a half-period of the voltage sinusoid.

The circuit of a D-TSC controller and its $V - I$ characteristic are presented on Figure 7.5b. The reactor has very small reactance compared to the capacitor, and

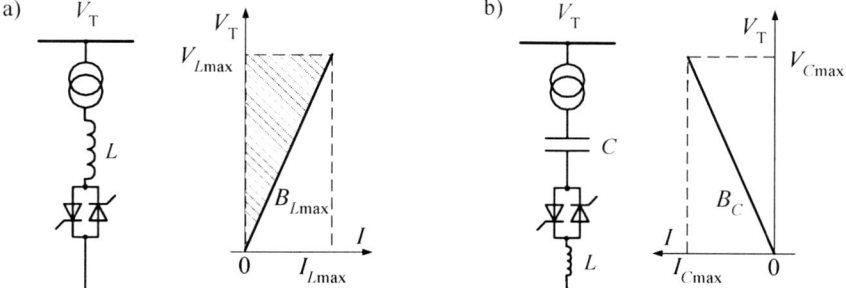

Figure 7.5. Basic circuits and $V-I$ characteristics of D-TCR/D-TSR (**a**) and D-TSC (**b**)

provides protection in the case of operation malfunction. In this type of shunt controller thyristors are switched on synchronously with system voltage. The instant of thyristor switch on after the turn off state is determined by the capacitor voltage. It is required that at the instant of turn on the instantaneous system voltage is equal to the capacitor voltage. This condition has to be fulfilled to not cause a short-circuit. The thyristor is turned off at the instant of current zero crossing, thus the capacitor voltage is equal to the positive or negative peak of the system voltage. In consequence, there is only one instant in every voltage period, when a thyristor can be turned on, so the response time of D-TSC is equal to one period of the system voltage sinusoid.

7.2.2 Combined D-SVC Controllers

There are two topologies of combined D-SVC controllers: fixed capacitor thyristor-controlled reactor (D-FC-TCR) and thyristor-switched capacitor thyristor-controlled reactor (D-TSC-TCR). A D-TSC-TCR controller circuit and $V-I$ characteristic are presented inn Figure 7.6.

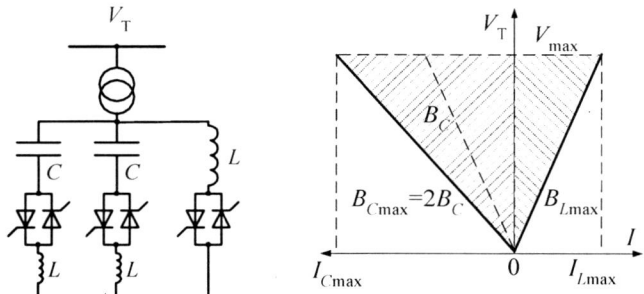

Figure 7.6. Circuit topology and $V-I$ characteristic of distribution TSC-TCR controller

Control of the D-TSC-TCR is realized by switching in an appropriate number of capacitors and varying continuously the delay angle in the TCR part. As a result, this controller is capable of both generating and absorbing reactive power. Additionally, harmonics distortion of injected currents is smaller than in D-TCR, because the power rating of the TCR part, which injects harmonics, is a fraction of the total rated power of the controller. The response time of D-TSC-TCR is one half of the system voltage period on decreasing its capacitive reactive power and can be up to one voltage period in the opposite case.

For custom power controllers it is necessary to precisely regulate the voltage supplied to the customer. D-SVC controllers inject harmonics (D-TCR) or are not capable of continuous control (D-TSR, D-TSC). In consequence, using only D-SVC it can be impossible to control the supply voltage at nominal level, sinusoidal and balanced. Additionally, the relative low response time of D-SVC deteriorates voltage controllability during a transients in distribution systems. Relatively fast, periodic disturbances (*i.e.* flicker of frequency above 15 Hz) also can not be effectively compensated using D-SVC. Although very important for FACTS, SVCs often have to be supplemented by D-STATCOM or active power filters when they are used in a distribution network as a custom power compensation controllers.

7.3 Distribution Static Synchronous Compensator (D-STATCOM)

D-STATCOM is the most important controller for distribution networks. It has been widely used since the 1990s to precisely regulate system voltage, improve voltage profile, reduce voltage harmonics, reduce transient voltage disturbances and load compensation. Rather than using conventional capacitors and inductors combined with fast switches, the D-STATCOM uses a power-electronics converter to synthesise the reactive power output. A D-STATCOM converter is controlled using pulse width modulation (PWM) or other voltage/current-shaping techniques [4-7]. D-STATCOMs are used more often than STATCOM controllers. Compared to STATCOM, D-STATCOMs have considerably lower rated power and, in consequence, faster power-electronics switches, thus the PWM carrier frequency used in a distribution controller can be much higher then in a FACTS controller. It has a substantial positive impact on the dynamics of the D-STACOM.

7.3.1 Topology

D-STATCOM controllers can be constructed based on both voltage source inverter (VSI) topology and current source inverter (CSI) topology (Figure 7.7). Regardless of topology, a controller is a compound of an array of semiconductor devices with turn off capability (*i.e.* IGBT, GTO, IGCT *etc.*) connected to the feeder via a relative small reactive filter. The VSI converter is connected to the feeder via reactor L_F and has a voltage source (capacitor C_D) on the DC side. On the other side, the CSI converter is connected on the AC side via capacitor C_F and has a current source (inductor L_D) on the DC side. In practice, CSI topology is not used for D-STATCOM. The reason for this is related to the higher losses on the DC reactor of CSI compared to the DC capacitor of VSI. Moreover, a CSI converter requires reverse-blocking semiconductor switches, which have higher loses than reverse-conducting switches of VSI. And finally, the VSI-based topology has the advantage because an inductance of a coupling transformer Tr (if present) can constitute, partially or completely, the inductance of AC filter. The following text will describe properties of VSI-topology based D-STATCOM only, but in many respects they are the same as for CSI-based controller.

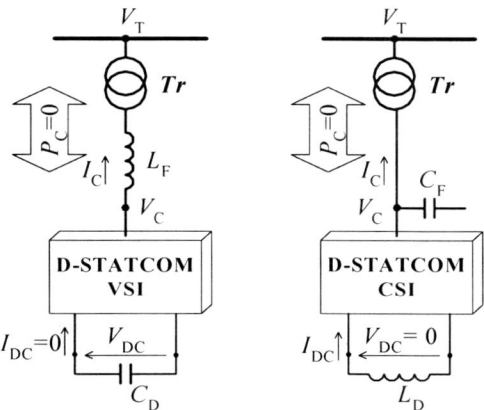

Figure 7.7. General topology of VSI-based and CSI-based D-STATCOM

The VSI converters for D-STATCOM are constructed based on multilevel topologies, with or without use of a transformer, which are described in Chap. 6. These solutions provide support for operation with a high level of terminal voltage. Additionally, D-STATCOM controllers can be a compound of several converters configured to various topologies, to achieve higher rated power, or lower PWM-related current ripples. The exemplary topologies are presented in Figure 7.8. In the parallel configuration (Figure 7.8a) converters are controlled to share the generated power equally, or at a given ratio, for example proportional to the rated power of the particular converter. In this solution it is necessary to provide inter converter communication at the control level, to distribute information about set controller power or currents. The cascade multiconverter topology (Figure 7.8b) is similar to the parallel configuration, but in this case the constituent converters do not share power equally, but successively, depending on the requirement. In this case, no communication between constituent converters is required, but on the other hand it is also not possible to use common PWM strategy. The converters in this case are exactly the same as for standalone operation. In Figure 7.8c and d are presented series and parallel master–slave topologies, respectively. The master–slave topologies require a high degree of integration between constituent converts, including a control system, and are treated and realised as a single, multiconverter controller. The master (called a "slow converter") converter has substantially higher rated power and, in consequence, considerably lower PWM carrier frequency than the slave converter (called a "fast converter"). The task of the master converter is to cover the requirements for power, while the slave has to compensate AC current/voltage ripples using serial superposition of voltages (Figure 7.8c) or parallel superposition of currents (Figure 7.8d).

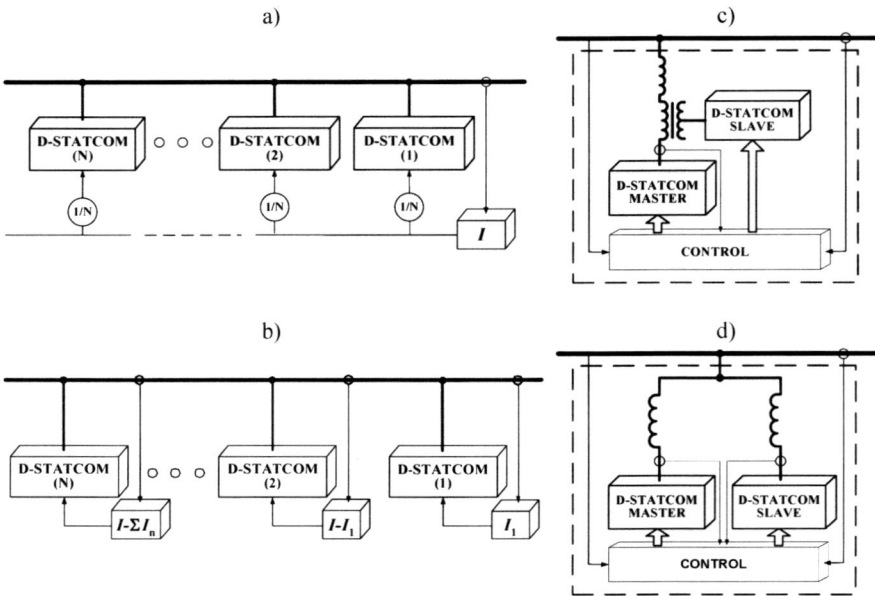

Figure 7.8. Multiconverter topologies of D-STATCOM controller: parallel (**a**), cascade (**b**), master–slave series (**c**), master–slave parallel (**d**)

8.3.2 Principle of Operation

For the operation analysis of the D-STATCOM converter, it is possible to represent its PWM-controlled VSI with an instantaneous (averaged for PWM period) voltage source. The principle of generating instantaneous active and reactive powers by D-STATCOM is shown in Figure 7.9. In this figure, voltages and currents are represented with instantaneous space vectors obtained using a power-invariant Clarke transform. In Figure 7.9 are presented three cases: the general one, for reactive power equal to zero and for active power equal to zero. From this figure it is clear that by generating an appropriate AC voltage it is possible to generate arbitrary instantaneous vectors of both active and reactive power. The real component of currents is related to the equivalent series resistance modelling losses on the AC side. The possible active and reactive powers that can be generated or absorbed by D-STATCOM are limited. This limitation is related to circuit parameters and maximum ratings of VSI components. In Figure 7.9 there is presented an exemplary limit for AC voltage, which depends on VSI DC voltage V_{DC}. This limit, together with filter inductance L_F and terminal voltage V_T, define the operating region of a D-STATCOM controller. The operating region of a two-level VSI-based controller is presented in Figure 7.10. In this figure, Y denotes the modulus of admittance on the AC side of VSI. In practice, the operating region does not limit the maximum ratings of VSI semiconductors, so the static $V - I$ characteristic of D-STATCOM reactive power is symmetrical (Figure 7.11).

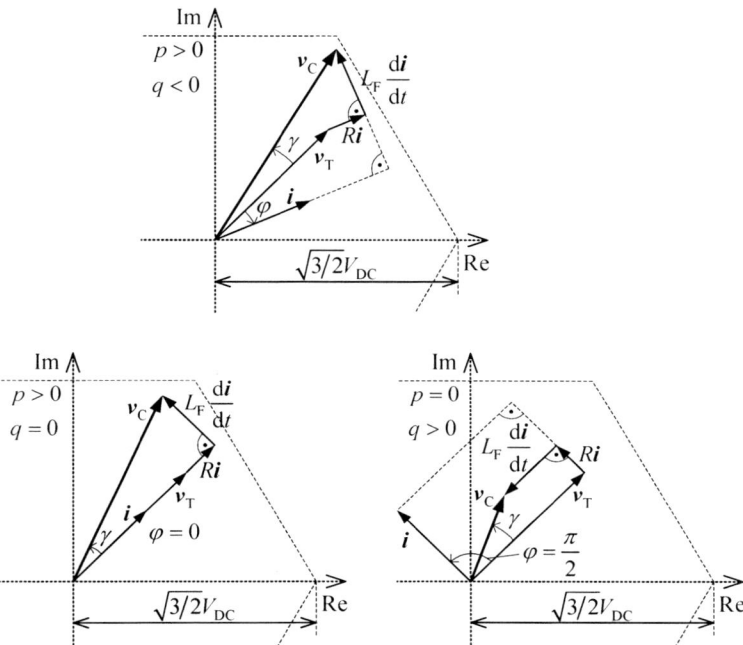

Figure 7.9. Principle of control of D-STATCOM instantaneous active and reactive power

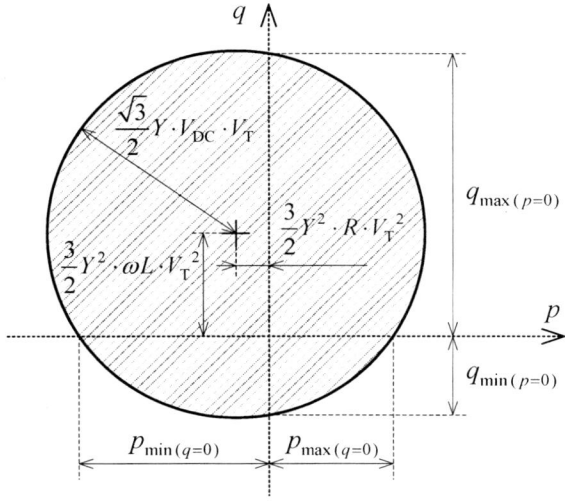

Figure 7.10. Operating region of two-level VSI-based D-STATCOM

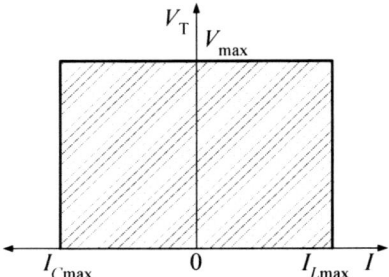

Figure 7.11. The $V - I$ characteristic of D-STATCOM

The active power is consumed by the D-STATCOM only to cover internal losses. Assuming lossless operation, the averaged (but not instantaneous) active power has to be zero. There are no similar limitations for reactive power, because it is only exchanged between phases, and is not converted between the AC and DC sides of D-STATCOM VSI.

There are two modes of D-STATCOM operation: load compensation in current-control mode and voltage regulation in voltage-control mode.

7.3.3 Load Compensation

In the load-compensation mode, D-STATCOM is controlled in current mode. The control system of D-STATCOM has to generate reference currents, compensating harmonic, unbalance and fundamental reactive components of nonlinear load supply currents. The required rated power of load-compensating D-STATCOM depends only on reactive power, harmonic distortion and power of the compensated load. In general, D-STATCOM is capable of compensating current disturbances from harmonics to long-duration effects, including active power transients. The possibility and effectiveness of compensation of a particular voltage-quality problem depends on the topology and rated power of the controller as well as on the capacity of the energy-storage system connected at the D-STATCOM DC side. This will be briefly described in Sect. 7.4.2.

Consider the block diagram of the system that is depicted in Figure 7.12. The instantaneous apparent power noted on that figure is defined as:

$$\underline{s}_{3f} = \underline{e}^* \underline{i} = p_{3f} + jq_{3f}, \text{ where } p_{3f} = \text{Re}(\underline{e}^* \underline{i}) \text{ and } q_{3f} = \text{Im}(\underline{e}^* \underline{i}). \quad (7.3)$$

Based on Equation 7.3, the instantaneous apparent power of a nonlinear load can be expressed as follows:

$$\underline{s}_{3f} = \overline{\underline{s}}_{3f} + \tilde{\underline{s}}_{3f} = \underline{s}_{e_1 i_1} + \sum_{n=-\infty}^{\infty} \underline{s}_{e_n i_n} + \sum_{n=-\infty}^{\infty} \tilde{\underline{s}}_{e_1 i_n} + \sum_{n=-\infty}^{\infty} \tilde{\underline{s}}_{e_n i_1} + \sum_{m=-\infty}^{\infty} \sum_{n=-\infty}^{\infty} \tilde{\underline{s}}_{e_m i_n}, \quad (7.4)$$

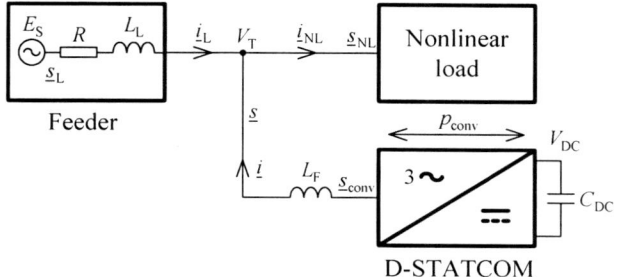

Figure 7.12. Configuration of system with D-STATCOM

where particular components are related with:

1. $\overline{\underline{s}}_{e_1 i_1}(t)$ — fundamental components of voltages and currents
2. $\sum \overline{\underline{s}}_{e_n i_n}(t)$ — higher, equal-order harmonics of voltages and currents
3. $\sum \overline{\underline{s}}_{e_1 i_n}(t)$ — higher-order harmonics of currents
4. $\sum \overline{\underline{s}}_{e_n i_1}(t)$ — higher-order harmonics of voltages
5. $\sum \overline{\underline{s}}_{e_m i_n}(t)$ — higher, not equal-order harmonics of voltages and currents

There are three possible compensation methods that ensure that feeder currents are:

- proportional to the fundamental, positive component of terminal voltage (the power components 1 and 4 are present);
- proportional to the terminal voltage V_T, which is equivalent to resistance (the power components 1, 2, 3 and 4 are present);
- corresponding to the constant instantaneous active power, and zero instantaneous reactive power (the power components 1 and 2 are present)

Among the possible strategies, the first one is usually considered as the best one. The reason is because it provides feeder currents that do not distort voltage at any point of the feeder. From the above considerations it is clear that load compensation provides also a reduction of the voltage distortion related with the feeder voltage drop. The level of distortion reduction depends on the configuration of the distribution network as well as the ratio between the power of the compensated nonlinear load and the feeder short-circuit power.

From the control viewpoint, the effectiveness of compensation realised by D-STATCOM operating in load-compensation mode depends on the reference-current calculation method and current controller (Figure 7.13). Among the most important properties of the reference current calculation algorithm are dynamics, exactness, possibility to selective compensation and computational effort. There are two major types of such algorithms: time based and frequency based. The exemplary reference calculation algorithms are presented in Figure 7.14. In the time-based algorithms the reference currents are derived directly from the transients of currents and voltages, thus, they provide an "instantaneous" result. The $p - q$, 3-wire theory is based on the power definitions given in Equation 7.3. The reference currents are determined using the following equation:

$$\underline{i} = \frac{\underline{s}}{\underline{e}^*} = \frac{1}{|\underline{e}|^2}(p\,\mathrm{Re}\,\underline{e} - q\,\mathrm{Im}\,\underline{e}) + \mathrm{j}\frac{1}{|\underline{e}|^2}(p\,\mathrm{Im}\,\underline{e} + q\,\mathrm{Re}\,\underline{e}). \qquad (7.5)$$

By selecting the appropriate instantaneous power components that are used in Equation 7.5 it is possible to calculate the desired reference currents according to the particular compensation method, with or without reactive power filtering. In a 4-wire system it is necessary to define independent zero sequence active power [8].

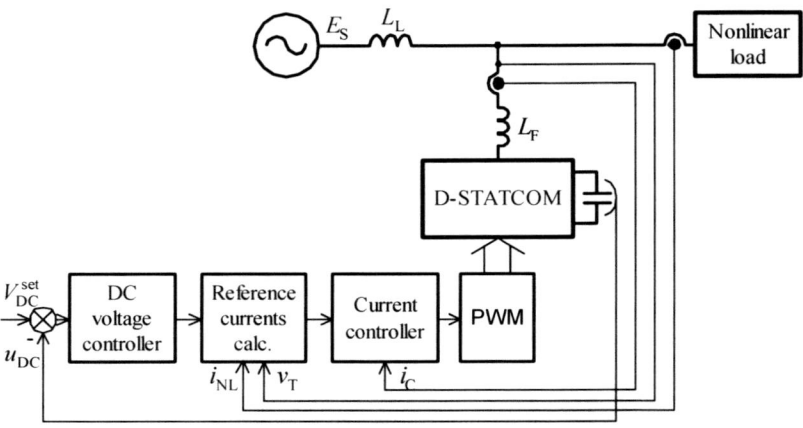

Figure 7.13. Block diagram of a control system for load compensating D-STATCOM

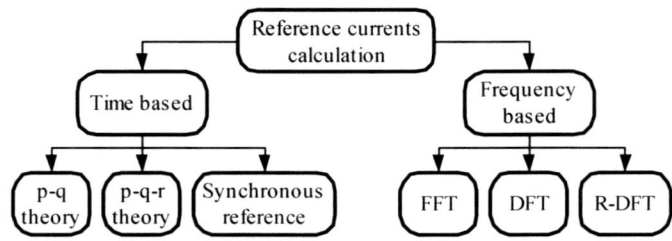

Figure 7.14. Selected reference current calculation algorithms

On the $p - q - r$ 4-wire method there are calculated three linear independent current components p, q and r, according to the equation:

$$\begin{bmatrix} i_p \\ i_q \\ i_r \end{bmatrix} = \begin{bmatrix} \cos\theta_2 & 0 & \sin\theta_2 \\ 0 & 1 & 0 \\ -\sin\theta_2 & 0 & \cos\theta_2 \end{bmatrix} \begin{bmatrix} i_\alpha \\ i_\beta \\ i_0 \end{bmatrix}, \quad \theta_2 = \tan^{-1}\left(\frac{e_0}{\sqrt{e_\alpha^2 + e_\beta^2}}\right), \tag{7.6}$$

where $\begin{bmatrix} i_\alpha \\ i_\beta \\ i_0 \end{bmatrix} = \begin{bmatrix} \cos\theta_1 & \sin\theta_1 & 0 \\ -\sin\theta_1 & \cos\theta_1 & 0 \\ 0 & 0 & 1 \end{bmatrix} \begin{bmatrix} i_\alpha \\ i_\beta \\ i_0 \end{bmatrix}, \quad \theta_1 = \tan^{-1}\left(\frac{e_\beta}{e_\alpha}\right).$ (7.7)

The synchronous reference method is the family of algorithms that is based on representation of the currents in the rotating, voltage synchronized reference frame. In that frame currents are filtered using a highpass filter to derive the desired compensating components. It is also possible to calculate particular, higher-order current harmonics using the multiple rotating reference frame method [8].

The frequency-based methods are the family of algorithms that use frequency representation of currents to separate compensating harmonics. The main difference between particular algorithms results from the method of frequency-domain calculation. Methods like fast fourier transform, discrete fourier transform, recursive discrete fourier transform and others, like kalman filtering or neural networks are used. These methods require higher computational effort and usually have lower response time compared to the time-based methods, but provide possibility to calculate reference currents compensating particular, selected harmonics of distorted load currents.

Among current-control techniques several methods, like PI controllers, sliding-mode controllers, predictive algorithms, dead-beat controllers, and hysteresis methods are used [6,9,10]. The most important property of such a controller is the dynamics, thus the predictive and dead-beat algorithms are usually preferred.

Despite current compensation, a D-STACOM controller can be used at the same time for AC/DC power conversion, for example providing a supply for a DC feeder or micro – DC distribution system, especially in distributed generation systems. In Figure 7.15 are presented transients of a shunt compensator providing AC/DC conversion. This solution can allow overall costs of D-STATCOM to be reduced in cases where it is necessary to also provide bidirectional AC/DC electrical power conversion.

7.3.4 Voltage Regulation

The idea of voltage regulation using D-STATCOM is consistent with the D-SVC described in Sect. 7.2 and presented in Figure 7.2. It is realised by compensating reactive power (*i.e.* by injecting or absorbing reactive power) [11,12]. The advantage of D-STATCOM over D-SVC is its also $V - I$ characteristics (Figure

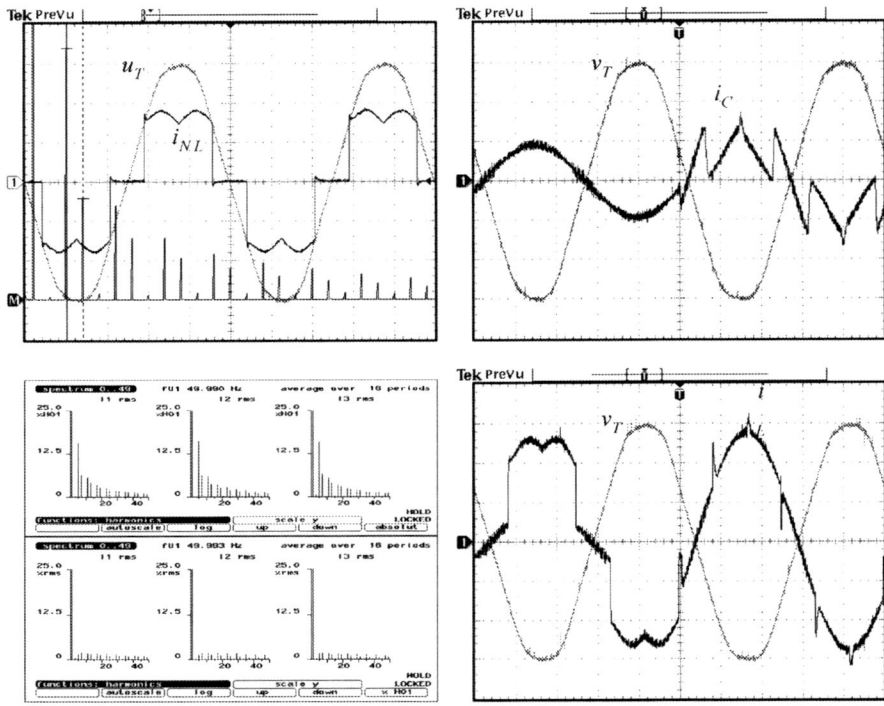

Figure 7.15. Transients and feeder-current harmonics without and with compensation for D-STATCOM compensating load and converting electrical energy from the AC to DC side. Activation of harmonic compensation

7.11) and dynamics, but this controller is more expensive. It is more important in the case of voltage-regulation mode because it requires, in general, higher compensating power than for load compensation.

The block diagram of a voltage-regulating D-STATCOM is presented in Figure 7.16. The reference AC voltage of VSI is calculated based on the output of a DC-voltage controller and terminal voltage controller. The DC-voltage controller defines the component of voltage across a filter inductor L_F that is perpendicular to the terminal voltage V_T, while the terminal-voltage controller determines the component of voltage at L_F that is parallel to V_T. Thus, the active power of D-STATCOM depends on the DC voltage error and the reactive power of D-STATCOM depends on the error of the terminal voltage V_T (see Figure 7.9). The dynamics of voltage regulation depends on terminal-voltage controller. State-feedback or output-feedback controllers can be applied for this purpose [2].

In fact, D-STATCOM in voltage-regulation mode compensates disturbances of voltage V_T taht are the result of all nonlinear or unbalanced loads connected to a distribution system, which have an influence on this voltage. Although shunt compensators are used more often than series compensators (see Chap. 8), the latter are better suited for voltage regulation from the viewpoint of the required

Static Shunt PE Voltage-quality Controllers 197

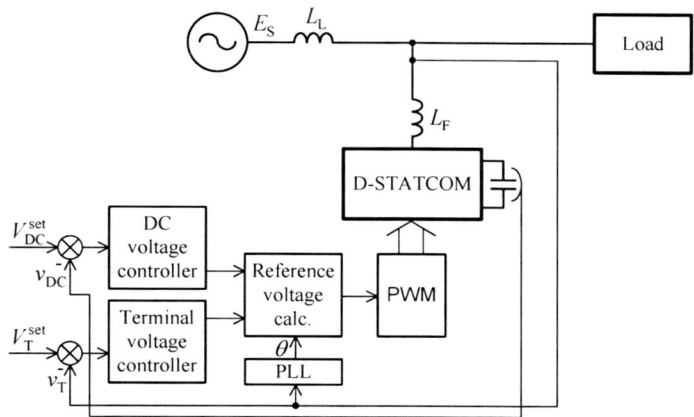

Figure 7.16. Block diagram of a control system for voltage-compensating D-STATCOM

rated power of the converter. The reason for using shunt controllers more often than series controllers is their immunity to feeder short circuit.

D-STATCOM controllers are especially useful for compensating voltage disturbances like flicker or voltage dips, caused by rapidly changing loads, like arc furnaces or high-power electrical machines during direct start up. In Figure 7.17 are presented the exemplary results of voltage-flicker compensation using a D-STATCOM controller.

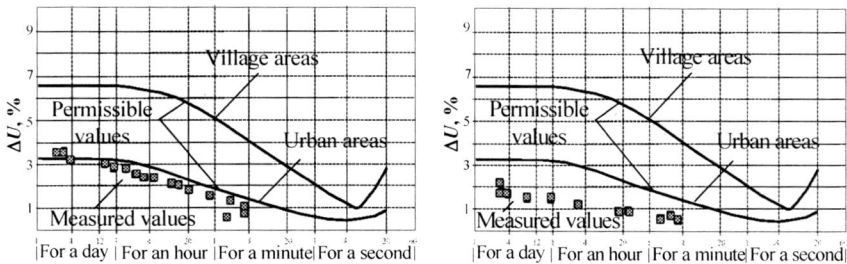

Figure 7.17. The example of voltage-flicker reduction caused by a sawmill using D-STATCOM installed near the industrial facilities. Flicker level without and with compensation

7.4 Other Shunt Controllers Based on D-STATCOM

The compensation capability of D-STATCOM controllers can be extended using combined topologies. In general, there are two major topologies extending the D-STATCOM compensation capability: hybrid arrangements and energy-storage system applications [13,14].

7.4.1 Hybrid Arrangements

The D-STATCOM-based hybrid arrangements can be used for both voltage-regulating and load-compensating controllers. D-STATCOMs are usually combined with D-SVCs or passive harmonics filters. The former are most often used for voltage regulation, while the latter are utilised for load compensation. Both provide improvement of compensation capabilities. Additionally, in hybrid topologies the rated power of D-STATCOM constitutes a part of a hybrid controller's rated power, thus they allow the installation costs to be reduced.

In Figure 7.18 to 7.20 are presented general topologies and $V - I$ characteristics of D-STATCOM D-SVC hybrid arrangements. The D-SVC part extends the operating region of D-STATCOM. The combination of D-STATCOM and D-TSC (Figure 7.18) extends the operating region towards the generation of reactive power (capacitive region). This property is important in practice, because it is often necessary in distribution systems to compensate inductive-type loads to provide terminal-voltage regulation. Extension of $V - I$ characteristics of D-STATCOM towards absorption of reactive power (inductive region) is possible by combining it with D-TSR (Figure 7.19). The symmetrical extension of $V - I$ characteristics is provided by the hybrid arrangement presented in Figure 7.20. Despite improvements of the $V - I$ characteristics, a hybrid arrangement can be used to optimize losses, cost and performance for a particular application.

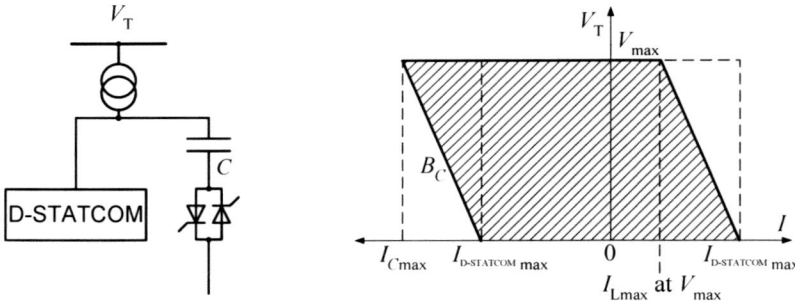

Figure 7.18. Hybrid D-STATCOM D-TSC controller and its $V - I$ characteristics

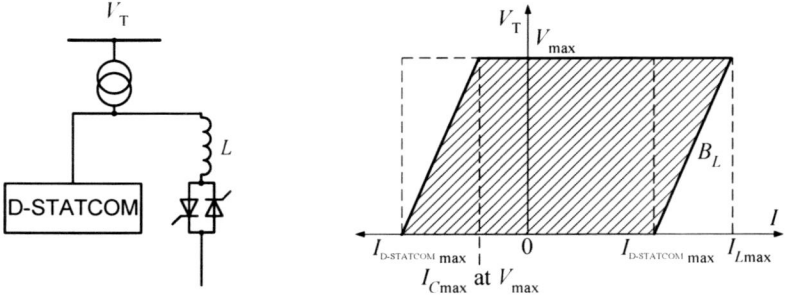

Figure 7.19. Hybrid D-STATCOM D-TSR controller and its $V - I$ characteristics

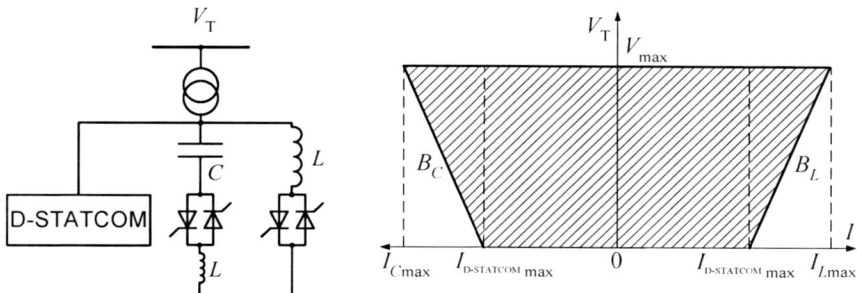

Figure 7.20. Hybrid D-STATCOM D-TSC D-TSR controller and its $V-I$ characteristics

In Figure 7.21 are presented hybrid-D-STATCOM based topologies of load-compensating controllers. Among the advantages of hybrid arrangements in this case are: better compensation performance and considerably lower rated power of D-STATCOM (lower voltages or/and currents). On the other hand, the disadvantages of hybrid topologies are their overall dimensions, longer response time and generating of reactive power. The last property, however, can be utilised for reactive power compensation in distribution systems.

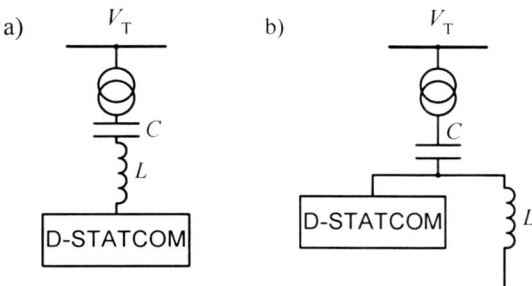

Figure 7.21. Hybrid D-STATCOM topologies intended for load compensation

In the topology presented in Figure 7.21a D-STATCOM is combined with a passive harmonic filter (or several filters tuned for particular harmonics) in series. The transients for this topology are presented in Figure 7.22. The notation is consistent with that used in Figure 7.12, and u denotes the AC voltage of D-STATCOM VSI. In this example the total harmonic distortion of the feeder current is equal to THD=1.87% and the power of D-SATCOM is equal to 8.2% of the compensated load power.

The hybrid topology presented in Figure 7.21b combines a passive harmonic filter with D-STATCOM connected in parallel with a passive filter inductor. The transients are presented in Figure 7.23. The feeder current THD=3.44% and D-STATCOM power is equal to 1.2% of the compensated load power.

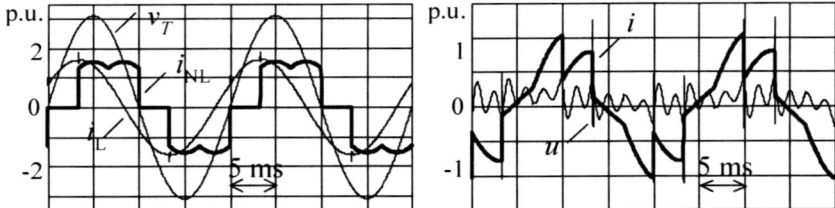

Figure 7.22. Exemplary transients of currents and voltages in steady state for hybrid topology presented in Figure 7.21a

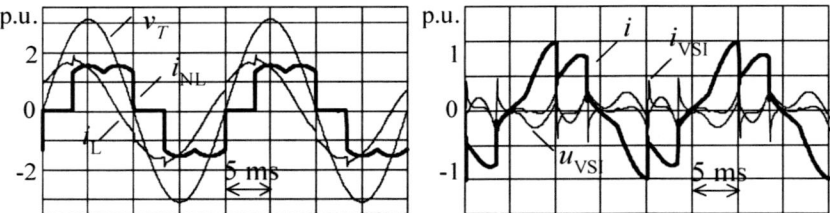

Figure 7.23. Exemplary transients of currents and voltages in steady state for hybrid topology presented in Figure 7.21b

7.4.2 Controllers with Energy-storage Systems

A D-STATCOM controller can be supplemented with an energy-storage (ES) system connected to the DC side of the converter to extend the compensation capabilities. Depending on its type, ES systems are connected by employing static converters or electromechanical converters in machine/generator (M/G) systems. For solutions with ES systems the averaged real power of D-STATCOM still needs to be zero, but the period of averaging can be considerably longer. This means that D-STATOM with ES can compensate (level) considerably slower fluctuations of feeder active power. In fact, a combined D-STATCOM/ES controller can be able to compensate disturbances of active power from harmonics to fluctuations up to several minutes long (for example feeder overloads).

There are many types of ES systems used. Among the most important are: supercapacitors energy storage (S-CES), superconducting magnetic-energy storage (SMES), flywheel-energy storage (FES) and battery-energy storage (BES). The description of these system is presented in [15-17]. Figure 7.24 summarises the power and energy capabilities of the selected ES systems.

The combined D-STATCOM ES systems can be controlled as an uninterruptible power supply (UPS) in the line interactive configuration [9,18] to provide high-quality terminal voltage V_T for critical loads, regardless of supply interrupts and the quality of the feeder voltage. The block diagram of such a system is depicted in Figure 7.25a. The reactor L_T has to be connected in series, to lower the short-circuit power of the feeder. This makes it possible to considerably lower the rated power of VSI and to reduce harmonic distortion of supply currents. In this application, the operation principle of D-STATCOM is the same as in voltage-regulation mode, and the control system is similar to that presented in Figure 7.16. The task of this UPS system is to regulate the terminal voltage to be nominal (or within a permissible range) and pure sinusoidal, regardless of the presence and quality of the source voltage V_S. In Figure 7.25b are presented voltage and current space vectors of the system. The averaged active power of D-STATCM has to be zero. In consequence, the controller has to ensure that the angle between the regulated terminal voltage and the source voltage corresponds to the active power of the load. This is realized by the D-STATCOM's DC voltage-control loop. To provide zero reactive power consumed from the feeder, it is necessary that the voltage across the reactor is perpendicular to the source voltage V_S. This means that the amplitude of the regulated terminal voltage V_T has to be greater than the amplitude of the source voltage V_S. In practice, the voltage has to be maintained within permissible limit and in consequence the region of operation with zero reactive power is limited, and depends on the reactor value, the amplitude of the supply voltage and the active power. In Figure 7.25b is presented the case for exceeded the limit of zero reactive power operation. As a result, the reactive power supplied from the feeder is not equal to zero, and there exists also a nonzero angle α between the supply voltage V_S and the supply current i_S.

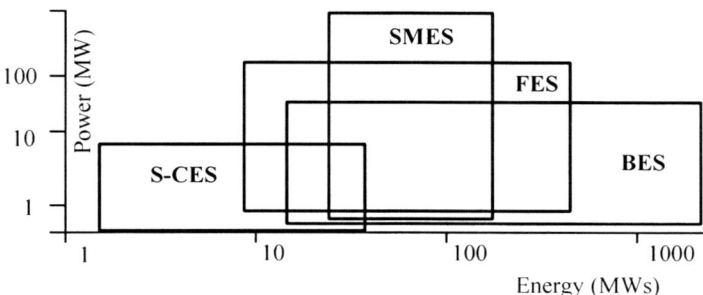

Figure 7.24. Power and energy ranges of the selected ES systems

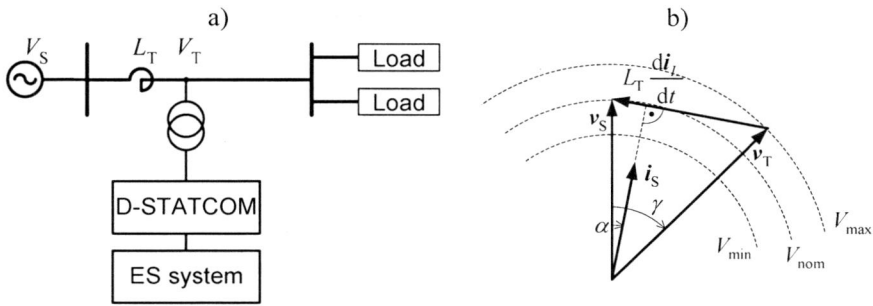

Figure 7.25. Block diagram (**a**) and space vectors (**b**) of line interactive UPS with D-STATCOM controller

From the principle of operation of a line-interactive UPS it is obvious that a distorted source voltage causes a distorted voltage across the reactor (for a sinusoidal, regulated terminal voltage) and, in consequence, a distorted supply current. Thus, it is not possible to regulate the terminal voltage and supply current at the same time. Figure 7.26 demonstrates the case for distorted supply voltage, which determines distorted voltage across reactor v_L. This simultaneous control of terminal voltage and supply current can be realised by a line-interactive delta UPS, but it is a combined series-shunt controller.

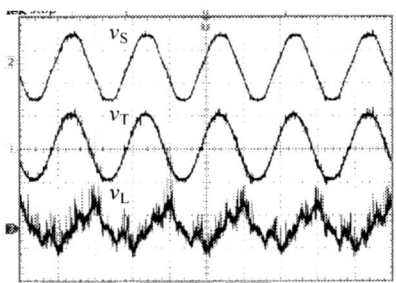

Figure 7.26. The source, terminal and reactor voltages of a line-interactive UPS controller

7.5 Summary

In this chapter, the possibilities and realization methods of shunt compensation in distribution systems for voltage-quality improvement were reviewed. This type of compensation is, in general, based on injecting or absorbing a reactive power to the system and, in consequence, regulating the terminal voltage. The two major types of compensators are passive-based (D-SVC) and active (D-STATCOM, combined). The most important differences between passive based and active distribution controllers are summarised in Table 7.1. For both, the active power has to be equal zero to (in practice equal to the power losses of the controller). The most important difference between them is that the active controllers, in opposition to passive ones, have the possibility to instantaneous generation of nonzero active power. Of course the averaged active power has to be still equal to zero, but the averaging period can be extended using energy storage systems that allow compensation of the power fluctuations and can even provide terminal voltage in the case of supply discontinuities in line-interactive UPS applications. The dynamics and transient behaviour are considerably better for active and combined controllers, which is beneficial for compensation of fast, periodic and non periodic disturbances. The passive (D-SVC) controllers have the lowest initial costs, and often supplement D-STATCOM controllers to optimise cost/performance for particular application.

At present, there is a tendency in new designs to utilise active PWM D-STATCOM controllers for voltage quality improvement because of their performance advantages and continuous lowering of installation costs due to development and cost reduction of power semiconductor switches.

Table 7.1. The summary of shunt controller properties

	D-SVC	D-STATCOM	Combined D-STATCOM with D-SVC	D-STATCOM with ES
Type	Passive based (controlled impedance)	Active (controlled voltage source)	Passive / active	Active (controlled voltage source)
$V-I$ characteristic	Dependent of voltage (Figure 7.5, 7.6)	Independent of voltage (Figure 7.11)	Dependent of voltage (Figure 7.18-7.20)	Independent of voltage (Figure 7.11)
Generation of instantaneous active power	No	Yes	Yes	Yes, averaging period up to several minutes
Maximum time response	Half to one cycle	Negligible	Negligible for limited power range	Negligible
Current harmonics introduction	D-TCR – Yes Other – Very low	Very low, higher order	Very low, higher order	Very low, higher order
Transient behaviour	D-TCR – Poor Other – neutral	Very good	Good	Excellent
Load compensation possibility	No	Yes	Limited	Yes
UPS operation possibility	No	No	No	Yes
Initial cost	Below medium	High	Medium	Very High

References

[1] Hingorani NG, Gyugyi L. Understanding FACTS. Concepts and Technology of Flexible AC Transmision Systems. New York: IEEE Press, 1999.
[2] Ghosh A, Ledwich G. Power Quality Enhancement Using Custom Power Devices. Bostom, Dordrecht, London: Kluwier Academic Publishers, 2002.
[3] Wong WK, Osborn DL, McAvoy JL. Application of Compact Static Var Compensators to Distribution Systems. IEEE Trans. Power Delivery, Vol. 5, No. 2, 1990; 1113–1120.
[4] Blazic B, Papic I. Improved D-STATCOM Control for Operation with Unbalanced Currents and Voltages. IEEE Trans. Power Delivery, Vol. 21, No. 1, 2006; 225–233.

[5] Escobar G, Stankovic AM, Mattavelli P. An Adaptive Controller in Stationary Reference Frame for D-STATCOM in Unbalanced Operation. IEEE Trans. Ind. Electr., Vol. 51, No. 2, 2004; 401–409.
[6] Moon G-W. Predictive Current Control of Distribution Static Compensator for Reactive Power Compensation. IEE Proc. Gener. Transm. Distrib., Vol. 146, No. 5, 1999; 515–520.
[7] Strzelecki R, Supronowicz H. Power Factor Correction in AC supply systems and improving methods. Warsaw: Warsaw University of Technology Publishing, 2000 (in Polish).
[8] Massoud AM, Finney SJ, Williams BW. Review of Harmonic Current Extraction Techniques for an Active Power Filter. 11th International Conference on Harmonics and Quality of Power, IEEE, 2004; 154–159.
[9] Emadi A, Nasiri A, Bekiarov SB. Uninterruptible Power Supplies and Active Filters. CRC Press, USA, 2005.
[10] Shin E-C, Park S-M, Oh W-H, Kim D-S, Lee S-B, Yoo J-Y. A Novel Hysteresis Current Controller to reduce the Switching Frequency and Current Error in D-STATCOM. The 30th Annual Conference of the IEEE Indust. Electr. Society, Busan, Korea, 2004; 1144–1149.
[11] Kincic S, Wan XT, McGillis DT, Chandra A, Ooi B-T, Galiana FD, Joos G. Voltage Support by Distributed Static VAr Systems (SVS). IEEE Trans. Power Delivery, Vol. 20, No. 2, 2005; 1541–1549.
[12] Haque MH. Compensation of Distribution System Voltage Sag by DVR And D-STATCOM. IEEE Porto Power Tech Conference, Porto, Portugal, 2001; CD-ROM.
[13] Lee S-Y, Wu C-J. Combined Compensation Structure of an SVC and an Active Filter for Unbalanced Three Phase Distribution Feeders with Harmonic Distortion. Proc. of the 4th Int. Conf. APSCOM-97, Hong Kong, 1997; 543–548.
[14] Akagi H. Active Filters and Energy Storage Systems for Power Conditioning in Japan. Proc. of First International Conference on Power Electronics Systems and Application, 2004; 80–88.
[15] Ribeiro PF, Johnson BK, Crow ML, Arsoy A, Liu Y. Energy Storage Systems for Advanced Power Applications. Proc. IEEE, Vol. 89, No. 12, 2001; 1744–1756.
[16] Saminemi S, Johnson BK, Hess HL, Law JD. Modeling and Analysis of a Flywheel Energy Storage System for Voltage Sag Correction. IEEE Trans. Industry Applications Vol 42, No. 1, 2006; 42–52.
[17] Schoenung SM, Burns C. Utility Energy Storage Applications Studies. IEEE Trans. Energy Conversion, Vol. 11, No. 3, 1996; 658–665.
[18] Oliveira da Silva SA, Donoso-Garcia PF, Cortizo PC, Seixas PF. A Three-Phase Line-Interactive UPS System Implementation With Series-Parallel Active Power-Line Conditioning Capabilities. IEEE Trans. Industry Applications Vol 38, No. 6, 2002; 1581–1590.

8

Static Series and Shunt-series PE Voltage-quality Controllers

Ryszard Strzelecki, Daniel Wojciechowski and Grzegorz Benysek*

Department of Electrical Engineering,
Gdynia Maritime University,
81–87 Morska Str, Gdynia, Poland
Emails: rstrzele;dwojc@am.gdynia.pl

Department of Electrical Engineering, Informatics and Telecommunications*
University of Zielona Góra,
50 Podgórna, Zielona Góra, Poland
Email: G.Benysek@iee.uz.zgora.pl

As presented in the Chap. 7 static shunt power electronics (PE) voltage-quality controllers protect the utility electrical system from the unfavorable impact of customer loads. Shunt controllers, as shown in Chap. 6, are recommended mainly for mitigation of the causes of disturbances, and not their effects in distanced nodes of a power-electronics system. In the case when reduction of disturbances effects is required, which leads to protection of sensitive loads from the deterioration in the supply-side voltage, we should rather apply dual arrangements – static series voltage PE quality controllers. Such controllers, operating as a synchronous voltage source, are included in the feeder in series between the supply voltage and the load, which is illustrated in Fig. 8.1 [1]. Series voltage controllers through inserting a voltage u_C of a required waveshape maintain the required quality in the load-side voltage u_L in case of various disturbances in terminal voltage u_T. The type and allowable value and duration of mitigated disturbances in voltage u_T, they depend on arrangement of the series voltage controller, its control system and capacitance of possible additional energy storage. In a distribution system we must maintain the energy balance, in which the following are taken into account: disturbance duration and instantaneous active powers p_T, p_C, p_L, absorbed from the power source, energy storage and by the load.

This chapter mainly discusses the most important responses and characteristics of static series controllers, designed on the basis of distribution static synchronous series compensator (D-SSSC) and AC/AC voltage regulators (VR). In addition, we also shall outline selected static shunt–series controllers, preserving a sensitive load from deterioration of supply voltage and at the same time protecting an electrical system from load impact.

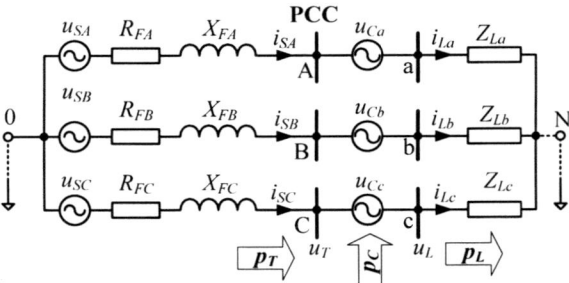

Figure 8.1. Connection of a series of voltage controllers in the distribution system

8.1 Distribution Static Synchronous Series Compensators

Figure 8.2 presents examples of fundamental structures of D-SSSC realised by transistorised 3-phase VSI topologies. We decided to pass over overvoltage protections, connected across the secondary winding of the transformer. From the viewpoint of the device the structure D-SSSC is similar to SSSC [2]. The difference between the two results from a place of installation (distribution network or transmission system) and rated power and control methodology. The SSSCs are flexible as transmission-system devices and are used to enhance controllability and increase power-transfer capability in transmission systems, the latter are custom power-compensating series controllers used in distribution systems.

The D-SSSC, which is presented in Fig. 8.2a is intended only for 3-wire supply systems, however, the other two D-SSSCs (Figures 8.2b,c) may be applied to 4-wire as well as to 3-wire systems. Among them, interesting characteristics present the structure D-SSSC, realised on the basis of 3-phase VSI topology with an additional fourth branch of switches (variant II – Fig. 8.2b). The additional branch of switches allows a significantly decreased capacity of capacitors when compared to the application of sectional capacitors (variant I – Fig. 8.2b). However, the structure D-SSSC shown in Fig. 8.2b allows application of unipolar PWM to the waveform of the required boosting voltage u_C. To this, the unipolar PWM allows application of a smaller filter $L_f - C_f$, preventing switching frequency harmonics from entering the power system, and thereby improve the transient and accuracy in output-voltage control.

The D-SSSC structures, which are presented in Fig. 8.2 obviously do not present all the possible arrangements. Sometimes, instead of the $L_f - C_f$ filter, we place capacitor C_f across the secondary winding of the transformer. The function of reactor L_f is then taken over by the leakage inductance of the transformer. This results not only in simplifiction of the arrangement, but also in significant increase of losses in the transformer, which further is a very significant disadvantage of the solution. In the feeders MV we can also apply D-SSSC with multilevel VSI (see Chap. 6). Basically, we could also apply D-SSSC realised on the basis of CSI topology (current source inverter), although industrial applications of these solutions are not known to the authors.

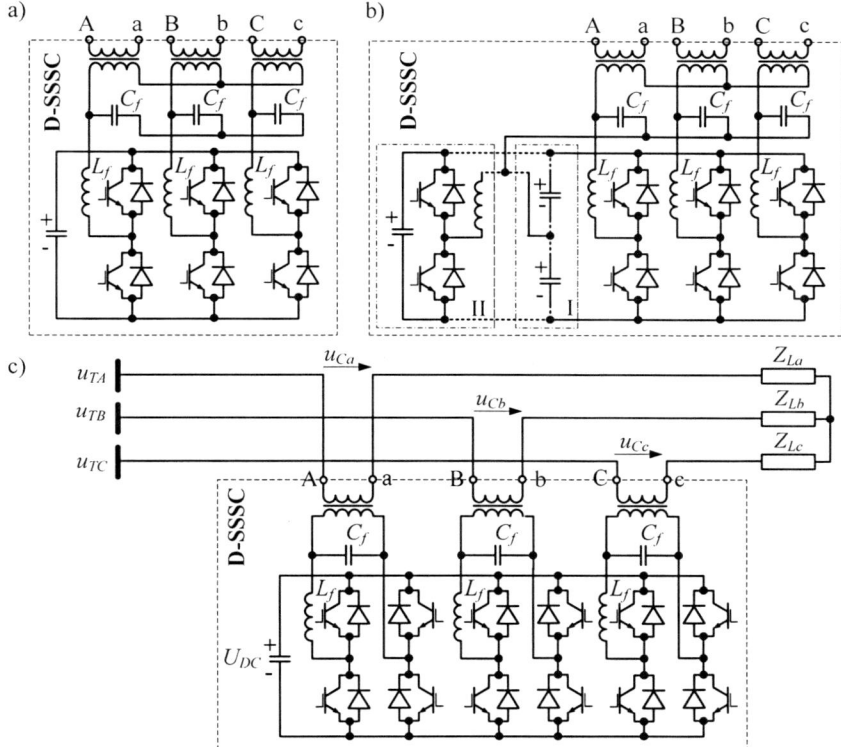

Figure 8.2. Examples of the D-SSSC structures realised by 3-phase VSI topologies

In the following part of this chapter we will consider D-SSSC as a controlled-voltage source without or with additional energy storage. We are allowed to do so because of the permissibility of continous models of PE converters with PWM in the area of their evaluation from a functional perspective (see Sect. 6.5) and the possibility to very careful wave-forming of the required output voltage, for example by application of output-voltage feedback control [3]. We will mainly concentrate on issues related to the indentification of compensated voltage components and determination of reference control signals in selected applications of the D-SSSC, and issues related to the energy balance between the power source, D-SSSC and load.

8.1.1 Identification of Separate Components of the Supply-terminal Voltage

Among various methods of identification of individual components of unbalanced and distorted supply voltage, one of the most important methods are the ones based on d–q transformation [4,5]. In many aspects they are convergent to similar methods used for controll of D-STATCOMs in shunt active and hybrid power-filter arrangements [6,7].

In Fig. 8.3 we present a block diagram of exemplary identification of seperated voltage and current componets in 3-wire systems, conducted on the basis of a d–q transformation. Initially we used (not marked in Fig. 8.3) space-vector transformation [8–10], investigating three-phase voltages and currents to an orthogonal system α–β on the basis of the equation:

$$\begin{bmatrix} u_\alpha + ju_\beta \\ i_\alpha - ji_\beta \end{bmatrix} = \sqrt{\frac{2}{3}} \begin{bmatrix} u_a & u_b & u_c \\ i_a & i_c & i_b \end{bmatrix} \cdot \begin{bmatrix} 1 \\ a \\ a^2 \end{bmatrix}, \tag{8.1}$$

where $a=exp(j2\pi/3)$, j – imaginary unit. After transformation of the vectors $u_\alpha+j\cdot u_\beta$ and $i_\alpha-j\cdot i_\beta$ to coordinates d–q by using operators $\exp(-j\omega_S t)$ and $\exp(j\omega_S t)$, where ω_S – fundamental angular frequency, and then filtration of the obtained output functions by using simple lowpass filters (LPF), and after application of inverse operators $\exp(j\omega_S t)$ and $\exp(-j\omega_S t)$ we obtained the current and voltage components of positive (+) and negative (–) sequence and fundamental frequency (^) within coordinates α–β. From here we can easily select higher harmonics (~).

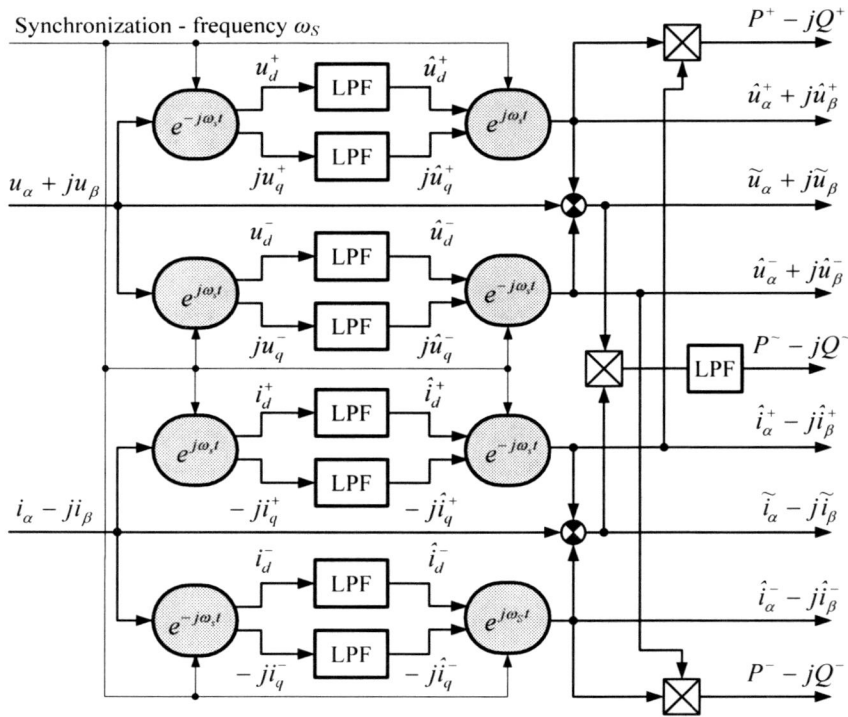

Figure 8.3. Block diagram of exemplary identification of separated components

Based on the conducted identification, as presented in Fig. 8.3, we can determine three average active power components $P = P^+ + P^- + P^\sim$ and three average reactive power components $Q = Q^+ + Q^- + Q^\sim$. These power components are directly related to the voltage and current components of positive (+) and negative (−) sequence and fundamental frequency (^), and also higher harmonics (~). As results from the energy balance [11], D-SSSC arrangement, which compensates unbalanced and/or distorted of the supply voltage, must insert active power $P_C = P^- + P^\sim$ into the distribution system. The average active power P_C obviously depends upon unbalanced and/or nonlinear load. If, after compensation, the load-side voltage U_L of positive sequence (+) is contained within allowable limits of the nominal voltage U_{nom} changes, then, a relatively small active power P_C can almost always be generated without an additional energy source, supplying D-SSSC. The voltage u_C, inserted by D-SSSC (see Fig. 8.1), must then include a positive sequence component $U^+ = U^P + jU^Q$ of the low real part U^P, which is illustrated in Fig. 8.4a. The above-presented opportunity is, however, very limited in situations where the load-side voltage U_L of the positive sequence is much lower than the voltage U_{nom} (Fig. 8.4b) and when required if restoration of voltage U_L to value U'_L located within the allowable limits of deviation from the rated voltage. In this case, the real part U^P of a component U^+ is most often high enough that D-SSSC must be supplied by additional energy source. The problem is discussed in greater detail in Sect. 8.2. We also recommend readers see references [12–15].

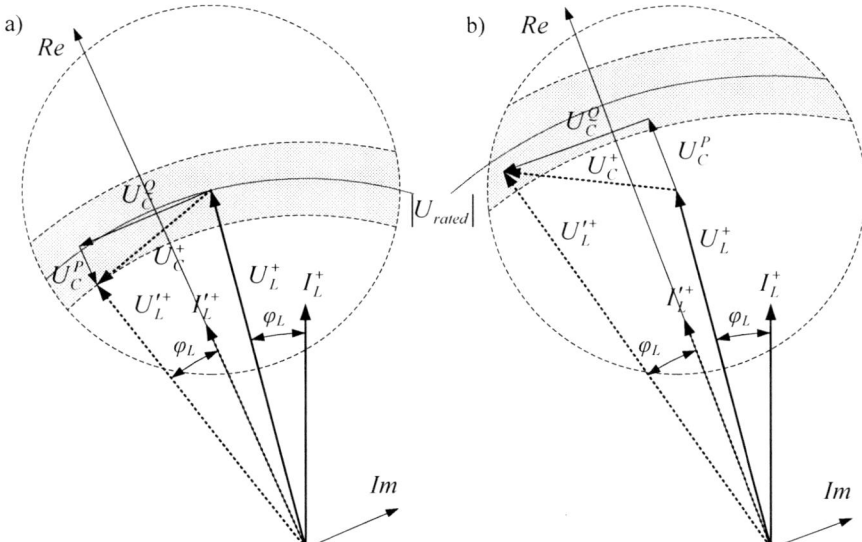

Figure 8.4. Phasor diagram of D-SSSC during an active power insert

We should note, that identification of separated components, the way it is presented in Fig. 8.3 as an example, does not take into account the zero-sequence voltage and zero-sequence current, and thereby is not sufficient in 4-wire 3-phase systems. However, in such systems, identification is not much harder and the process not more complicated. We could easily prove it, for example, on the basis of the instantaneous PQ theory [16] or other similar theories [17,18].

8.1.2 Harmonic Filtration and Balancing of the Voltage in 3-wire Systems

To 3-phase 3-wire distribution systems we can apply simpler D-SSSC structures than those presented in Fig. 8.2. In these systems the D-SSSC agreement can be, for example, realised on the basis of two independent 1-phase H-Bridge VSI topologies [19]. Another example of a simpler structure D-SSSC is shown in Fig. 8.5.

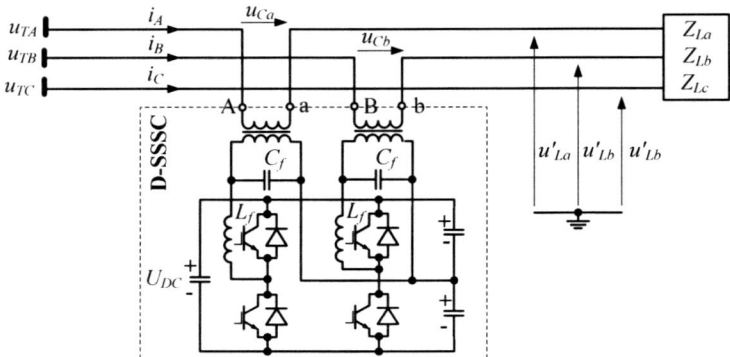

Figure 8.5. Simple structure D-SSSC for 3-phase 3-wire system

8.1.2.1 Harmonic Voltage-supply Filtration
The D-SSSC structure presented in Fig. 8.5 can be used for filtration of voltage harmonics in a 3-wire system, since the sum of line-to-line voltage is always equal to zero, that is:

$$u_{TA-TB} - u_{TB-TC} - u_{TC-TA} = (u_{TA} - u_{TB}) + (u_{TB} - u_{TC}) + (u_{TC} - u_{TA}) = 0. \quad (8.2)$$

When required, instantaneous values of inserted voltages u_{Ca} and u_{Cb}, compensating components $\tilde{u}_{TA}$, $\tilde{u}_{TB}$ and $\tilde{u}_{TC}$ of higher harmonics of phase voltages u_{TA}, u_{TB} and u_{TC}, must satisfy equations:

$$u_{Ca} = \tilde{u}_{TC} - \tilde{u}_{TA} \quad \text{and} \quad u_{Cb} = -\tilde{u}_{TB} + \tilde{u}_{TC}, \quad (8.3)$$

to which:

$$u_{TA} = \hat{u}_{TA} + \tilde{u}_{TA}, \quad u_{TB} = \hat{u}_{TB} + \tilde{u}_{TB}, \quad u_{TB} = \hat{u}_{TB} + \tilde{u}_{TB}, \quad (8.4)$$

where $\hat{u}_{TA}$, $\hat{u}_{TB}$, $\hat{u}_{TC}$ – components of the phase voltage of the fundamental frequency.

Taking into account Eq. 8.4, phase voltages on the load terminals after compensation are described by the equations:

$$u'_{La} = \hat{u}_{TA} + \tilde{u}_{TA} + u_{Ca}, \quad u'_{Lb} = \hat{u}_{TB} + \tilde{u}_{TB} + u_{Cb}, \quad u'_{Lc} = u_{TC} = \hat{u}_{TC} + \tilde{u}_{TC}. \quad (8.5)$$

Therefore, the line-to line voltages on the load terminals are:

$$u'_{La-Lb} = u'_{La} - u'_{Lb} = (\hat{u}_{TA} + \tilde{u}_{TA} + u_{Ca}) - (\hat{u}_{TB} + \tilde{u}_{TB} + u_{Cb}),$$

$$u'_{Lb-Lc} = u'_{Lb} - u'_{Lc} = (\hat{u}_{TB} + \tilde{u}_{TB} + u_{Cb}) - (\hat{u}_{TC} + \tilde{u}_{TC}), \quad (8.6)$$

$$u'_{Lc-La} = u'_{Lc} - u'_{La} = (\hat{u}_{TC} + \tilde{u}_{TC}) - (\hat{u}_{TA} + \tilde{u}_{TA} + u_{Ca}).$$

Following Eqs. 8.6, we can see that satisfying Eq. 8.3 is completely sufficient for full filtration of load-side voltage harmonics. The most fundamental case is accurate determination of the instantaneous values of inserting voltages) u_{Ca} and u_{Cb}. To this end we can use the identification method presented in Fig. 8.3, however, the difference is that Eq. 8.1, on which basis we determine components within the orthogonal coordinates α–β, is changed to the following equation [20]:

$$\begin{bmatrix} u_\alpha + ju_\beta \\ i_\alpha - ji_\beta \end{bmatrix} = \sqrt{\frac{2}{3}} \begin{bmatrix} u_{TA} - u_{TB} & u_{TB} - u_{TC} & u_{TC} - u_{TA} \\ i_A - i_B & i_C - i_A & i_B - i_C \end{bmatrix} \cdot \begin{bmatrix} 1 \\ a \\ a^2 \end{bmatrix}. \quad (8.7)$$

On the basis of identified $\tilde{u}_\alpha$, $\tilde{u}_\beta$, components and in accordance with the equation:

$$\begin{bmatrix} u_{Cb} - u_{Ca} \\ -u_{Cb} \\ u_{Ca} \end{bmatrix} = \begin{bmatrix} \tilde{u}_{TA} - \tilde{u}_{TB} \\ \tilde{u}_{TB} - \tilde{u}_{TC} \\ \tilde{u}_{TC} - \tilde{u}_{TA} \end{bmatrix} = \sqrt{\frac{2}{3}} \cdot \begin{bmatrix} 1 & 0 \\ 1/2 & \sqrt{3}/2 \\ 1/2 & -\sqrt{3}/2 \end{bmatrix} \cdot \begin{bmatrix} \tilde{u}_\alpha \\ \tilde{u}_\beta \end{bmatrix} \quad (8.8)$$

instantaneous values of the voltage u_{Ca} and u_{Cb} are determined. As results from the comparison of Eq. 8.3 and Eq. 8.8, determination of the inserting voltages completely compensate the higher harmonics of terminal-supply voltages under balance and unbalance conditions in 3-wire systems.

8.1.2.2 Balancing and Regulation of the Load-side Voltage

Let us consider application of a D-SSSC structure, which is presented in Fig. 8.5, to balance and regulate of load-side voltages. On the basis of transformation:

$$u_T^+ = \left[(u_{TA} + u_{Ca}) + a \cdot (u_{TB} + u_{Cb}) + a^2 \cdot u_{TC} \right]/3, \quad (8.9a)$$

$$u_T^- = \left[(u_{TA} + u_{Ca}) + a^2 \cdot (u_{TB} + u_{Cb}) + a \cdot u_{TC} \right]/3, \quad (8.9b)$$

we identified positive (+) and negative (−) sequences of the terminal-supply voltages: u_{TA}, u_{TB}, u_{TC}. From Eq.8.9b it results also that the negative sequence may be compensated by inserting a voltage u_{Ca}, determined on the basis of the equation:

$$u_{Ca} = -u_{TA} - a^2 \cdot (u_{TB} + u_{Cb}) - a \cdot u_{TC} \ . \tag{8.10}$$

After inserting a voltage u_{Ca}, positive and negative sequences of load-side voltages: u'_{La}, u'_{Lb}, u'_{Lc} take the following values:

$$u_L'^{-} = \{u_{TA} + \underbrace{[-u_{TA} - a^2 \cdot (u_{TB} + u_{Cb}) - au_{TC}]}_{u_{Ca}} + a^2 \cdot (u_{TB} + u_{Cb}) + a \cdot u_{TC}\}/3 = 0, \tag{8.11a}$$

$$u_L'^{+} = \{u_{TA} + \underbrace{[-u_{TA} - a^2 \cdot (u_{TB} + u_{Cb}) - au_{TC}]}_{u_{Ca}} + a \cdot (u_{TB} + u_{Cb}) + a^2 \cdot u_{TC}\}/3$$
$$= (a - a^2) \cdot (u_{TB} + u_{Cb} - u_{TC})/3 \tag{8.11b}$$

As one can see from Eq.8.11b, the positive sequence of a load-side voltages may be changed by inserting a voltages u_{Cb} into the terminal voltage u_{Tb}. In this manner, by the simple structure of the D-SSSC, presented in Fig. 8.5 we can not only filter higher harmonics (see Sect. 8.1.2.1), but also balance and regulate load-side voltages. Balancing and regulation principles, described by Eqs.8.10 and 8.11, illustrate voltage phasors, presented in Fig. 8.6. In practical applications, we stress again D-SSSC (including the structure as in Fig. 8.5), which are dedicated to balancing and regulation of the voltage in the broad sense and independently from a load type, must be additionally power supplied. Owing to the following functional opportunities, these arrangements perform their function under the generally accepted name dynamic voltage restorer (DVR).

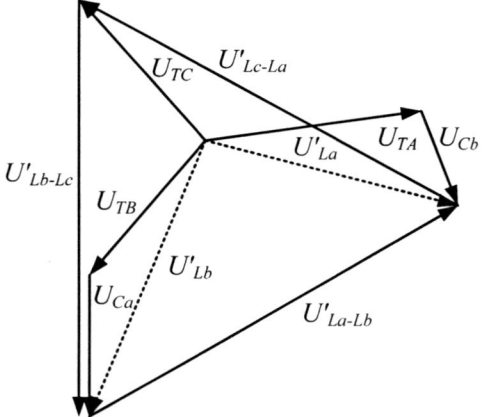

Figure 8.6. Voltage phasors of a D-SSSC (Fig. 8.5) during balancing and regulation

8.2 Dynamic Voltage Restorer (DVR)

8.2.1 What is a DVR?

A dynamic voltage restorer is a PE converter-based D-SSSC, which can protect sensitive loads from all supply-side disturbances other than outages. It is connected in series with a distribution feeder and also is capable of generating or absorbing real and reactive power at its AC terminals. The basic principle of a DVR is simple: by inserting a voltage of the required magnitude and frequency, the DVR can restore the load-side voltage up to the desired amplitude and waveform even when the source voltage is either unbalanced or distorted. Usually, a DVR, as a cost-effective solution when compared to very costly UPS agreements, is connected in order to protect loads and can be implemented at both a LV level and a MV level; which gives an opportunity to protect high-power applications from voltage sags during faults in the supply system. A typical location in the distribution system and the operation principle of the DVR is shown in Fig. 8.7 [21], where U_T – terminal supply voltage, U'_L – the load side voltage after restore; U_C – the inserted voltage by the DVR, I_L and I_S are the load and feeder currents, P_C – the real power generated or absorbed by the D-SSSC.

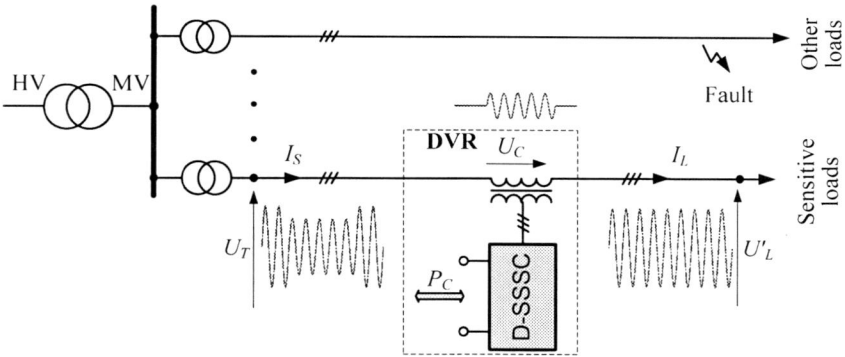

Figure 8.7. A typical location and operation principle of the DVR

DVR was commercially introduced in 1994 for the first time, and its first important installation was in North Carolina, for the rug manufacturing industry [22]. Since then, the number of installed DVR has increased continuously. Obviously, it is implemented especially in those industry branches where supply-side disturbances can lead to dangerous situations for personel or serious production losses.

8.2.2 Control Strategies of the DVR Arrangements

The following DVR control strategies can be applied:
- "presag" strategy, where the load-side voltage U'_L after restoration is assumed to be in phase with this presag voltage [23];

- "inphase" strategy, where the voltage U'_L after restoration is in phase with terminal-supply voltage U_T during sag [24];
- "minimal energy" strategy, where the phase angle α of the voltage U'_L after restoration is determined on the basis of requirements $P_C=0$ or $dP_C/d\alpha = 0$ [25].

These strategies and differences among them are presented in Fig. 8.8, where dashed-line phasors are related to the presag condition, while solid-line phasors relate to restoration.

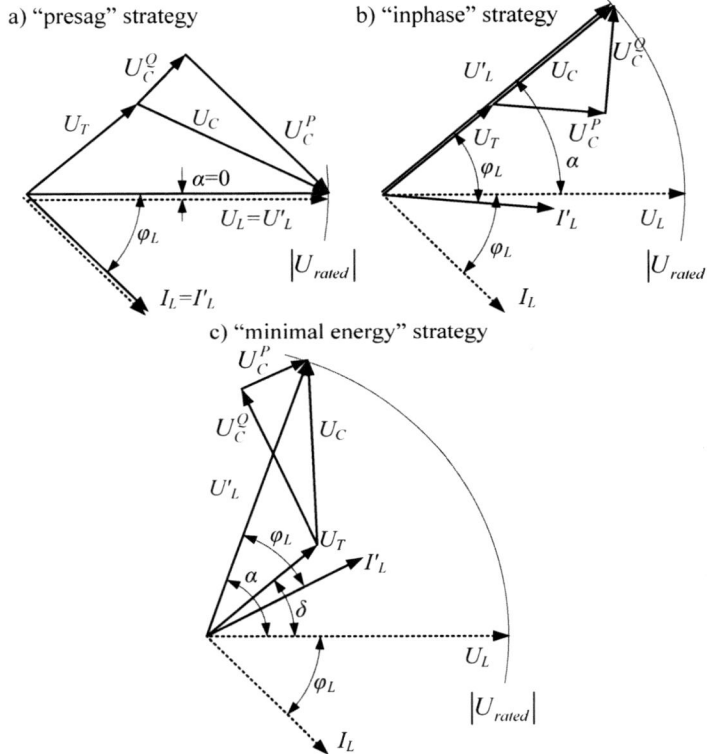

Figure 8.8. Phasor diagrams; illustration of the DVR control strategies

The "presag" strategy's characteristic (Fig. 8.8b) is a significant magnitude of a inserted voltage U_C. Furthermore, a DVR controlled on the basis of this particular strategy should allow injection of heavy active power P_C, which especially impacts the required capacity of the energy storage or energy absorbed from the grid supply during voltage sags (see Sect. 8.2.3). Therefore, the "presag" strategy is applied only in cases where the load is exceptionally sensitive to changes of the supply-voltage phase angle α (for example some applications of thyristor converters). If the load is not sensitive to the changes of the supply-voltage phase angle α, then the most often applied strategy is the "inphase" strategy (Fig. 8.8b). The main advantage of this solution is that the magnitude of an inserted voltage $U_C=U'_L-U_T$ is minimal. If the magnitude of the terminal-supply voltage is disturbed, the DVR generates the same voltage as the voltage drop. Therefore, the rated power of the DVR is minimized for the existing current and terminal-supply voltage [24]. In addition, application of the "inphase" strategy decreases the demand for injection of active power P_C, however, it does not minimize its value.

When the value of injected active power P_C is critical because of the capacity of the energy storage, it is advised to apply a "minimal energy" strategy (Figure 8.8c) instead. For this we distinguish two possible variations of this strategy. Its first variation refers to voltage sags, during which the magnitude of the terminal-supply voltage U_T is not dropped below the value $U'_L \cdot cos\varphi_L$, where φ_L is the load power factor angle. Then, suitably selecting the angle $\alpha=\alpha_{opt}$ we have the possibility of restoration of the load-side voltage U'_L at active power $P_C=0$. In the second case, when $U_T<U'_L \cdot cos\varphi_L$, angle $\alpha=\alpha_{opt}$ should be determined on the basis of $dP_C/d\alpha =0$. In this case the active power P_C takes a minimum value, however, different from "zero". Since both cases can be combined to determine an optimum angle $\alpha=\alpha_{opt}$, the "minimal energy" strategy for any given disturbance is realised as follows:

if $U_T \geq U'_L \cdot cos\varphi_L$,

then $\alpha_{opt} = \varphi_L + \delta - arccos\left(\dfrac{U'_L}{U_T} \cdot cos\varphi_L \right)$, (8.12a)

else $\alpha_{opt} = \varphi_L + \delta$. (8.12b)

The corresponding DVR insertion voltage U_C and injection active power P_C requirement under the α_{opt} control strategy are:

$$U_C = \sqrt{U'^2_L + U^2_T - 2 \cdot U'_L \cdot U_T \cdot cos(\alpha_{opt} - \delta)},$$ (8.13)

$$P_C = S_L \cdot cos\varphi_L - S_L \cdot (U_T/U'_L) \cdot cos(\varphi_L - \alpha_{opt} + \delta).$$ (8.14)

Equations 8.12 to 8.14, which point out the essence of the "minimal energy" strategy, can be directly applied to control and evaluation of DVR characteristics only in the case of 1-phase distribution systems or balanced 3-phase systems, with the assumption that only balanced voltage disturbances occur. Other cases of the "minimal energy" strategy, for example during single-phase or two-phase sags, with or without phase deviation, are presented in the literature [25].

Figure 8.9 shows the characteristics of DVR controlled on the basis of "inphase" and "minimal energy" strategies, in the case of restoration of the load-side voltage during single-phase voltage sags. As we can see, the injection active power P_C is lower by about 20% when applying the "minimal energy" strategy. However, at the start and end points of time interval Δt_{sag} of the sag, also 15–25%

short-time overvoltages occur. This, however, is not advisable for numerous sensitive loads. Therefore, "minimal energy" strategy is sporadically realised in commercial DVRs.

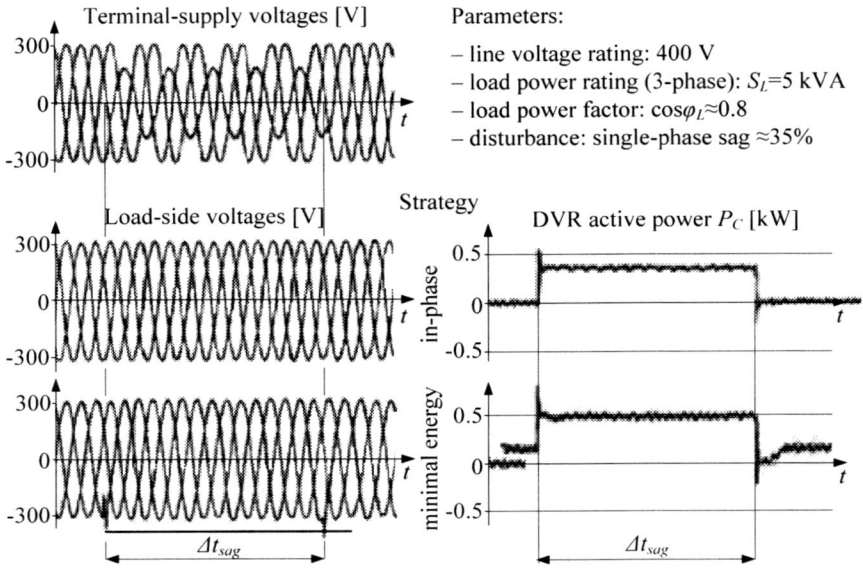

Figure 8.9. Comparison of the „inphase" and „minimal energy" control strategies

8.2.3 Comparison of the DVR Types

Installed DVR, in general, can be divided into two main groups with and without energy storage (ES), including 2 topologies, as is illustrated in Fig. 8.10. Topologies 1 and 2 (Fig. 8.10a) use stored energy to supply the delivered power. The stored energy can be delivered from various ES systems such as capacitors, batteries, flywheel, or supermagnetic energy storage. For topologies 2 and 3 (Fig. 8.10b), the DVR has essentially no internal energy-storage capacity and instead the energy is taken from the supply grid during disturbances.

8.2.3.1 DVR Types with Energy Storage
Electrical ES are relatively expensive, but for certain voltage disturbances the DVR types may be necessary. This refers especially to the cases when terminal-supply voltage sag crossed 40–50%. If, added to that, the voltage sag is short lived, usually enough it is to apply a DVR arrangement of topology 1 (Fig. 8.10a), for which a variable DC-link voltage is characteristic. In this case, the restoration of the load-side voltage may take as long as energy ΔE, delivered by DVR to the load, does not cause a change of the voltage U_{DC} in capacitor C_{DC}, below assumed bottom level $U_{DC(min)}$. Below this level, the D-SSSC is not able to insert the required voltage U_C. Thus, the energy ΔE is limited by the inequality:

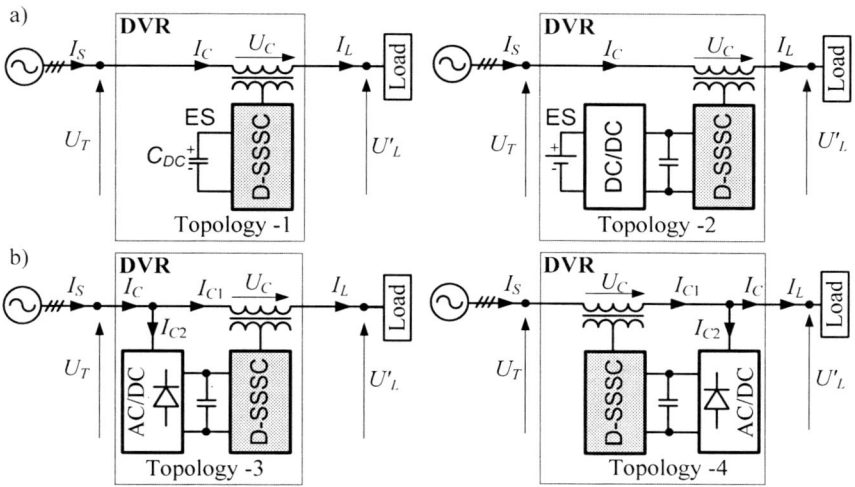

Figure 8.10. Topologies of the DVR with (**a**) and without (**b**) energy storage

$$\Delta E \leq C_{DC} \cdot (1/2) \cdot \left(U_{DC(0)}^2 - U_{DC(\min)}^2 \right),$$

where $U_{DC(0)}$ is the rated DC-link voltage. If we assume terminal-supply voltage sags, which take longer, then it is advisable to apply topology 2 (Fig. 8.10a). A separated ES system and an additional DC/DC converter maintain a constant DC-link voltage. During the restoring process, energy is transferred from a large ES to a smaller rated DC-link storage using a DC/DC converter.

The undoubted advantage of DVR of topology 1 is the lowest power rating in the PE converters. On defining the reduction factor of the terminal-supply voltage as $\xi = (U_{rated} - U_T)/U_{rated}$ and considering that $U_{rated} = U'_L$, and because the only PE converter is D-SSSC, this power rating can by calculated as:

$$S_{DVR(1)} = S_{D-SSSC} = U_C \cdot I_C = \xi \cdot S_L, \tag{8.15}$$

where $U_C = \xi \cdot U_{rated} = \xi \cdot U'_L$ and $I_C = I_L = I_S$. However, for topology 2 we obtain:

$$S_{DVR(2)} = S_{D-SSSC} + S_{DC/DC} = \xi \cdot S_L + S_{DC/DC} = 2 \cdot \xi \cdot S_L, \tag{8.16}$$

where the power rating of the additional DC/DC converter $S_{DC/DC} = S_{D-SSSC} = \xi \cdot S_L$ is calculated for the "worst" case – resistance load. Then, the lowest possible value of S_{D-SSSC} equals the active power P_C, delivered from ES to the load through a DC/DC converter and a D-SSSC arrangement.

8.2.3.2 DVR Types without Energy Storage

DVR topologies without energy storage (Fig. 8.10b) include an additional shunt AC/DC converter, as a residual supply source. It is usually a diode rectifier, since only unidirectional power flow is assumed necessary and it is a cheap solution. In these topologies, saving is obtained on the energy-storage system, and the ability exists to compensate very much longer sags, which constitutes its great advantage. The main disadvantage is, however, that they draw more current from the feeder during the terminal-supply voltage drop. Therefore, DVRs without energy storage and with a shunt diode rectifier are not fit for soft-grid cases of deep voltage sags.

The limitation for maximal values of voltage sags refers especially to topology 3 (Fig. 8.10b), being characterized by a supply-side-connected diode rectifier. For this topology, the uncontrollable DC-link voltage U_{DC} is proportional to the terminal-supply voltage. Hence, when load-side voltage $U'_L = U_{rated}$, we can conclude that:

$$U_{DC} = k_1 \cdot (1-\xi) \cdot U_{rated} \quad \text{and} \quad U_C = \xi \cdot U_{rated}, \tag{8.17}$$

whereas the following inequality must be always satisfied:

$$U_C \leq n \cdot k_2 \cdot U_{DC}, \tag{8.18}$$

where k_1 and k_2 are coefficients depending on the structure of the diode rectifier and the D-SSSC, and n is the transformer ratio. From Eq.8.17 and Eq.8.18 we conclude that in order to fully restore the load-side voltage, the transformer ratio must be selected in accordance with the following formula:

$$n \geq k_1 \cdot k_2 \cdot \xi/(1-\xi). \tag{8.19}$$

For example, assuming $k_1 \cdot k_2 = 1$, for $\xi = 0.5$ we obtain $n = 1/2$. Such a transformer ratio, which is typical for commercial DVR, limits their ability for full restoration of the load-side voltage to the changes of factor ξ within the limits 0–0.3, which is shown in Fig. 8.11 [26].

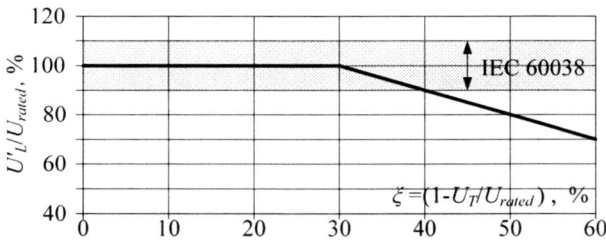

Figure 8.11. Typical output-voltage characteristic of a commercial DVR

As we can see, on the basis of the schemes presented in Fig. 8.10b, current I_{C1} through D-SSSC in cases of DVR of topology 3 equals the load current $I_{C1} = I_L$, however, in cases of DVR of topology 4, being characterised by a load-side-connected diode rectifier, this current equals $I_{C1} = I_L/(1-\xi)$. Thus, since the inserted voltages by the DVRs $U_C=\xi \cdot U'_L$, the rated power of the D-SSSC structures are different and are given by:

– for topology 3: $S_{D-SSSC} = \xi \cdot S_L$, (8.20)

– for topology 4: $S_{D-SSSC} = \xi \cdot S_L /(1-\xi)$. (8.21)

However, the rated powers of the shunt AC/DC converters (diode rectifiers), which are determined on the basis of the maximal values of currents and voltages, are the same and amount to [27]:

$$S_{AC/DC} = \xi \cdot S_L /(1-\xi). \quad (8.22)$$

Taking into account Eqs. 8.20 to 8.22, the overall power ratings of the DVR topologies without energy storage $S_{DVR}= S_{D-SSSC}+S_{(AC/DC)}$ are calculated as:

– for topology 3: $S_{DVR(3)} = (2-\xi) \cdot \xi \cdot S_L /(1-\xi)$, (8.23)

– for topology 4: $S_{DVR(4)} = 2 \cdot \xi \cdot S_L /(1-\xi)$. (8.24)

Figure 8.12 shows how the overall converter power rating varies with a change of factor "ξ" for the four DVR topologies considered. The DVR of topology 4, which is characterised by a load-side-connected AC/DC converter, requires the highest power rating when compared to other topologies DVR. Th significant disadvantage of this topology is also larger currents to be handled by the D-SSSC. In addition, the load can be disturbed by the nonlinear currents drawn by the shunt diode rectifier (AC/DC converter). However, a DVR using the topology 4 may be an efficient solution in terms of the shunt converter design, since the DC-link voltage can be held constant.

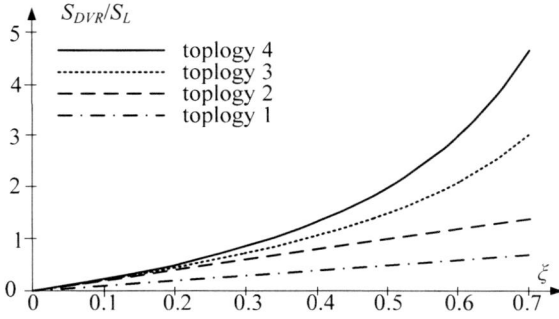

Figure 8.12. Overall power rating of the DVR topologies (only converters) versus factor ξ

It should be stressed that the ability to hold a constant DC voltage is also a characteristic of DVR of topology 3, where in the role of shunt AC/DC converter we apply an active front-end (AFE) rectifier. Moreover, the AFE rectifier can serve the function of a shunt active compensator (see Chap. 7). The same applies to DVR of topology 4, however, in this case we can unload the D-SSSC structure and series transformer from nonactive components of the load current. For this way DVR structures presented in Fig. 8.10b, with a shunt APF rectifier instead of a diode rectifier, can also serve the function of shunt-series or series-shunt compensators, popularly called active power line conditioner (APLC) [28–30], or unified power-quality conditioner (UPQC) [31,32]. Figure 8.13 presents three different topologies UPQC with a load-side-connected shunt compensator.

a) Typical UPQC topology

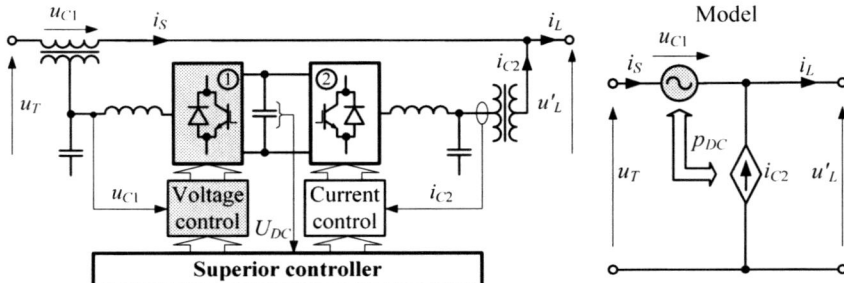

b) UPQC topology with inverse inserted sources

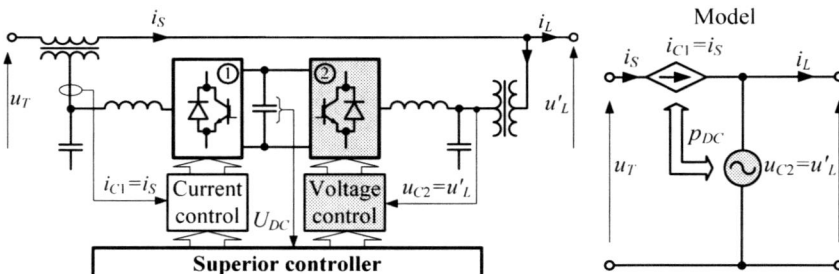

c) UPQC topology without DC-link

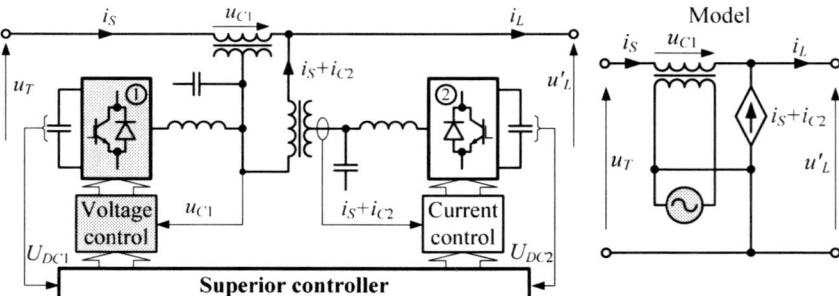

Figure 8.13. Basic UPQC arrangement with load-side-connected shunt compensator

The most often researched and discussed in the literature is UPQC of the first topology – typical (Fig. 8.13a), for example [33–35]. In UPQC of the second topology (Fig. 8.13b), convergent with the topology delta conversion UPS (see Sec. 6.4.4)[36], a series compensator forces a sinusoidal line current i_S, and a shunt compensator provides a sinusoidal load-side voltage u'_L. Therefore, since the waveforms of the reference signals are also sinusoidal, the current control and voltage control of the instantaneous voltages and currents may be realised with very high accuracy [37], and without the need for identification of separate components (see Sec. 8.1.1). However, an advantage of the third topology UPQC (Fig. 8.14c) is the lack of DC-link. This is very important when applying shunt and series compensators realised on the basis of multilevel inverters of topology HB or FC [38] (see Sec. 6.5.3). The UPQC arrangements are discussed in greater detail in Chap. 9.

8.3 AC/AC Voltage Regulators

Most voltage sags on the supply system have a significant retained voltage, so that energy is still available, but at too low a voltage to be useful to the load. In these cases, besides DVR arrangements without energy storage, which are discussed in Sect. 8.2, as series boosters for restoration of the load-side voltage, we can also apply an AC/AC voltage regulator (VR) [39,40]. The overall structure and principle of operation of such 3-phase VRs is presented in Fig. 8.14, where the inserted voltages U'_C are inphase, and the inserted voltages U''_C are in quadrature with the terminal-supply voltage U_T during sags. Also possible are relations of VRs, with a load-side connected primary winding of the transformers, which we do not focus on. The two inserted voltages U'_C and U''_C, are also needed only when the phase angles of the voltages U_T during sags are not balanced, which is actually a requirement for load-side voltages U'_L, or when the "presag" strategy is applied (see Sect. 8.2.2).

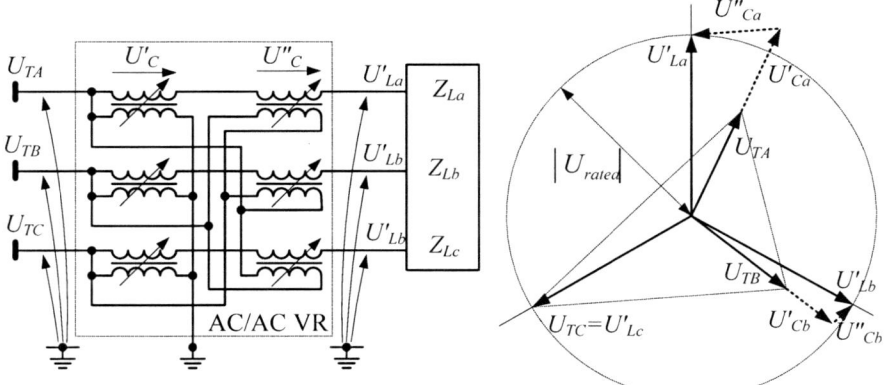

Figure 8.14. Overall structure and principle of operation of AC/AC voltage regulators

In practice, AC/AC VRs usually allow only insertion of the voltage U'_C inphase with the terminal-supply voltage U_T. If regulation of the magnitude and phase of the load-side voltage is required, then most often, even if considering the number of necessary series transformers (Fig. 8.14), then first of all DVR (Fig. 8.10b) or UPQC (Fig. 8.13) arrangements are applied. Alternative future for them may be VRs with matrix converters [41,42], however, so far solutions of this type have never been applied in practice. Further more, in this section we describe only three types of AC/AC VRs allowing only regulation of the magnitude of the load-side voltage, classified as follows:

- electromechanical voltage regulators (as archetype modern solutions);
- step voltage regulators (with a multitaps transformers or autotransformers);
- continuous voltage regulators (on the base PWM AC choppers).

An important point to note in the selection of a VR is that the chosen solution must solve the particular problem without creating additional problems.

8.3.1 Electromechanical Voltage Regulators

The principle of this type of VR is to automatically control an internal variable autotransformer to compensate for the variation of the magnitude of the terminal-supply voltage. The output of the autotransformer feeds the primary winding of a buck/boost transformer, of which the secondary winding is connected in series between the supply terminals and load terminals to inject an adding or opposing inserting voltage U_C into the supply line as shown in Fig. 8.15.

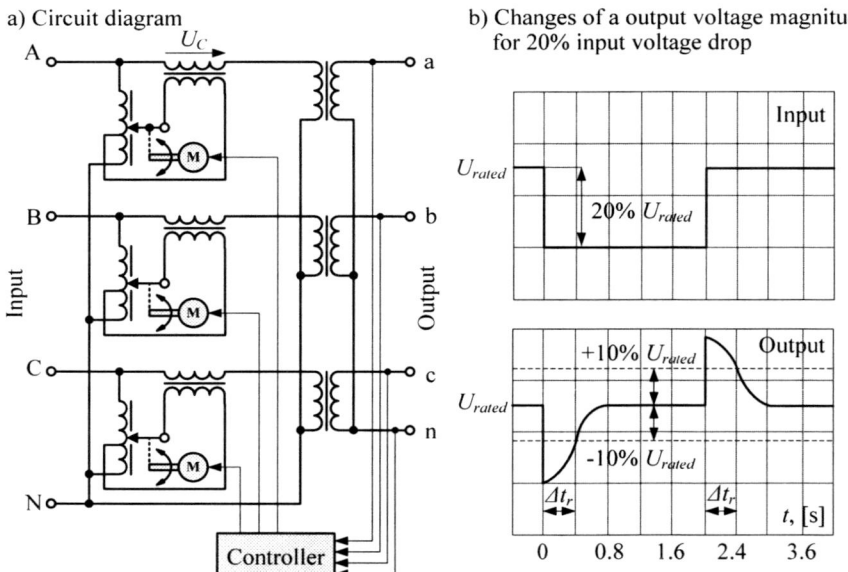

Figure 8.15. A typical electromechanical voltage regulator and its dynamic property

The 3-phase electromechanical VRs are offered either as arrangements of operation common to each of the three phases of the feeder or as arrangements allowing individual regulation of the voltage magnitude independently in every phase. In the first arrangements one servomotor (M) common to the three autotransformers is applied, however, in the second, each autotransformer must have its own servomotor, controlled independently. As presented in Fig. 8.15a an electromechanical VR realises individual regulation. The load-side voltage of the electromechanical VR is monitored by individual controllers in each line. If this voltage deviates from the preset value due to a change in the terminal-supply voltage or the load current, the controller will drive an adequate servomotor, which then rotates the brush arm of the variable autotransformer in the required direction to boost or buck the input (terminal-supply) voltage until the correct preset value of the output (load-side) voltage is restored. This VR does not produce harmonics and therefore does not inject distortion into the incoming voltage supply.

The main advantages of the electromechanical VRs are: simple design, relatively low cost and noise immunity. The most important disadvantages, which show to some extent technical obsoleteness of these VRs, are: moving parts, relatively large overall dimensions and weight, and what is more important, low regulation dynamics. For example, the dimensions, weight and regulation rate of a commercial 3-phase electromechanical VR with individual regulation of power rating 45 kVA are: 134×93×63 cm, 340 kg and 45 ms/volt. As calculated during experimental research the response time Δt_r of this VR, in the case of a 20% voltage change (U_{rated}=220 V) is approximately 400 ms (see Fig. 8.15b). It is slower than step VRs and much slower than continuous VRs. Therefore, it is obvious that electromechanical VRs do not allow mitigation of rapid and short-time voltage sags and swells.

8.3.2 Step Voltage Regulators

Step-voltage regulators, also called static tap changers series boosters (STCSB) operate by selecting separate taps of the autotransformer or transformer. In Fig. 8.16 are presented structures of two basic types of 1-phase STCSB (structures of 3-phase STCSB are similar). A change of the tap causes changes of the inserted voltage U_C, and thus regulation of the output (load-side) voltage. The output voltage changes in steps, and therefore, the more tap points, the more precise can be the regulation of the output voltage.

Tap selection in the STCSBs is performed by fully bidirectional switches – contactors or power-electronics semiconductor devices such as an antiparallel connection of the SCR. Variations in the input (terminal-supply) voltage are monitored by an electronic controller that in turn automatically selects the appropriate tap using a switch, thus maintaining the required output voltage. The instant of tap changing is phased by the controller to occur very near the zero crossing of the input voltage thus ensuring that any RF interference or switching transients are reduced to a minimum. If, additionally, as SCR switches are applied, then the response time of the STCSB with variations of the input voltage usually does not exceed one cycle (20 ms for the fundamental frequency, 50 Hz), and when applying special control algorithms, the time may be reduced to even about one half-cycle.

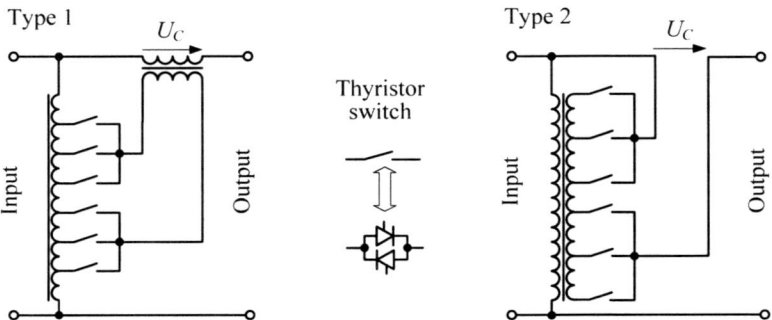

Figure 8.16. Basic types of static tap changer series boosters

It should be stressed that an additional series-boost transformer, necessary in STCSB of type 1 (with autotransformer), is often used in STCSB of type 2 (Fig. 8.16). In this case tap-switching devices are insulated by a supply source and load, and this especially facilitates the working current and working voltage to fit the load parameters. Additionally, series-boost transformers are used in most of the improved precision of output-voltage regulation STCSBs arrangements. For example in [43,44,49], STCSBs are used, among the others, for rapid voltage-magnitude regulation, *e.g.* voltage-sag compensation. Modified STCSBs are also used as static phase shifters rated among FACTS (flexible AC transmission system) arrangements [45,46].

When summarising STCSB arrangements, on the one hand we can see their advantages such as:
- very high efficiency
- relatively small size and weight, and low cost;
- relatively fast response, typically 1–1.5 cycles (20–30 ms).

On the other hand they have some limitations in their application, resulting from the following characteristics:
- the voltage regulation (stabilization) is in steps;
- the output voltage tolerance is normally only ±3–5 %.

These limitations are not valid for continuous-voltage regulators [47,48,49], which in addition characterise even faster response to input voltage changes.

8.3.3 Continuous-voltage Regulators

AC/AC continuous voltage regulators (CVR), also called PWM (or self-commutated) AC voltage controllers (or stabilizers) or PWM AC boosters, are very fast and very tolerant power-electronics arrangements without moving parts and no need for tap changing. The main component of CVRs is PWM AC chopper [50,51] (see also Sect. 6.5.2.2). The PWM AC chopper supplies a voltage with regulated magnitude to the primary winding of a series-boost transformer. The secondary winding of this transformer, similarly to former types of VRs, is connected in series between the supply terminals and load terminals. The PWM AC chopper can thus add (or subtract) an inserting voltage to the input voltage, stabilising the magnitude of the output voltage. The CVRs are also very similar to modern static phase shifters, for example [52].

Figure 8.17 shows an exemplary solution of a 3-phase CVR arrangement, where the applied PWM AC chopper is the same as for a transistorised reactive current regulator (see Fig. 6.48). Since the AC chopper realises only unipolar PWM, the inserted voltage U_{C1} can be only added to the input voltage (or only deducted). The two series-boost transformers Tr_1 and Tr_2 applied to this CVR arrangement allow for regulation of the output voltage up and down. If we also take into consideration the transformer voltage ratio $1/n_1$ – for Tr_1 and $1/n_2$ – for Tr_2, it is not difficult to show that the output voltage can be expressed as follows:

$$U'_L = U_T(1 - 1/n_2 + \delta/n_1), \tag{8.25}$$

where δ is the controllable duty cycle ($0 \leq \delta \leq 1$) of the switches S_{Aa}, S_{Bb}, and S_{Cc}. Hence, the control range of this CVR is the following:

$$1 - 1/n_1 \leq U'_L/U_T \leq 1 - 1/n_1 + 1/n_2 . \tag{8.26}$$

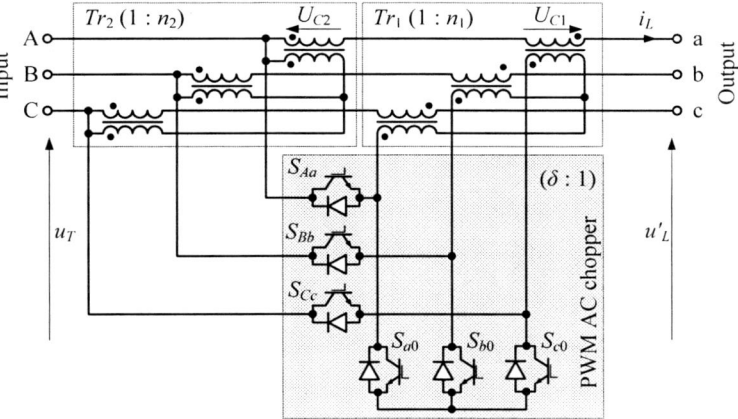

Figure 8.17. The structure of an example of the AC/AC transistored CVR

Figure 8.18 shows exemplary input (terminal-supply) and output (load-side) voltages and load-current waveforms for the VCR (Fig. 8.17), in the case where the full control range of the output voltage is $0.7 \leq U'_L/U_T \leq 1.3$, i.e. when the transformer voltage ratio are: $1/n_1=0.6$ and $1/n_2=0.3$. The presented waveforms apply to two values of the duty cycle: $\delta=0.2$ and $\delta=0.8$, with the switching frequency of the switches S_{Aa}, S_{Bb}, S_{Cc}, and S_{a0}, S_{b0}, S_{c0} only $f_i=2$ kHz. As regards high-harmonics disturbance, the switching frequency should be the highest possible.

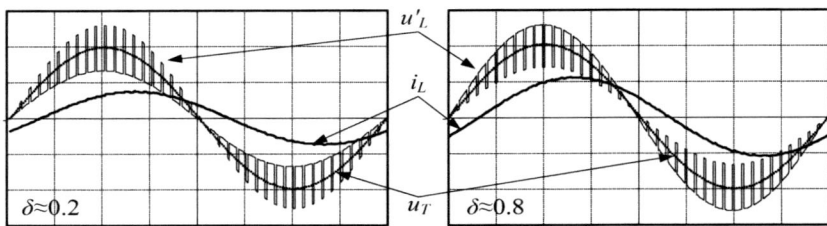

Figure 8.18. Input and output voltage and load-current waveform for the presented VCR

The significant disadvantage of VCR of the structure shown in Fig. 8.17 is the inability of this arrangement to independently regulate the load-side phase voltages. This inability is associated with realisation of three-phase switches, and thereby with the principle of operation applied to a PWM AC chopper [53,54]. However, such a type of regulation allow VCR topologies with somewhat different PWM AC choppers applied, among them are also topologies presented in Fig. 8.19. In these VCRs, AC choppers are realised on the basis of single fully bidirectional switches with turnoff capability (see Fig. 6.45). We should note that a VCR of the topology shown in Fig. 8.19a (without an additional series-boost transformer – see Fig. 8.17) allows regulation of the phase output voltage only up and down, and a VCR of the topology shown in Fig. 8.19b in both up as well as down. It depends on whether the 3-phase PWM AC chopper allows either unipolar PWM or bipolar PWM [37,48] to be applied.

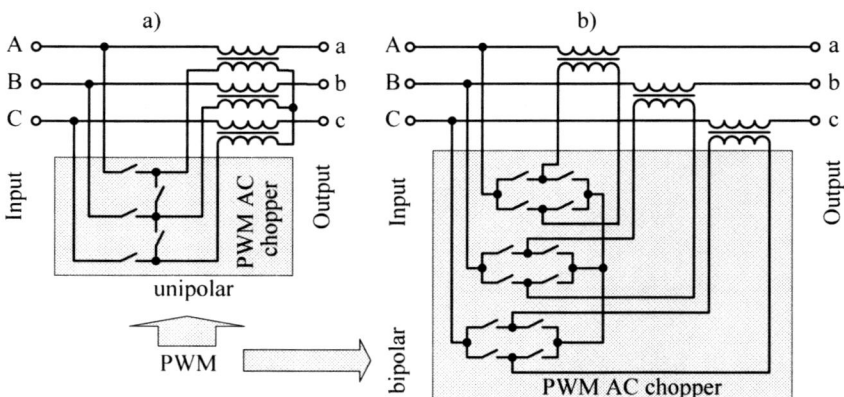

Figure 8.19. Examples of topology VCRs, allowing independent regulation in phases

Table 8.1 shows a simple comparison of AC/AC voltage-regulation techniques, however, evaluation of STCBS depends on the number of taps – key $^{(*)}$. As we can see, the VCRs on the base of AC PWM choppers are the most effective, but also more expensive than the other two VR arrangements. Currently, it seems to be a serious shortcoming, however, the future development and availability of higher-power turnoff semiconductor devices and their inevitable reduction in cost bodes well for the future of the VCRs as the fastest, most efficient and cost-effective arrangements of stabilising the voltage supplying sensitive loads.

Table 8.1. Comparison of the AC/AC voltage-regulation techniques (1 – bad, 10 – excellent)

VR technique	Input range	Smoothness of control	Speed of response	Stabilization accuracy	Size per kVA	Cost per kVA	Total out of 60
Electromech.	10	10	3	10	6	8	47
STCSB	7(*)	6 (*)	8	6 (*)	9	10	48
VCR	10	10	10	10	10	6	56

8.4 Summary

The content of this chapter focuses on the problems of mitigation of the voltage disturbance by means of different arrangements, connection in series between a supply source and load. In this context, Sect. 8.1, is only an introduction to the topic, focusing mainly on typical D-SSSC structures, being realised by VSI topologies and an outline for identification of the separate components of the disturbed voltages. At the same time, with regard to functional limitation, D-SSSC without additional residual supply source are applied only sporadically.

For the purpose of mitigation of the voltage disturbance, in distribution systems usually arrangements of type DVR or AC/AC VR are installed, discussed in Sects. 8.2 and 8.3. Thus, an answer to the question about the particular arrangement we should install, can be given only on the basis of complete technicaleconomic analyses, which consider possible voltage disturbances and their impact on sensitive load, detailed characteristics and features of these arrangements, the capital costs and operating costs. Evaluation of functional characteristics supports DVRs. And when comparing a simple DVR presented in Fig. 8.10b to the VCR presented in Fig. 8.19b (of similar characteristics), we notice that DCR includes: first of all – more and additionally fully bidirectional switches, and secondly – without additional development only an "inphase" strategy can be realized. Furthermore, DCR does not include any small capacitor for energy storage and therefore can not absorb an instantaneous active power pulsation [13,14]. As a consequence, under certain conditions a result may be additional distortion of the feeder current. The authors believe that in the future, the leading solution of voltage-quality controllers in distribution systems will become the most solid DVRs, in terms of their functionality, with active front-end rectifiers (or UPQC), and dependent on the need also with energy storages connected to a DC-link.

References

[1] Ghosh A., Jindal AK., Jodhi A. Design of a capacitor-supported dynamic voltage restorer (DVR) for unbalanced and distorted loads. IEEE Trans. Power Delivery, Vol. 19, No. 1, 2004; 405–413.
[2] Song YH., Johns AT. Flexible AC Transmission Systems (FACTS). London, UK: The Institution of Electrical Engineers, 1999.
[3] Tanaka T., Wada K., Akagi H. A new control scheme of series active filters. IPEC'95 Conf., Yokohama, Japan, 1995; 376–381.
[4] Blaschke F., The principal of field orientation as applied to the new trans-vector close-loop control system for rotating-field machines. Siemens Rev., Vol. 34, 1972; 217–220.
[5] Park RH. Two reaction theory of synchronous machines. AIEE Trans., Vol.48, 1929; 716–730.
[6] Soares V., Verdelho P. Instantaneous active and reactive current id-iq calculator suitable to active power filters. 8th Int. PEMC'98 Conf., Vol. 8.,1998; 111–114.
[7] Strzelecki R., Frąckowiak L.,Benysek G. Hybrid filtration in conditions of asymmetric nonlinear load current pulsation. 7th Europan Power Electronics and Applications Conf. – EPE'97, Trondheim/Norway, 1997; 1.453–1.458.
[8] Aller JM., Bueno A., Paga T. Power system analysis using space-vector transformation. IEEE Trans. Power Systems. Vol. 17, No. 4, 2002; 957–965.
[9] Fortescue CL. Method of symmetrical coordinates applied to the solution of polyphase networks. Trans. AIEE, Vol.37; 1027–1140.
[10] Clarke E. Circuit Analysis of AC Power Systems. New York, Wiley, 1943.
[11] Ferrero A. Leva S., Morando AP. A systematic, mathematically and physically sound approach to the energy balance in three-wire, three-phase systems. L'Energia Elettrica, Vol.81, 2004; 51–56.
[12] Ferrero A. Leva S., Morando AP. About the role of the Park imaginary power on the three-phase line voltage drop. ETEP, Vol. 10, No. 5, 2000, 287–286
[13] Akagi H., Kanazawa Y,. Nabae A., Instantaneous reactive power compensators comprising switching devices without energy storage components. IEEE Trans. Ind. Appl. Vol. IA_20, No. 3, 1984; 625–630.
[14] Peng FZ., Lai JS., Generalized instantaneous reactive power theory for three-phase power systems. IEEE Trans Inst. Meas., Vol. 45, No. 1, 1996; 293–297.
[15] Vilathgamuwa DM., Perera AADR., Choi SS. Voltage sag compensation with energy optimized dynamic voltage restorer. IEEE Trans. Power Delivery, Vol. 18, No. 3, 2003; 928–936.
[16] Akagi H., Ogasawara S., Kim H. The theory of instantaneous power in three-phase four-wire systems. A comprehensive approach. IEEE IAC Conf., Vol. 1, 1999; 431–439.
[17] Cardenas V., Moran L., Bahamondes A., Dixon J., Comparative analysis of real reference generation techniques for four-wire shunt active power filters. 34th IEEE Power Electronics Specialist Conf. – PECS'03, Acapulco, Mexico, 2003; 791–796.
[18] Kim H. Blaabjerg F., Bak-Jensen B., Choi J. Instantaneous power compensation in three-phase systems by using p-q-r theory. IEEE Trans. Power Electron., Vol. 17, No. 5, 2002; 701–710.
[19] Bhavaraju VB., Enjeti PN., An active line conditioner to balance voltages in a three-phase system. IEEE Trans. Ind. Appl., Vol. 32, No. 2, 1996; 287–292.
[20] Strzelecki R., Supronowicz H. Filtration of the harmonic in AC supply systems. Toruń, Poland: Adam Marszałek Publishing House, 1997/1999 (in Polish).

[21] Fitzer C., Arulampalm A., Barnes M., Zurowski R. Mitigation of saturation in dynamic voltage restorers connection transformers. IEEE Trans. Power Electron., Vol. 17, No. 6, 2002; 1058–1066.
[22] Woodley NH., Morgan L., Sundaram A. Experience with an inverter-base dynamic voltage restorer, IEEE Trans. Power Delivery, Vol. 14, 1999; 1181–1185.
[23] Jauch T., Kara A, Rahmani M., Westermann D., Power quality ensured by dynamic voltage correction. ABB Rev., Vol. 4, 1998.
[24] Kim H., Minimal Energy Control for a Dynamic Voltage Restorer. IEEE-PCC'02 Conf., Osaka, Japan, Vol. 2, 2002; CD-ROM.
[25] Vilathgamuwa DM., Perera AADR., Choi SS. Voltage sag compensation with energy optimized dynamic voltage restorer. IEEE Trans. Power Delivery, Vol. 18, No. 3, 2003; 928–936.
[26] Didden M. Techno-economic analysis of methods to reduce damage due to voltage dips. Ph.D. Thesis, Leuven, Belgium: Catholic University of Leuven, 2003.
[27] Nilsen JG., Blaabjerg F. A detailed comparison of system topologies for dynamic voltage restorers. IEEE Trans. Ind. Appl., Vol. 41, No. 5, 2005; 1272–1280.
[28] Aredes M., Heumann K., Watanabe E. H. A universal active power line conditioner, IEEE Trans. Power Delivery, Vol. 13, No. 2, 1998: 545–551.
[29] Strzelecki R., Klytta M., Frąckowiak L., Rusiński J. Power flow in APLC topologies. Proc. 5th Int. EPQU '99 Conf., Crakow, Poland, 1999; 391–398.
[30] Strzelecki R., Kukluk J., Suproniwicz H., Tunia H. A universal symmetrical topologies for active power line conditioners. 8th EPE '99 Conf., Lausanne, Switzerland, 1999, CD-ROM.
[31] Fujita H., Akagi H. The Unified Power Quality Conditioner: The integration of series- and shunt-active filters. IEEE Trans. Power Electron., Vol. 13, No. 2, 1998; 315–322.
[32] Emadi A., Nasiri A., Bekiarov S.B. Uninterruptible power supplies and active filters. Boca Raton, USA, CRC Press, 2005.
[33] Farrukh Kamran F., Habetler TG., Combined deadbeat control of a series-parallel converter combination used as a universal power filter. IEEE Trans. Power Electron., Vol. 13, No. 1, 1998; 160–168.
[34] Strzelecki R., Benysek G., Rusiński G., Dębicki H. Modeling and experimental investigation of the small UPQC systems. IEEE Conf. Compatibility in Power Electron. –CPE'2005, Gdańsk, Poland, 2005, CD-ROM.
[35] Han B., Bae B., Baek S., Jan G. New configuration of UPQC for medium-voltage application. IEEE Trans. Power Delivery. Vol. 21, No. 3, 2006; 1438–1444.
[36] Oliveira da Silva SA., Donoso-Garcia PF., Cortizo PC., Seixas PF. A three-phase line-interactive UPS system implementation with series-parallel active power-line conditioning capabilities. IEEE Trans. Ind. Appl., Vol. 38, No. 6, 2002; 1581–1590.
[37] Kaźmierkowski M.P., Krishnan R., Blaabjerg F. Control in power converters. Selected problems. San Diego, USA, Academic Press, 2002.
[38] Jin Wang J., Peng FZ., Unified Power Flow Controller using the cascade multilevel inverter. IEEE Trans. Power Electron., Vol. 19, No. 4, 2004; 1077–1084.
[39] Dmowski A. AC voltage regulation. Selected systems. WNT, Warsaw, Poland, 1983 (in Polish)
[40] Montenero-Hernandez OC., Enjeti PN., Application of a boost AC-AC converter to compensate for voltage sags in electric power distribution systems. 31st Power Electronics Specialists Conference PESC'00. Galway, Ireland, Vol.1, 2000; 470–475.
[41] Gyugyi L., Pelly BR., Static Frequency Changers. New York: John Wiley, 1976.
[42] Strzelecki R., Noculak A., Tunia H., Sozański K., Fedyczak Z.UPFC with matrix converter", Conference EPE'2001, Graz, Austria, 2001; CD-ROM.

[43] Demiric O., Torrey DA., Degeneff RC., Schaeffer FK., Frazer RH. A new approach to solid-state on load tap changing transformer. IEEE Trans Power Delivery. Vol. 13, No. 3, 1988; 952–961.
[44] Lipkowski KA. Transformer-switches performance topologies of the AC/AC converters. Naukova dumka, Kiev, Ukraine, 1983 (in Russian).
[45] Hingorani NG, Gyugyi L. Understanding FACTS. Concepts and Technology of Flexible AC Transmision Systems. New York: IEEE Press, 1999.
[46] Iravani MR., Maratukulam D. Review of semiconductor controlled (static) phase shifters for power systems applications. IEEE Trans Power Systems, Vol. 9, No. 4, 1994; 1833–1839.
[47] Mozdzer AJ., Bose BK., Three-phase AC power control using power transistors. IEEE Trans. Ind. Appl., Vol. IA-12, 1976; 499–505.
[48] Hamed SA. Modelling and design of transistor-controlled AC voltage regulators. Int. Journal Electronics, Vol. 69, No. 3, 1990; 421–434
[49] Kwon BH., Jeong GY., Han SH., Lee DH., Novel line conditioner with voltage up/down capability. IEEE Trans. Indust. Electronics, Vol. 49, No. 5, 2002; 1110–1119.
[50] Fedyczak Z., Strzelecki R. Power electronics agreement for AC power control. Torun, Poland: Adam Marszalek Publishing House, 1997 (in Polish) .
[51] Fedyczak Z. Strzelecki R., Benysek G. Single phase PWM AC/AC semiconductor transformer topologies and applications. 33rd Power Electronics Specialists Conference PESC'02. Cairns, Australia, Vol. 2, 2002; 1048–1053
[52] Lopes LAC., Jóos G., Ooi BT., A high-power PWM quadrature booster phase shifter based on a multimodule AC controller. IEEE Trans. Power Electronics, Vol. 13, No. 2, 1998; 357–365.
[53] Vincenti D., Jin H., Ziogas P. Design and Implementation of a 25-kVA three-phase PWM AC Line Conditioner. IEEE Trans. Power Electronics, Vol. 9, No. 4, 1994; 384–389.
[54] Strzelecki R., Fedyczak Z.Properties and structures of three-phase PWM AC power controllers. 27th Power Electronics Specialists Conference PESC'96. Baveno, Italy, Vol.1, 1996; 740–746.

9

Active Power-line Conditioners

Patricio Salmerón and Jesús R. Vázquez

Department of Electrical Engineering
Escuela Politécnica Superior
University of Huelva
Ctra/ Palos de la Frontera, s/n
21819, Palos de la Frontera
Huelva, Spain
Emails: patricio@uhu.es; vazquez@uhu.es

Nowadays, the active power filters, APFs, can be used as a practical solution to solve the problems caused by the lack of electric power quality, EPQ. The emerging technology of power-electronic devices and the new developments in digital signal processing, DSP, have made possible its practical use. These power filters can fully compensate the nonlinear loads of electrical power systems: harmonics, reactive power, unbalances, *etc.* So, they can be called active power-line conditioners (APLCs). There are many configurations of APLCs, from shunt and series connection to hybrid passive-active filters. The target is to optimize the design using the advantages of each filter with the different load configurations.

In this chapter, the more common APLC configurations will be presented. The usual power blocks, the control strategy and the modulation control method will be shown. In particular, a shunt APLC design will be detailed. In this case, the goal is to inject, in parallel with the load, a compensation current to get, *e.g.*, a sinusoidal source current.

At the end of the chapter, practical design considerations will be presented. The APLC parameters will be justified. The proposed filter will be issued in a simulation platform to adjust the component values. Finally, a laboratory prototype will be probed to contrast the final design.

9.1 Introduction

In recent years, the rated power and the switching speeds of electronic devices (IGBTs, GTOs,...) have increased. On the other hand, there are new possibilities of digital signal processing (DSP boards). An old idea, active power filters, is now possible from a practical point of view. The APF concept is to use a DC/AC converter to inject currents or voltages harmonic components to cancel the load harmonic components [1–4]. The more usual configuration is a shunt APF to inject into the system the current harmonics. Besides, the control strategy can include other targets to compensate the reactive power or to balance the asymmetrical load currents.

An APF can be installed in a low-voltage power system to compensate one or more loads; thus, it avoids the propagation of current harmonics along the system. The developments of power and control stages of APFs are made possible to compensate the reactive power, the negative and zero sequences. For this, a new term is used to name the APFs: active power-line conditioners, APLCs. There are shunt, series, combined shunt series APFs, and some hybrid passive and active configurations. Two usual hybrid topologies are series active-shunt passive filter, and shunt active-shunt passive filter. In the second topology, the active filter can compensate current harmonics and reactive power eventually, whereas in the first one, the active filter target is to improve the performance of the passive filter. In this case, the active filter insulates the harmonics between the source and the load [5,6].

The first APLC for harmonic compensation was installed in 1982. From this date, many high-power devices have been installed thought the world, principally in Japan. The experimental and industrial advances have improves the APLC utilities. So, from 1993, some studies on unified power quality conditioners (UPQCs) have appeared. The UPQCs integrate a combination of series and shunt active filters. The target is to compensate flickers and voltage unbalances in three-phase systems, besides the reactive power, unbalanced currents or load harmonics currents. This scheme is a new step to improve the power quality in electrical installations [7,8].

The three main aspects of an active power conditioner are:

- the configuration of power converter (the scheme and the topology of converter, and the electronics device used);
- the control strategy (the calculation of APLC control reference signals);
- the control method used (how the power inverter follows the control reference, usually through the pulse width modulation of the switch device trigger signals, that is, the PWM method).

So, the design of the control strategy (according to the application), and the election and implementation of the control method are key to the APLC design [9,10].

The structure of this chapter is:

- In Section 9.2, the APLC's possibilities to improve the electrical power quality are shown.

- In Section 9.3, is presented the APLC power stage. The different topologies and its performance are detailed.
- The main shunt APLC control strategies to compensate nonlinear loads are presented in Section 9.4.
- The basic rules to choose the passive components of a converter network connection are presented in Section 9.5. This component design will be contrast helped by MATLAB® and Simulink® simulations. Then, a laboratory prototype will allow checking of the practical performance of a shunt APLC.
- Finally, the state-of-the-art of commercial APLCs in electrical power systems will be analysed in Section 9.6.

9.2 Power Quality and Active Power Filters

The electrical power quality (EPQ) is associated with alterations and disturbances of the electrical supply that may generate electrical and electronic systems performance failure and malfunctions. In particular, the number of commercial and industrial equipment based on power electronics is increasing, which are the origin of the harmonic distortion. The traditional solutions proposed to eliminate the harmonic currents include: electrical equipment over dimensioning, three-phase transformer special connections to eliminate the third and its proportional harmonics, and the connection of passive elements, but they have more disadvantages than advantages. The technical evolution in rated power and switching speed of electronic devices, allows nowadays the application of the active power filters, APFs, [11-15].

An APF is an electronic converter that produces and injects into the system the necessary harmonic components to cancel the harmonics of load current. An APF can be installed in the point of common coupling (PCC) of an AC system to compensate one or several loads. Once installed, the current harmonic circulation to the system is limited. Nowadays, APF development allows its application to compensate the reactive power, the negative sequence currents, and the harmonic currents. So, the APFs are generically named active power-line conditioners (APLC). Besides, hybrid systems with active and passive filters have been proposed, and since 1995, several studies on unified power-quality conditioners (UPQC) have appeared. The UPQCs include series and shunt active filters in the same module. The general targets are the correction of flicker and three-phase voltage unbalances, and the compensation of reactive power, harmonic currents and current unbalances. This is the next step to correct the power quality with active power conditioners [16-19].

The parameters to define an APLC are the circuit configuration of a power converter (its scheme, the power inverter topology, and the electronic devices used), the control method (PWM modulation), and the control strategy (the way to obtain the reference signal). In this section, the current APLC configurations and the basic performance will be presented.

9.2.1 Distribution Static Compensator, DSTATCOM

The more usual APLC configuration is the shunt or parallel connection, [20]. Figure 9.1 shows the basic scheme of the connection, where an IGBT switching device represents the APLC power block. The loads with current harmonics can be compensated by this APLC configuration. A typical example of a current-source load is a rectifier with an inductive branch in dc side.

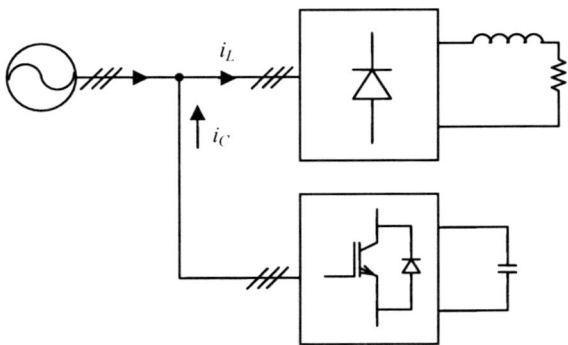

Figure 9.1. A shunt APLC scheme

Figure 9.2 shows the basic performance of a shunt APLC. The general aim is that the shunt APLC will inject into the system a compensation current, i_C, to cancel the harmonic component of the load current, i_L. The source current i_S becomes sinusoidal after the compensation.

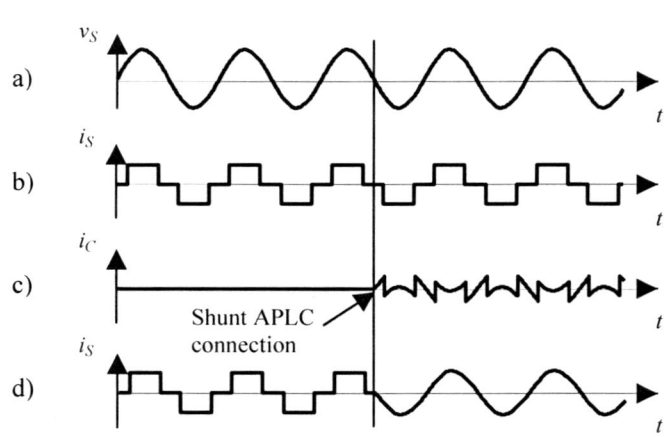

Figure 9.2. Performance scheme of a shunt APLC

The current waveform of a nonlinear load, a three-phase diode rectifier with a highly inductive DC branch, is shown in Figure 9.2b. Alter the shunt APLC connection, this injects a compensation current, Figure 9.2c, in parallel with the load. Figure 9.2d shows the source current of the system. Before the compensation is equal to the current load, and after it is sinusoidal. In this example, the source voltage is sinusoidal, Figure 9.2a.

9.2.2 Series Active Filters

Figure 9.3 shows the connection scheme of a series APLC. It is connected to the system through a coupling transformer. The compensation voltage, v_C, is used to cancel the voltage harmonics of the load, *e.g.* diode rectifiers with high capacitance in the DC side.

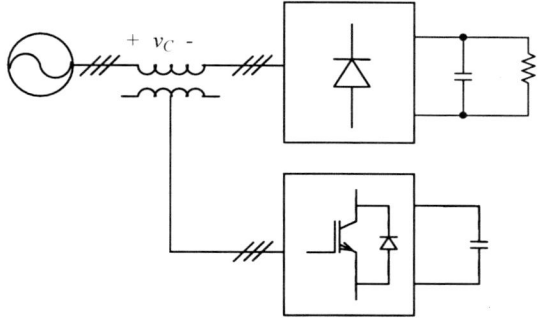

Figure 9.3. A series APLC scheme

The performance scheme of a series APLC is shown in Figure 9.4. Different control targets are possible, [21–24]. In this case, the APLC supplies a compensation voltage, Figure 9.4b, when it is connected. These harmonic components cancel the voltage harmonics of the load, Figure 9.4a. So, after the compensation, the source voltage will be sinusoidal, Figure 9.4c.

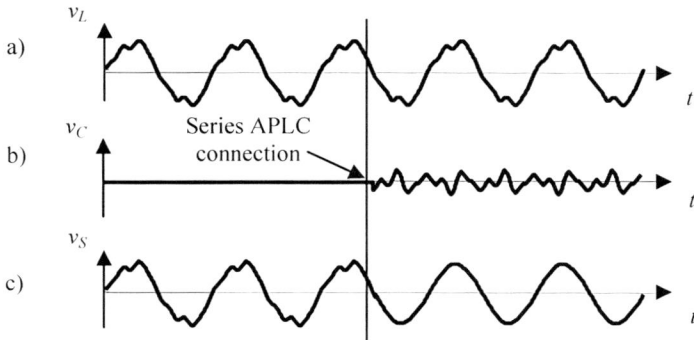

Figure 9.4. Performance scheme of a series APLC

9.2.3 Hybrid Filters

The next figures show some hybrid passive and active filters. The basic aim of these combinations is to reduce the cost of the static compensation. The passive filters are used to cancel the most relevant harmonics of the load, and the active filter is dedicated to improving the performance of passive filters or to cancel other harmonics components. As result, the power of the active filter is reduced, and the passive filter problems (*e.g.* resonances with the source impedance) are mitigated. In summary, the total cost decreases without reduction of the efficiency [25–29].

Figures 9.5, 9.6 and 9.7 show the more usual hybrid topologies.

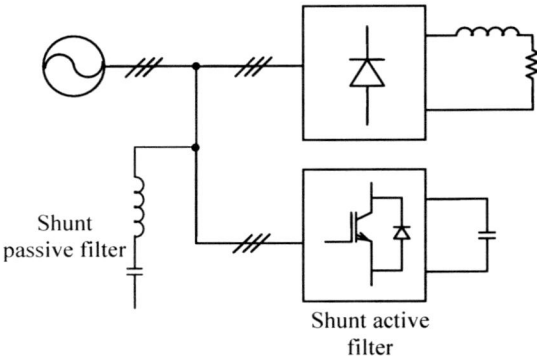

Figure 9.5. Hybrid filter with a shunt passive filter and a shunt active filter

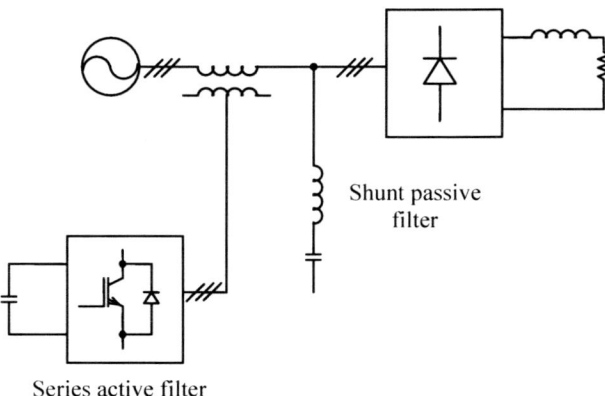

Figure 9.6. Hybrid filter with a shunt passive filter and a series active filter

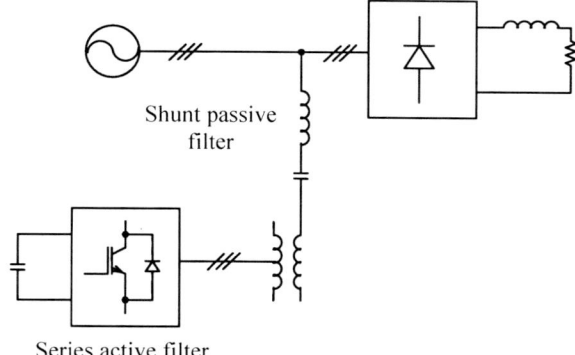

Figure 9.7. A shunt passive filter and an active filter in series with it

The passive filter is designed with some LC branches resonant to some harmonics or adjusted as high-pass filter. The main characteristics of the presented hybrid topologies are summarized in Table 9.1.

Table 9.1. Comparison between the different hybrid topologies

	Shunt active filter + shunt passive filter	Series active filter + shunt active filter	Active filter in series with a shunt passive filter
Power circuit of active filter	PWM inverter with closed-loop current control	PWM inverter without current control	PWM inverter with or without closed-loop current control
Main aim of active filter	Current-harmonic compensation	Harmonic insolate and voltage-harmonic compensation	Harmonic compensation or to improve the passive filter
Advantages	Reactive power regulation Commercial active filters	There is no harmonic current in active filter Commercial passive filters	Low protection of active filter Commercial passive filters
Disadvantages	Compensation intervals	Over-currents. There is no reactive power control	There is no reactive power control

9.2.4 Unified Power-quality Conditioner

It is possible to design unified configurations of series and shunt active filters, [30–33]. The basic scheme of a global or unified power-quality conditioner is shown in Figure 9.8.

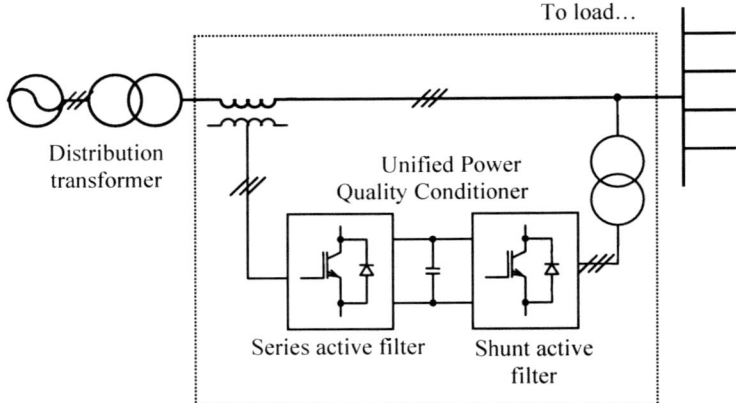

Figure 9.8. General scheme of a unified power quality conditioner

The series filter function is to isolate the voltage harmonics between the source and the load. In addition, it regulates the voltage and compensates the flicker and the PCC (point of common coupling) voltage unbalances. The shunt filter aim is to compensate the load-current harmonics, the reactive current and the unbalanced currents, Table 9.1.

9.3 Power-electronic Inverters in APLCs

There are two kinds of power circuits in an APLC, [34–36]:

- Voltage source inverter (VSI). It is a DC/AC inverter with a capacitor in the DC side. It works as a voltage source.
- Current source inverter (CSI). It is an inverter with an inductance in DC side. It works as a current source.

The corresponding three-phase power circuits are shown in Figures 9.9 and 9.10 respectively.

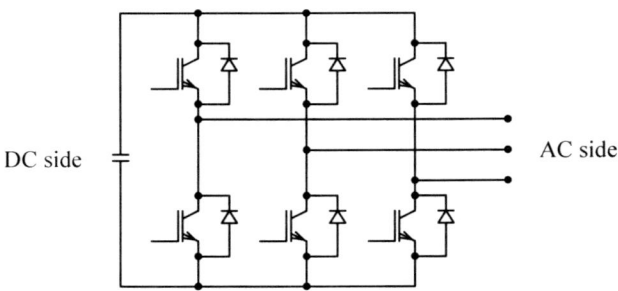

Figure 9.9. Power circuit of a VSI inverter

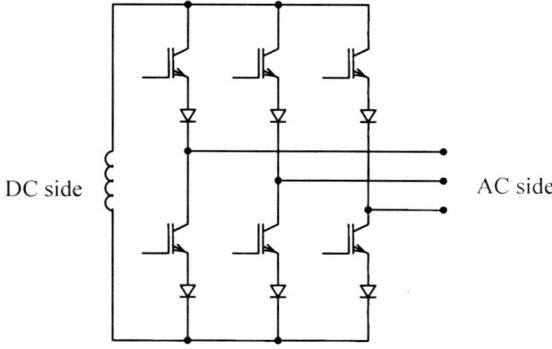

Figure 9.10. Power circuit of a CSI inverter

The usual converter switching devices are insulated gate bipolar transistors (IGBTs). In the DC side, a capacitor (in the VSI configuration) or an inductance (in the VSI configuration) is the energy stored. The inverter does not need an additional energy source in the DC side. So, the APLC control will enable the electrical system to provide the converter energy losses. The VSI topology cost is lower and it is the usual extended configuration.

Table 9.2 compares the main power-device characteristics. These values were presented by Akagi, [2].

Table 9.2. Power-device characteristics.

Switching devices	Power of active filter	Voltage and current limits	Power-circuit topology	Switching frequency	Year
GTO	20 MVA	4500 V 3000 A	VSI	300 Hz	90
SI	200 MVA	1200 V 300 A	VSI	5 kHz	88
BJT	500 MVA	1200 V 300 A	VSI	1.3 kHz	87
IGBT	100 MVA	1000 V 100 A	VSI	8 kHz	88

Nowadays, the performance limits have been augmented, but the comparison shows than the IGBT works with a switching frequency over the rest (about 25 kHz in actuality). Besides, its energy losses are low. For this, the IGBT is the more usual switching device. The GTOs are used in some applications where it is necessary to work with a very high voltage or current.

9.3.1 Voltage-source Inverter Topologies

As was mentioned above, an active filter uses a DC/AC inverter as a power circuit. A DC voltage value is connected to the load in an alternative way to synthesise a desired waveform in the AC side, [37].

The shunt APLC use a VSI inverter, because it is possible to obtain a high efficiency with a low initial cost. In the next section, the basic performances of the VSI converter are described. Single-phase and three-phase configurations are presented.

9.3.1.1 Single-phase Full-bridge Inverters

Figure 9.11 shows two possible schemes of a single-phase inverter. In this analysis, ideal switching devices and a constant-voltage source in the DC side have been considered. Figures 9.11a, b present a half-bridge and a full-bridge configuration, respectively.

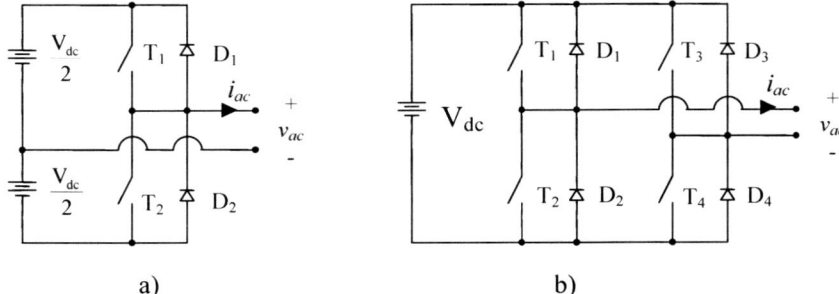

Figure 9.11. Basic schemes of an ideal inverter

The adequate ON/OFF connection of ideal switches allows a rectangular waveform in the AC side. In an half-bridge scheme, if T1 is ON and T2 is OFF, the output voltage will be $+V_{dc}/2$. If T1 is OFF and T2 is ON, the output voltage will be $-V_{dc}/2$. In the full-bridge scheme, if T1 and T4 are ON and T2 and T3 are OFF, the output voltage is V_{dc}. When T2 and T3 are OFF and T1 and T4 are ON, the output voltage is $-V_{dc}$.

When an inductive load is connected in the AC side, it is necessary to include shunt diodes with each switch. The inductive current is delayed with regard to the voltage. If the switches change the position, e.g. T1 and T4 in Figure 9.11b, the current needs a freewheeling way through diodes D2 and D3. The current can flow in the same way in the AC side. In this period, the load returns the energy to the DC source. For this reason, the diodes are called energy-recover or freewheeling diodes. If the load is resistive, the diodes will not be necessary.

In Figure 9.12, a sinusoidal current has been built with a half-bridge inverter. It is a basic commutation scheme of a rectangular waveform. The desired waveform is the fundamental component, also included in Figure 9.12.

Active Power-line Conditioners 241

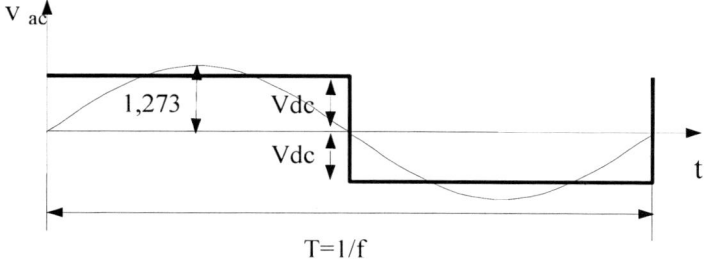

Figure 9.12. Sinusoidal waveform from a single-phase inverter, with a quadrangular switching pulse

In effect, the Fourier decomposition allows the first harmonics of a rectangular waveform to be obtained:

$$V_{ac,1} = \frac{4}{\pi} V_{dc} = 1,273 V_{dc}, \qquad (9.1)$$

Figure 9.13 shows the waveform harmonic spectrum.

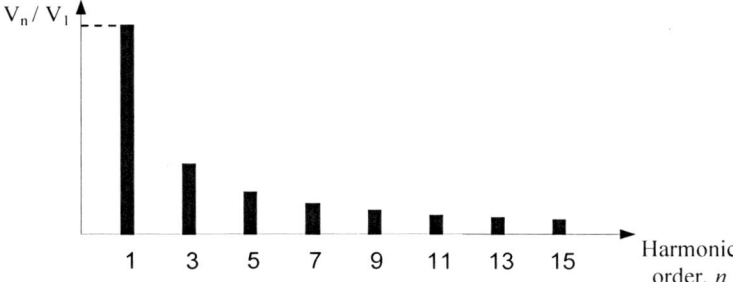

Figure 9.13. Harmonic spectrum of Figure 9.12 waveform

The output signal frequency can be modified by the switch commutations. The amplitude can be adjusted by the DC voltage value or by the switch pulse modulation. This is the more practical and usual procedure. There are two possibilities. The first one is modulating constant-width pulses and the second one is a sinusoidal modulation in a period.

The output voltage is an alternated waveform. It includes the desired sinusoidal output voltage and other non desired frequencies as a result of device commutations. The current waveform depends on the output load. If the load is resistive, the current waveform is the same as the voltage waveform. If the load is inductive, the current waveform is smoother than the voltage waveform, and their non desired harmonic components will be reduced. For this reason, an output inductance will be fixed in the AC side. This inductance is not enough to filter all no desired frequencies of output voltage, and it is convenient, in a general case, to add a LC passive filter.

The usual electronic switches are BJTs, MOSFETs or IGBTs. Figure 9.14 shows the previous inverter schemes designed by IGBTs, the more extended device.

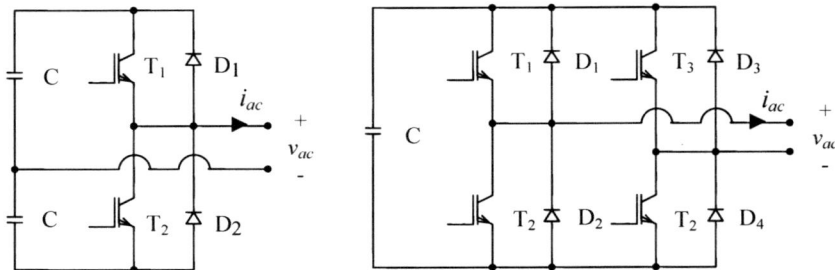

Figure 9.14. VSI single-phase inverter schemes with IGBTs, (a) a half-bridge, (b) full-bridge

An IGBT is a hybrid combination of a BJT and a FET power device. Its performance is similar to a BJT, but the gate current is very low, as in a FET. The commercial power transistors include the freewheeling diodes. The dc voltage is fixed by one or two capacitors, according to the configuration.

9.3.1.2 Three-phase Full-bridge Inverters

A three-phase inverter can be built with three single-phase inverters connected to the same DC source. Also it is possible to design a three-phase configuration as is presented in Figure 9.15 [38].

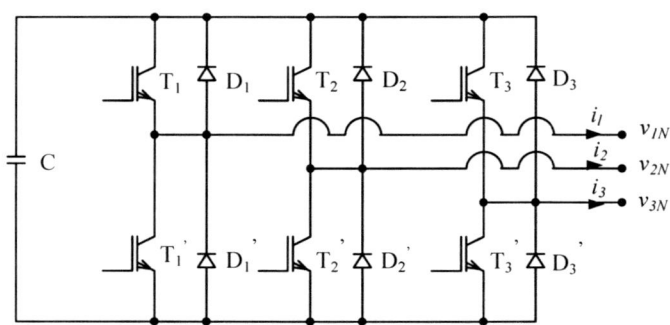

Figure 9.15. VSI three-phase full-bridge inverter with IGBTs

Figure 9.16 shows the three-phase inverter performance in generating sinusoidal three-phase voltages, v_{1N}, v_{2N} y v_{3N}. It is considered as only a switching pulse along the conduction period. In this configuration, it is necessary to apply complementary pulses in the top and bottom IGBT of each branch. There are some periods where T1-T2-T3 are ON and periods where T1'-T2'-T3' are ON. The three output line-to-neutral voltages and a line-to-line voltage, with their fundamental component are presented.

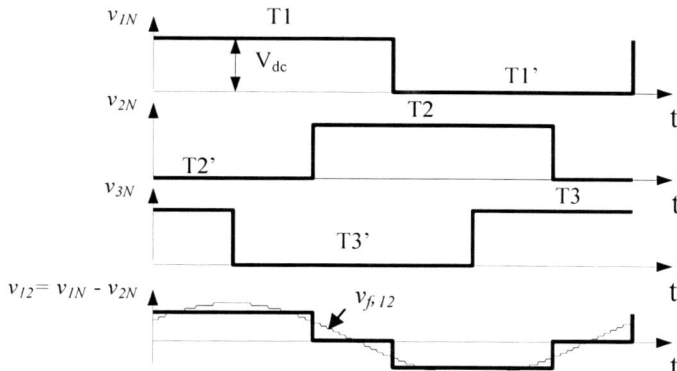

Figure 9.16. Output voltages in a three-phase inverter with square-wave switching

Figure 9.17 shows the voltage-harmonic spectrum, where the harmonic amplitude values are relative to the fundamental component. The use of polyphase systems allows a reduction in the presence of harmonics. The PWM techniques will reduce the harmonic distortion significantly.

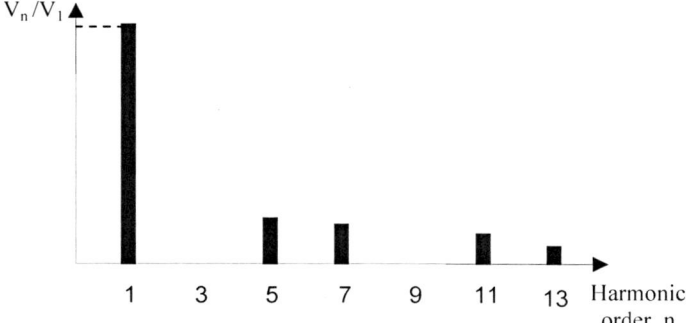

Figure 9.17. Harmonic spectrum of line-to-line voltage v_{12}

It is possible to use different power circuits in shunt three-phase APLCs. In low-power applications, it is usually a three-phase scheme. Figure 9.18 shows a three-phase three-wire system compensated by a shunt APLC. The power circuit has three IGBTs branches with a capacitor in the DC side.

In this simple scheme, the connection between the APLC and the system is realised by means of an output inductance. When the APLC output voltages are positives, the output currents increase. When the voltages are negatives, the currents decrease. The quadrangular output voltages will produce triangular output currents. A correct design of output inductances (and *a posteriori* passive filter) is very important to get an adequate active conditioner answer. The increment of compensation current will allow, or not, to follow the reference of control.

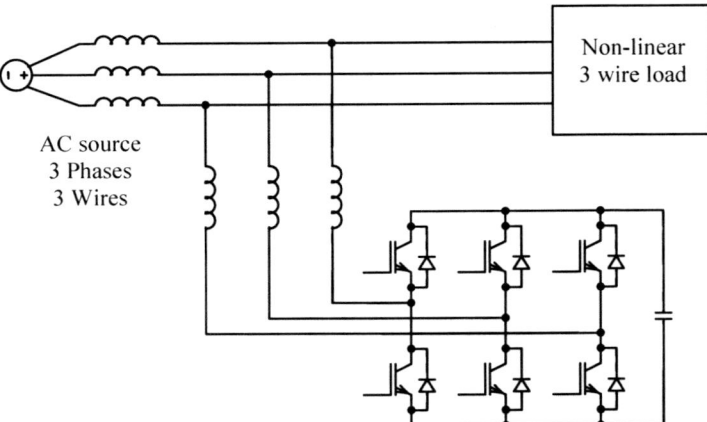

Figure 9.18. Three-phase three-wire system scheme, with a nonlinear load compensated by a shunt APLC using a VSI three-phase inverter

In three-phase four-wire systems with unbalanced loads, it is possible to use three single-phase inverters as an APLC power circuit. The aim is to compensate phase by phase. Figure 9.19 shows a three-phase four-wire system with a nonlinear load compensated by a shunt APLC. The network connection needs a coupling transformer, because there is no a common neutral point.

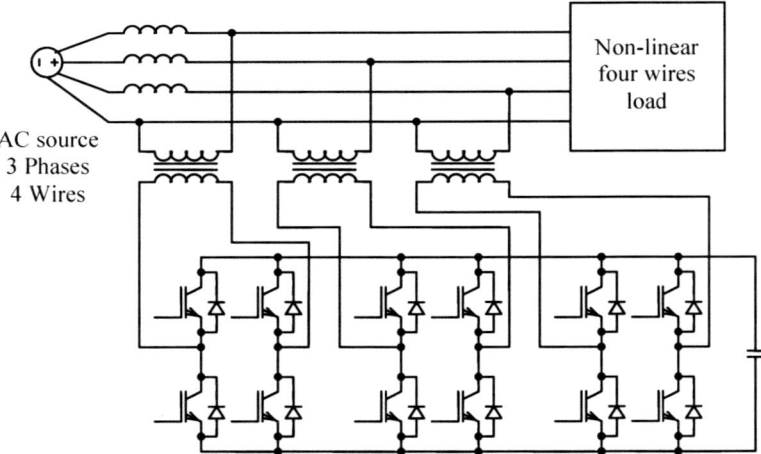

Figure 9.19. Three-phase, four-wire system scheme, with a nonlinear load compensated by a shunt APLC using three single-phases as an APLC power circuit

In general, in four-wire power systems, it is usual to use APLCs with three-phase configurations [39]. In this case, a split capacitor is necessary in the DC side. The middle point is connected to the neutral wire.

Active Power-line Conditioners 245

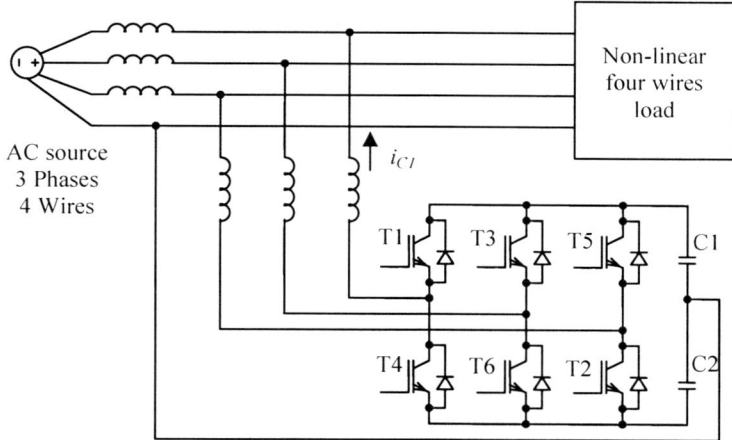

Figure 9.20. Three-phase, four-wire system scheme, with a nonlinear load compensated by a shunt APLC using a three-phase three-leg APLC power circuit

This scheme has a problem when the compensation currents have sequence-zero components. In this case, these current will flow through one only DC capacitor. As a result, the DC capacitor voltage will be unbalanced. This implies a malfunction in the inverter performance, and some solutions have been proposed [34]. Figure 9.21 summarises, *e.g.*, all the situations when the APLC follows a sinusoidal reference.

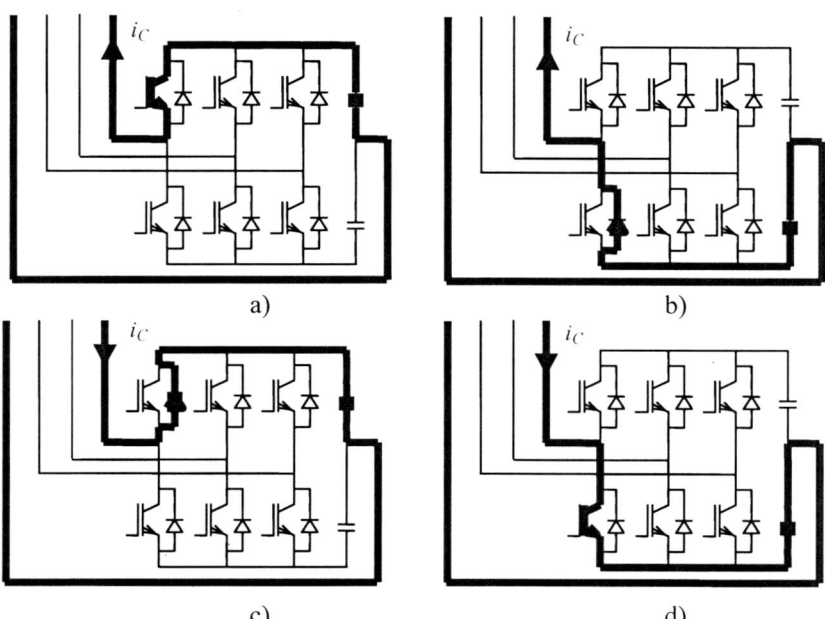

Figure 9.21. Neutral compensation current in a three-phase inverter with a DC split capacitor

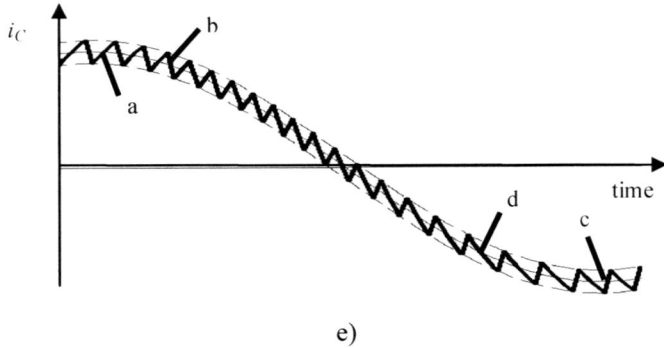

e)

Figure 9.21. Neutral compensation current in a three-phase inverter with a DC split capacitor (contd.)

In this example, the inverter uses a hysteresis band PWM control to follow the sinusoidal reference, (Section 9.3.2). If the compensation current i_C is increasing and positive ($i_C > 0$, $di_C/dt > 0$), the current will flow through T1 and C1, and the C1 voltage will decrease. Figure 9.21 shows other possible situations. Table 9.3 summarises all the voltage variations.

Table 9.3. Voltage variations of DC capacitors in a three-branch, four-wire inverter

cases	i_C	di_C/dt	Voltage variations
a	$i_C > 0$	$\dfrac{di_C}{dt} > 0$	C1 voltage decreases
b	$i_C > 0$	$\dfrac{di_C}{dt} < 0$	C2 voltage increases
c	$i_C < 0$	$\dfrac{di_C}{dt} > 0$	C1 voltage increases
d	$i_C < 0$	$\dfrac{di_C}{dt} < 0$	C2 voltage decreases

When the compensation current does not present a zero-sequence component, the voltage unbalance is not important. Anyway, it is possible to implement a proportional or proportional-integral control block to adjust the signal reference using the capacitor voltage difference. This allows balancing of their voltages.

Other inverter configurations are possible to avoid the problem mentioned above. Figure 9.22 shows a three-phase inverter with four IGBT branches.

Active Power-line Conditioners 247

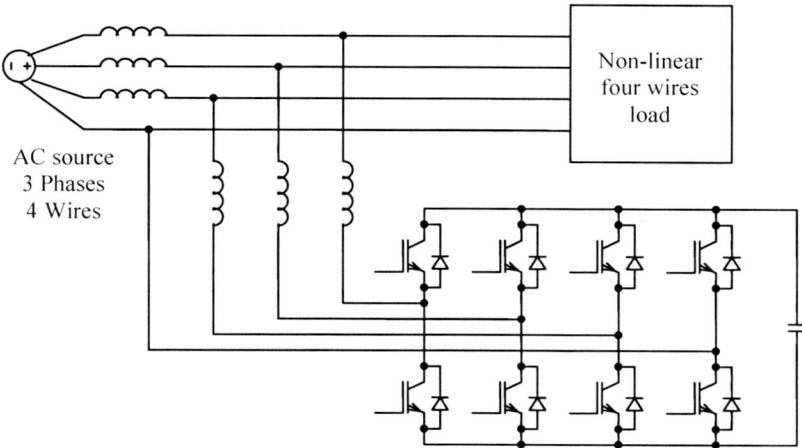

Figure 9.22. Three-phase four-wire system with a nonlinear load compensated by a shunt APLC, using a three-phase, four-arm APLC power circuit

In this case, the middle point of the fourth arm is connected to the neutral wire. The control of these IGBTs allows fixing of the neutral voltage and it avoids voltage unbalance. Anyway, this scheme is not extended because it uses more devices.

9.3.1.3 Multilevel Inverters

The basic inverter presented up to now usually presents a switching frequency, a maximum value of the output voltage and a rated power that becomes limited by the characteristics of the power-electronic devices that constitute it. Some of these limitations can be overcome by means of the employment of a higher number of electronic devices with an appropriate topology. In this area, a solution that recently has received great attention, are multilevel voltage source inverters [40–41].

The inverter's configuration shown in Figure 9.15 belongs together with the basic topology of an inverter of two levels, since each output can be connected as much to the positive terminal as to the negative terminal of the DC side, and the voltage output takes the values $+V$ and 0, Figure 9.16, for a capacitor voltage value of V volts. However, there is a class of inverters that can take more than two values in their voltage output; they are denominated multilevel inverters. The multilevel inverters receive that name according to the different voltage values that can be obtained in their output. Thus, a 3-level inverter can produce voltage levels of $+V$, 0, and $-V$, where V represents the DC side voltage. In the same way they obtain four different levels of voltage, $+V$, $+V/3$, $-V/3$, $-V$, in a 4-level inverter, five different values, $+V$, $+V/2$, 0, $-V/2$, $-V$, in a 5-level inverter, and so forth. Thus, higher values of voltage are obtained in the output than in the classic inverter, and currents of higher quality occur although at the expense of a higher number of devices.

The more common multilevel inverter is the 3-level inverter that is presented in Figure 9.23. Each side of the inverter is made up of for four IGBTs, T1 to T4, with the freewheeling diodes, D1 to D4 and two clamping diodes, D5 and D6 that avoid the DC side capacitors short-circuited. The DC side includes two capacitors, Ci, that makeup a capacitive divider. The circuit topology contains twelve power switches however, which implies a high number of possible inverter states, in practice twenty-seven states only are used for this configuration.

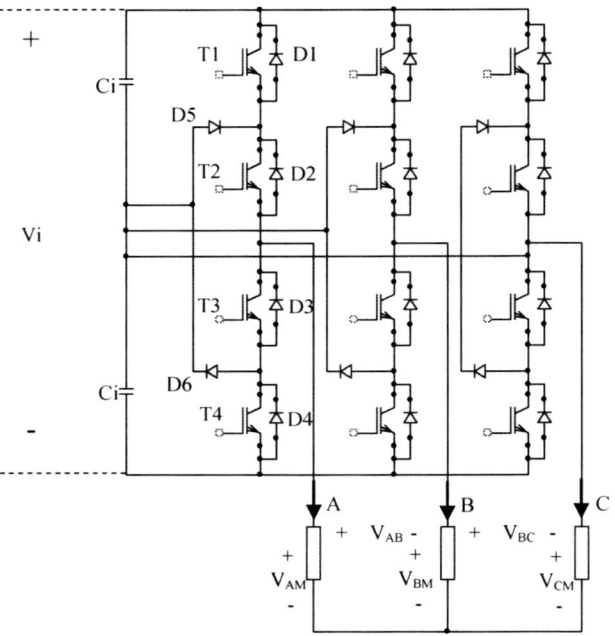

Figure 9.23. Three-level inverter

In short, the inverter leg only adopts one of the three following states:

- T1 and T2 on, T3 and T4 off;
- T2 and T3 on, T1 and T4 off;
- T3 and T4 on, T1 and T2 off.

The DC voltage, Vi, is always applied to a couple of series-connected switches. Therefore the voltage obtained by a 3-level inverter can be double that of the rated voltage of the switches; this corroborates our statement at the beginning of this section.

Active Power-line Conditioners 249

In what follows the techniques of synthesis in voltage waveforms will be discussed by means of the use of multilevel inverters. In the same way that selective harmonic elimination with 2-level inverters has been used, it is possible to extend that technique to multilevel inverters. Here the case of a 5-level inverter will be considered.

Figure 9.24 shows half of the period in the 5-level inverter voltage waveform. The negative half-period is the mirror image of the positive haft-period.

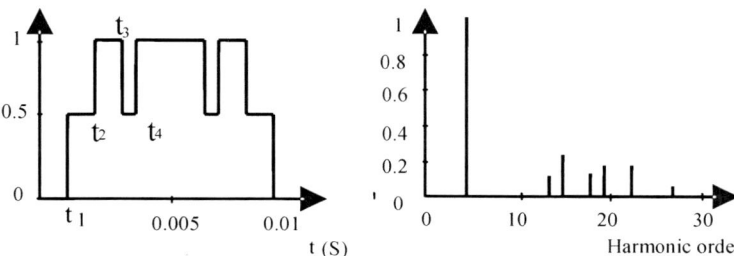

Figure 9.24. Output voltage of a five-level inverter and its harmonic spectrum

In a 5-level inverter, each leg is constituted by eight switching electronic devices. The output-voltage waveform is reconstructed by the driving turn on of the corresponding switches in the specified instants. Figure 9.24 indicates the times t_1, t_2, t_3, and t_4 corresponding to the changes of the inverters state. The waveform presents the symmetry of a half-wave, which means that the component DC and all their even harmonics are null. Besides, the waveform presents odd symmetry, as its development in Fourier expansions will only include terms of sine type.

For all that indicated previously, the harmonic of order tension h is given by,

$$b_h = \frac{2}{\pi} \int_0^{\pi} V_{AH} \sin(h\omega t) d(\omega t), \qquad (9.2)$$

This equation gives as a result for odd h,

$$b_h = \frac{2}{h\pi} \left[-\cos h\omega t_1 - \cos h\omega t_2 + \cos h\omega t_3 - \cos h\omega t_4 \right], \qquad (9.3)$$

For simplicity, it has taken the unit as the DC voltage value. To eliminate the order harmonic 3, 5, 7 and 9, four equations are obtained when substituting the harmonic orders in the previous equation. As a result the following solutions are obtained: $\omega t_1 = 14.57°$; $\omega t_2 = 45.43°$; $\omega t_3 = 57.43°$; $\omega t_4 = 62.57°$. The harmonic spectrum normalised regarding the fundamental component of the waveform synthesised in Figure 9.24 is shown; a fundamental frequency of 50 Hz has been supposed. One can observe all their harmonics have an order above eleven.

The main disadvantage of this inverter type is the high number of capacitors that are included in its configuration. The switching process can cause an unequal capacitor charge taking place an unbalanced output voltage. The multilevel inverters can also be operated by means of PWM control as will be seen later.

9.3.2 Control of Voltage-source Inverters

The main advantages of modern power-electronic converters are being achieved through the use of the so-called switch-mode operation, in which power semiconductor devices are controlled in an on/off fashion, with no operations in the active region, [42].

9.3.2.1 Switched Waveform

Usually, to explain the inverter control principles, the named switching variables are usually introduced [43]. In the case of a 2-level inverter they are necessary to help the three binary switching variables *a, b, c*, one for each one of the phases. Each inverters leg is make up by two switches that cannot be simultaneously in the on state, since in that case the voltage of the DC side would be in short-circuit. Therefore, in an ideal condition, a switch of each phase will be in the on and the other one in the off state; each leg can assume in this way only two states, from which the number of the inverter states is $2^3 = 8$.

If we refer to phase 1 of Figure 9.15, the switching variable *a* is defined such that it takes the value 1 if the transistor T1 is in the on state and the transistor T'1 is in the off state. If, inversely, T1 is in the off state and T'1 is in the on state, the switching variable *a* takes the value 0. The two remaining variables *b* and *c*, are defined in a similar way. Thus, the state of a 2-level inverter is established by the number binary *abc*. For example, with $a=1$, $b=1$, $c=0$, it is said that the inverter is in the state 6, since $110_2 = 6_9$.

The output line-to-line voltage of the inverter v_{12}, v_{23}, v_{31} are given by

$$\begin{bmatrix} v_{12} \\ v_{23} \\ v_{31} \end{bmatrix} = V_i \begin{bmatrix} 1 & -1 & 0 \\ 0 & 1 & -1 \\ -1 & 0 & 1 \end{bmatrix} \begin{bmatrix} a \\ b \\ c \end{bmatrix}, \qquad (9.4)$$

where V_i it is the voltage value in the DC side.

When the output terminals are connected to a load in Y of three conductors it is verified that for the line-to-neutral voltages,

$$v_{1N} + v_{2N} + v_{3N} = 0, \qquad (9.5)$$

On the other hand, it is also verified that,

$$v_{1N} - v_{2N} = v_{12} ; \quad v_{2N} - v_{3N} = v_{23} , \tag{9.6}$$

The last three equations allow the line-to-neutral to voltages be expressed as a function of the line-to-line voltages,

$$\begin{bmatrix} v_{1N} \\ v_{2N} \\ v_{3N} \end{bmatrix} = \frac{1}{3} \begin{bmatrix} 1 & 0 & -1 \\ -1 & 1 & 0 \\ 0 & -1 & 1 \end{bmatrix} \begin{bmatrix} v_{12} \\ v_{23} \\ v_{31} \end{bmatrix} , \tag{9.7}$$

that combined with Equation 9.4, the following holds

$$\begin{bmatrix} v_{1N} \\ v_{2N} \\ v_{3N} \end{bmatrix} = \frac{V_i}{3} \begin{bmatrix} 2 & -1 & -1 \\ -1 & 2 & -1 \\ -1 & -1 & 2 \end{bmatrix} \begin{bmatrix} a \\ b \\ c \end{bmatrix} . \tag{9.8}$$

The simplest control strategy of an inverter consists of imposing the state sequence 5-4-6-2-3-1. The result is the square wave or six-step operation that is shown in Figure 9.25. There, the line-to-line voltage waveforms are represented and one of the line-to-neutral voltages, v_{1N}, where N is the neutral of the load in star configuration.

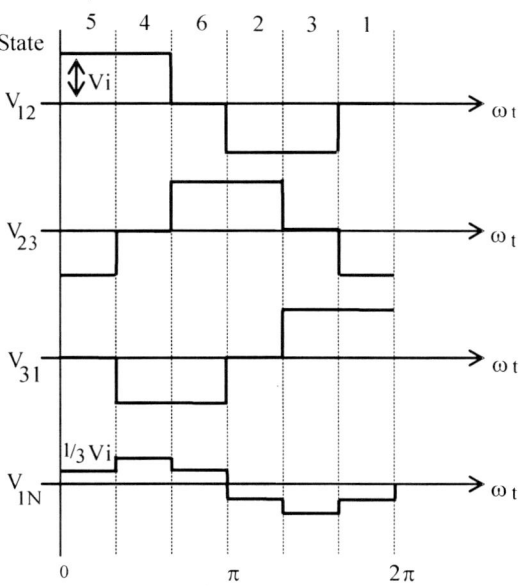

Figure 9.25. Line-to-line and line-to-neutral output voltage waveforms in a voltage source inverter in the square-wave mode

In the square-wave operation mode, each transistor is only turned on and off once per cycle, and the peak value of the fundamental component of the output line-to-line voltage is $1.1 V_i$. In fact, the expansion in Fourier series is,

$$v_{12} = \frac{2\sqrt{3}}{\pi} V_i \left(\sin\omega t - \frac{1}{5}\sin 5\omega t - \frac{1}{7}\sin 7\omega t + \frac{1}{11}\sin 11\omega t + \ldots \right), \qquad (9.9)$$

The fundamental harmonic has a peak value,

$$V_p = \frac{2\sqrt{3}}{\pi} V_i = 1.1 V_i, \qquad (9.10)$$

Thus, in the square-wave operation mode the possible maximum output voltage is obtained via the PWM technique, as will be seen later. Nevertheless, this output voltage value cannot be controlled, which is one of the main disadvantages of this operation mode.

In the power inverters the most convenient form of synthesising the desired output voltage waveform, is pulse width modulation (PWM). The modulation in power electronics is a process that allows a switched representation of some waveform to be obtained. For this, the period of the output voltage is split into the so-called switching intervals. A typical PWM waveform appears in Figure 9.26. There, the voltage period has been split into 12 switching intervals. A pulse of each switching variable appears in each switching interval, and it is the width of the pulse that adjusts its duration. The number N of switching intervals per period is given by,

$$N = \frac{f_{sw}}{f}, \qquad (9.11)$$

where f_{sw} corresponds to the switching frequency and f is the fundamental frequency of the inverter output voltage. Usually, the switching frequency is constant; thus N depends only on the output frequency, and it is not necessarily a whole number. The harmonic content in the voltage waveforms are distributed around the orders of the multiple switching frequencies. In this way, when the PWM inverter has connected inductive loads in the AC side, the current waveforms come closer to sinusoidal waveforms due to the low-pass action of the inductances.

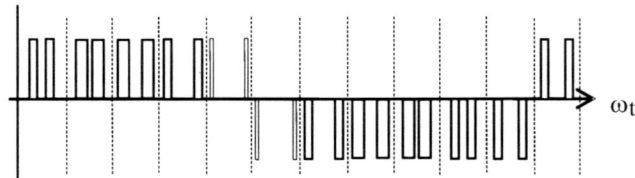

Figure 9.26. Typical waveform in the PWM mode

The most commonly used PWM methods for three-phase converters impress either the voltages or the currents into the AC side. Two basic schemes of PWM are utilized; the open-loop PWM voltage-control technique, Figure 9.27a, and the closed-loop current-control technique, Figure 9.27b.

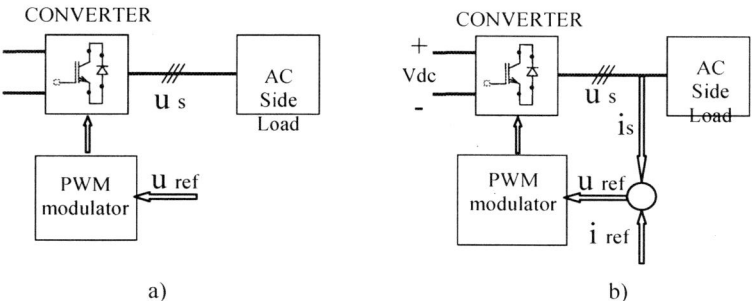

Figure 9.27. Open-loop voltage control and closed-loop current-control

In the approach of open-loop voltage control the well-known modulation method is the triangular carrier-based (CB) sinusoidal PWM. This method is based on equalling the average value from the switched waveform to the average value of the modulating signal in a switching interval. The scheme for the generation of a PWM control signal is shown in Figure 9.28.

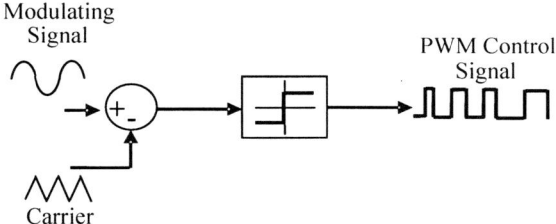

Figure 9.28. Basic principles of PWM control scheme

This is composed of a carrier signal and a modulating signal. The magnitudes of both signals are compared; the PWM control signal takes a high value when the modulating signal has higher numeric values than the carrier signal and it takes a low value when it is the carrier signal that has high numeric values. If this PWM control signal is applied to the inverter of Figure 9.11b, the switches T1 and T4 are closed when the modulating signal is higher than the carrier, and the switches T2 and T3 are closed when the carrier is higher than the modulating signal. The inverter output voltage is $\pm V_{DC}$, this constitutes a bipolar voltage switching.

Many applications of three-phase voltage-source PWM inverters, among them active power filters, have a control structure that contains an internal current feedback loop, Figure 9.29. In comparison to the conventional inverter of open-loop voltage PWM, the current-controlled PWM (CC-PWM) has the following advantages:

- control of high accuracy in the current instantaneous waveform;
- protection against overcurrents;
- very good dynamic operation;
- compensation of the effects due to changes in the parameters of the load, voltage drop in the switches, and voltage fluctuations in the DC and AC sides.

Figure 9.29 presents the basic block diagrams of a current-controlled PWM inverter. The control system forces the load currents to follow the current references.

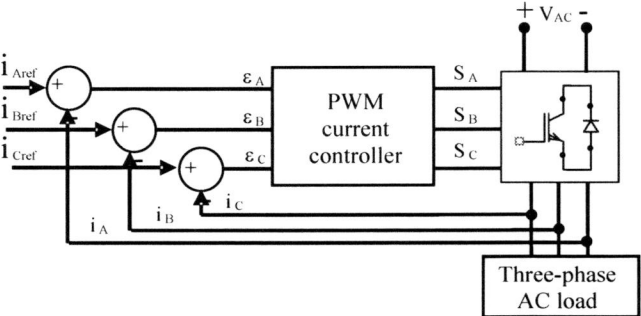

Figure 9.29. Current-controlled PWM inverter

For comparison of the reference current instantaneous values i_{Aref}, i_{Bref}, i_{Cref}, with the measured currents i_A, i_B, i_C, of each one of the phases, the current-controlled inverter generates the switching states S_A, S_B, S_C, for the inverters electronic devices that decrease the errors of the currents ε_A, ε_B, ε_C. Therefore, the CC carries out two tasks: compensation of the error (decrease of ε_A, ε_B, ε_C) and modulation (determination of the switching states S_A, S_B, S_C).

In the following will be presented in more detail, on the one hand an open-loop PWM voltage-control technique, the denominated space vector modulation (SVM), and on the other hand, three closed-loop current-control techniques; periodical sampling, hysteresis band and the triangular carrier technique [44].

9.3.2.2 Space Vector Modulation
In the inverter of Figure 9.30, e.g., there are 8 different states possible.

The instantaneous positive sequence is defined in Equation 9.12, where $a = 1 \angle 120°$ and v_{AN}, v_{BN} and v_{CN} are the output terminal voltages of the inverter.

$$v^+ = \frac{1}{\sqrt{3}}(v_{AN} + a \cdot v_{BN} + a^2 \cdot v_{CN}), \qquad (9.12)$$

Active Power-line Conditioners 255

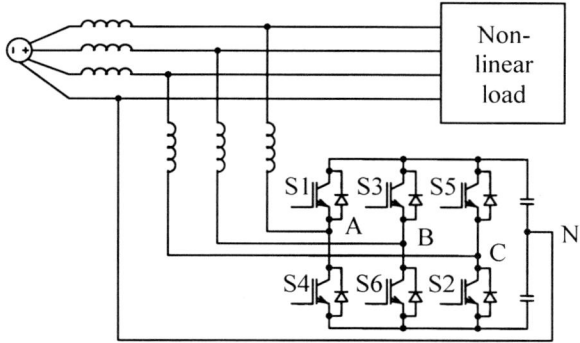

Figure 9.30. Inverter of 2 levels that makes up an APLC

The 8 states are listed in Table 9.4, in pu values. Two states mean a null v^+ vector. Figure 9.31 presents a graphical representation, where the complex plane is divided into 6 triangular regions between adjacent vectors.

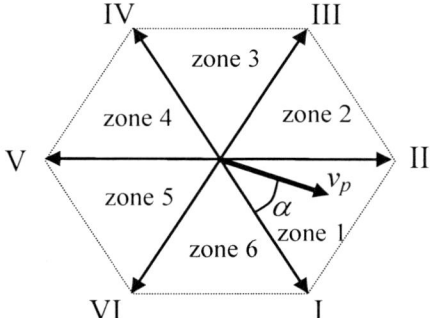

Figure 9.31. Inverter space vector

At a given instant, a certain vector v_P can be obtained with the average values of the states nearby (I and II) and the zero vector (VII or VIII place in the plane origin). This average value is made over a time interval t_a previously selected. In Equation 9.13, v_I, v_{II} and v_{VII} are the inverter output voltages in the states I, II and VII, and t_1 and t_2 are the time that the inverter operates in the states I or II.

$$v_P t_a = v_I t_1 + v_{II} t_2 + v_{VII}(t_a - t_1 - t_2) \qquad (9.13)$$

For a vector v_P, with modulus V_{Pm} and angle α, and with the values of v_I, v_{II} and v_{VII} from Table 1.4, it can be deduced that:

$$V_{Pm} t_a \cos(60 - \alpha) = 0.5774\, t_1 + 1.1546\, t_2 ,$$
$$V_{Pm} t_a \sin(60 - \alpha) = t_1 . \qquad (9.14)$$

Table 9.4 is generated for $V=1.0$ pu. For any value V, it can be written as shown,

$$\frac{t_1}{t_a} = \frac{V_{Pm}}{V}\sin(60-\alpha),$$
$$\frac{t_2}{t_a} = \frac{V_{Pm}}{V}\sin\alpha.$$
(9.15)

Thus, the inverter must operate in state I during the time t_1, in state II during t_2, and the rest of the interval $t_a - t_1 - t_2$ in the zero state (VII or VIII).

Table 9.4. Inverter switch states and output voltages in pu values

State No.	Switch Status						Output Voltages			v^+
	S_1	S_2	S_3	S_4	S_5	S_6	v_{AN}	v_{BN}	v_{CN}	
I	On	Off	Off	Off	On	On	+1	−1	+1	0.577 − j1.0
II	On	On	Off	Off	Off	On	+1	−1	−1	1.1547+j0.0
III	On	On	On	Off	Off	Off	+1	+1	−1	0.5774+j1.0
IV	Off	On	On	On	Off	Off	−1	+1	−1	− 0.577+j1.0
V	Off	Off	On	On	On	Off	−1	+1	+1	− 1.154+j0.0
VI	Off	Off	Off	On	On	On	−1	-1	+1	− 0.5774 − j1.0
VII	On	Off	On	Off	On	Off	+1	+1	+1	0
VIII	Off	On	Off	On	Off	On	−1	−1	−1	0

The maximum output voltage value one will have when vectors different from zero only are used, this is when it is verified that,

$$t_{VII} = t_{VIII} = 0,$$
(9.16)

where t_{VII} and t_{VIII} represent the time that the inverter is located in states VII and VIII, respectively. In this case, the ends of the inverter output voltage vectors are located in a radio circle V_{max}. Let $t_a = T_{sw} = 1/f_{sw}$ the switching period, that is, the duration of the switching interval. Let us denote the quotients t_X/t_a and t_Y/t_a for d_X and d_Y, that are called state duty ratios of the inverters generic states X and Y (I, II, ... VI). Having presented Figure 9.31, any maximum vector will be expressed as a function of two components

$$v_{p\max} = d_X v_X + d_Y v_Y,$$
(9.17)

That vector will be located in one of the six regions of Figure 9.31 and together with their components they will determine a triangle in which it is possible to apply the cosine theorem. Since v_X and v_Y form 60°, the v_{pmax} modulus is

$$V_{max}^2 = d_X^2 + d_Y^2 + d_X d_Y, \quad (9.18)$$

and together with the constraint

$$d_X + d_Y = 1, \quad (9.19)$$

it can determine the V_{max} value from the Lagrange multiplier technique. The result is $V_{max} = \sqrt{3}\, V_i/2$, which means that the maximum peak value available of the fundamental component of the output line-to-neutral voltage of an PWM inverter is $V_i/\sqrt{3}$, and the fundamental component of the line-to-line voltage is V_i; that is, the DC side voltage. There are to think that for a sinusoidal balanced voltages set, the resulting voltage space vector has a magnitude similar to 3/2 times the peak value of each one of those voltages, as it is proven applying Equation 9.12. As was already indicated, the fundamental output voltage in the square-wave mode is 1.1 V_i.

Once the durations of the individual states of the inverter are been obtained, it will be to specify their sequence inside the switching interval. In practice, two such sequences, a high-quality sequence and a high-efficiency sequence are usually used. The first one is the sequence $X - Y - Z_1$, $Y - X - Z_2$, ..., where the states zero, Z_1 and Z_2, are chosen so that the transition of states supposes switching in a single phase; that is, only an inverter leg makes a switching of its electronic devices. For example, in zone 1, where $X = I$ and $Y = II$, the high-quality sequence is I - II - VIII; II - I - VII, see Table 9.4. This state sequence that for a given number N of switching intervals for output voltage cycle is the output current of the best quality, $N/2$ switching pulses take place per switch and per cycle.

The switching number can be reduced at the expense of increasing the distortion in the current waveforms, when the high-efficiency state sequence, $X - Y - Z$; $Z - Y - X$ is used. Here, $Z = VIII$ in even zones, and $Z = VII$ in odd zones are used. For the previous example the high-efficiency sequence is I – II – VII; VII – II – I. Note that the branch of the left constituted by the switches $S1$ and $S4$ non-switching in the whole zone in question see Table 9.4. With this state sequence, the switching number pulses per switch and per cycle is $N/3 + 1$, very much lower than the high-quality sequence; that is a large reduction of the switching losses in the inverter.

As was mentioned above, Section 9.3.1.3, a multilevel inverter can be controlled with a PMW mode. In a 3-level inverter, three states are used for each leg. So, the switching variables are:

$$a = \begin{cases} 0 \text{ if } S_1, S_2 \text{ are off and } S_3, S_4 \text{ are on} \\ 1 \text{ if } S_1, S_4 \text{ are off and } S_2, S_3 \text{ are on} \\ 2 \text{ if } S_1, S_2 \text{ are on and } S_3, S_4 \text{ are off} \end{cases}.$$

The switching variables b and c can be defined in a similar way. The line-to-line output voltages are:

$$\begin{bmatrix} v_{AB} \\ v_{BC} \\ v_{CA} \end{bmatrix} = \frac{V_i}{2} \begin{bmatrix} 1 & -1 & 0 \\ 0 & 1 & -1 \\ -1 & 0 & 1 \end{bmatrix} \begin{bmatrix} a \\ b \\ c \end{bmatrix}, \qquad (9.20)$$

and the line-to-neutral voltages:

$$\begin{bmatrix} v_{AN} \\ v_{BN} \\ v_{CN} \end{bmatrix} = \frac{V_i}{6} \begin{bmatrix} 2 & -1 & -1 \\ -1 & 2 & -1 \\ -1 & -1 & 2 \end{bmatrix} \begin{bmatrix} a \\ b \\ c \end{bmatrix}, \qquad (9.21)$$

The notation *abc* is corresponding now to a base 3 number that can be used to represent the 27 inverter states. In Figure 9.32, the line-to-neutral voltage space vectors, in pu values, are shown. The states 0, 13 and 26 produce null vectors.

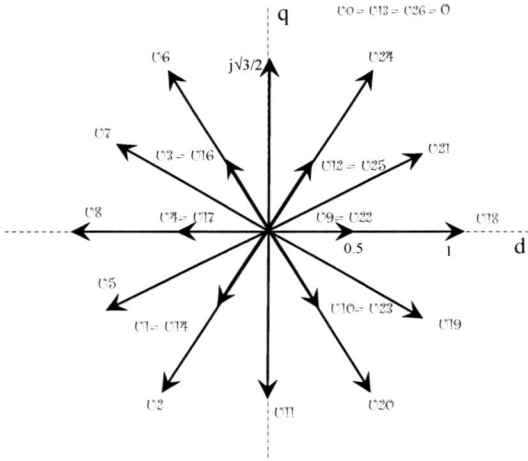

Figure 9.32. Space vectors in the three-level voltage-source inverter

The number of states in a multilevel inverter allows current of higher quantity to be obtained than that obtained with a 2-level inverter. With a PWM mode, a 3-level inverter can operate with a very low switching frequency, and it can generate a current with a quality similar to a 2-level inverter.

9.3.2.3 Techniques of Closed-loop PWM Current Control

In this section, the main closed-loop current techniques used in shunt APLC control will be presented. In the APLC control block, the inputs are the load voltage and current signals, and the outputs are the compensation current references, depending on the control strategy. The APLC power circuit will inject the compensation currents using a closed-loop PWM current control [35,45–46].

For this, APLC output currents are measured and compared with their references. A current controller uses the error signals to generate the IGBT trigger signals. These pulses allow to be changed the electronic switch states. The goal is to reduce the current errors, Figure 9.29.

The main closed-loop current controls through the pulse-width modulation are:

- periodical sampling, PS;
- hysteresis band, HB;
- triangular carrier, TC.

In the periodical sampling method, the switching devices change with a periodical switch bipolar wave. At these times, the output signal is compared with their reference. If it is bigger, the switches change and the output voltage becomes negative. So, the real signal will decrease. On the other hand, if it is less, the switches change to obtain a positive output voltage. Then, the real signal will increase. A general control scheme is presented in Figure 9.33. In this case, the reference signal is sinusoidal. Figure 9.33 shows that the error signals allow to be obtained the trigger pulses to control all the IGBT gates.

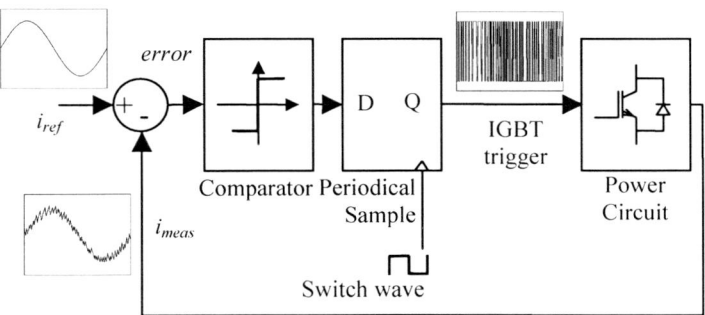

Figure 9.33. Control scheme of periodical sampling

Figure 9.34 presents a period of different waveforms: the reference current, the output current and the trigger signal. If the power circuit is a single-phase half-bridge inverter, Figure 9.14a, these trigger signals will be the upper IGBT gate pulses. The complementary signal will be the bottom IGBT gate pulses. It is easy to implement this control method, but the average error is not null.

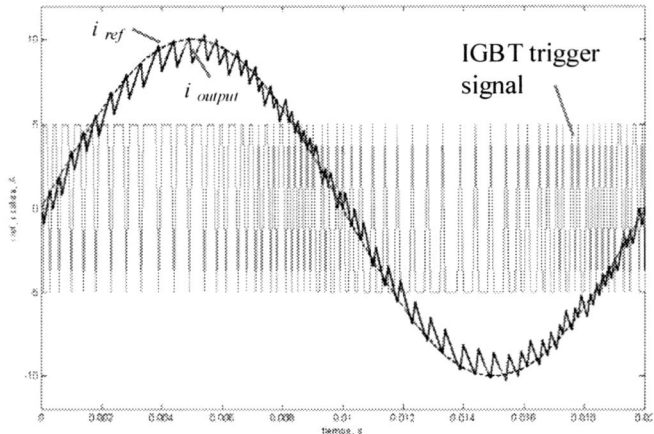

Figure 9.34. Waveforms in a periodical sampling control method

The hysteresis band control is presented in Figure 9.25. A single-phase inverter and a sinusoidal reference are considered. When the measured current crosses a band around the reference current, the switch devices change (it turns ON or OFF) and the current returns to the band.

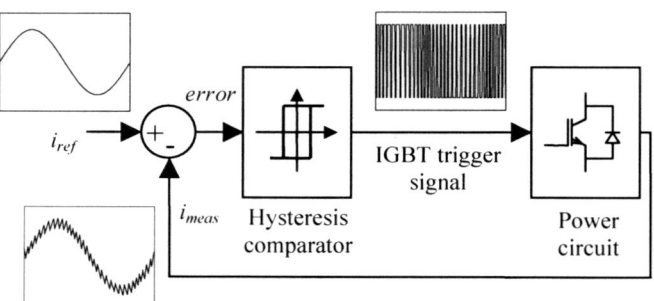

Figure 9.35. Control scheme of hysteresis-band method

The control performance is next: the currents are measured and compared with the reference ones; the error signals are the inputs of a hysteresis comparator; their outputs are the IGBT trigger pulses. If the measured current is bigger (a half of the band value) than the reference one, it is necessary to commute the corresponding switch devices to get a negative inverter output voltage; this voltage allows to be decreased the output current, and it goes to the reference current. On the other hand, if the measured current is less (a half of the band value) than the reference one, the switch devices commute to obtain a positive inverter output voltage; the output current increases, and it goes to the reference current. As a result, the output current will be in a band around the reference current.

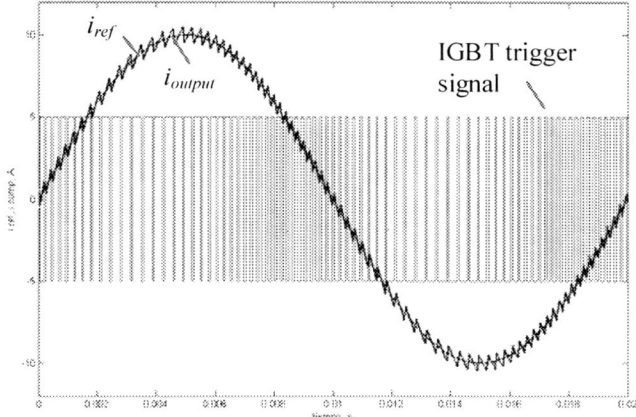

Figure 9.36. The main waveforms in a hysteresis-band control method

The implementation of the hysteresis-band control is not expensive and the dynamical answer is excellent. It allows a fast current control. Therefore, in this control it is not possible to fix the commutation frequency. This disadvantage is not ever critical, and this method is one of the more extended closed-loop current controls.

In the triangular-carrier method, the pulse sequences of the PWM control are calculated by comparing the current error signal with a triangular carrier, as Figure 9.37 shows. It is usual to include, e.g., a proportional integral gain (PI control) to process the error signal. The k_p and k_i parameters determine the stationary error and the dynamical answer of the control.

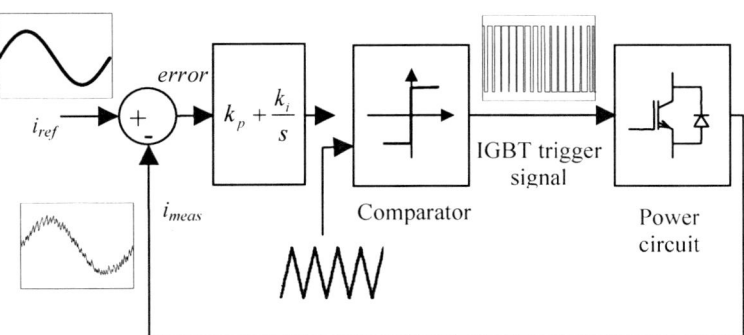

Figure 9.37. Triangular carrier PWM control of a single-phase inverter

Figure 9.38 shows the reference current, the output current and the IGBT trigger pulses with a single-phase inverter controlled by this method.

This method uses a triangular wave of a fixed frequency. So, the converter device commutation frequency is constant. In contrast to this advantage, there are some disadvantages. There are amplitude and phase errors in the output current, and there are some intervals where the converter state corresponds to a null output voltage.

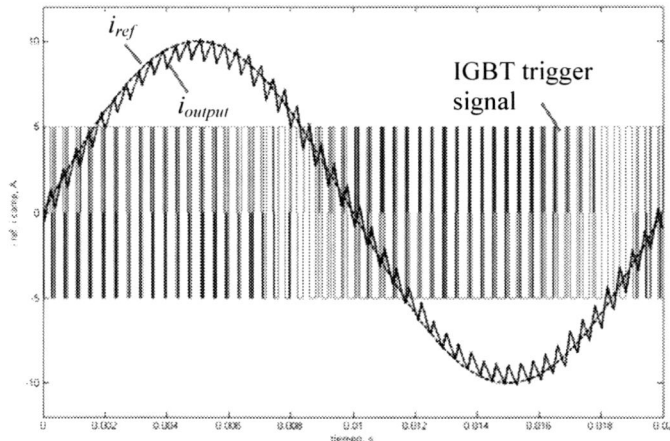

Figure 9.38. The main waveforms in the triangular-carrier current-control method

In any modulation method, the APLC output current will include no desired high frequencies, because the switch device's commutation frequency is not high enough. So, it is necessary to design an APLC output passive filter to eliminate these high-frequency components. This will be considered in Section 9.5.

9.4 Strategies for Load Static Compensation

The election of strategy control is essential to get the desired compensation aim. Many strategies in the time domain and in the frequency domain have been proposed. Some of these are modifications of the Akagi $p-q$ theory, [34, 47–50]. This theory is based on a coordinates transformation, from $a-b-c$ (or 1–2–3) axes to new $\alpha-\beta-0$ axes. In these coordinates, the compensation current references are calculated, as Section 9.4.1 will describe. Other control formulations used are the modified $p-q$ and the $d-q$ theory. The latter will be presented in Section 9.4.2.

9.4.1 Instantaneous Reactive Power Theory

In 1983, Akagi, Kanazawa and Nabae introduced the instantaneous reactive power theory, the so-called $p-q$ theory, [51–56]. The aim was to find an effective strategy to compensate nonlinear loads using active power filters. Initially, it was applied to balanced three-phase, three-wire systems, and then it was extended to unbalanced four-wire systems.

This theory defines the power terms in new $\alpha-\beta-0$ axes, where $\alpha\beta$ is coplanar to 123 axes and the 0 axis is orthogonal to both. In $\alpha\beta$ coordinates, the power terms are the instantaneous real and imaginary powers, respectively. The load current can be expressed in these real and imaginary components and the source and compensation currents too. The aim of p–q control strategy is to obtain a constant source power after the compensation with APLCs.

Besides, it is necessary to introduce the 0-axis to consider four-wire systems. The new power term is the instantaneous zero-sequence real power. So, an additional compensation target is to null the neutral current. A mandatory constraint of the static compensation is that the active power of the conditioner will be null.

The zero-sequence voltage and current components are:

$$e_0 = \frac{u_1 + u_2 + u_3}{\sqrt{3}}, \quad (9.22)$$

$$i_0 = \frac{i_1 + i_2 + i_3}{\sqrt{3}}. \quad (9.23)$$

The 123 system is transformed to $0\alpha\beta$ coordinates as follows:

$$\begin{bmatrix} e_0 \\ e_\alpha \\ e_\beta \end{bmatrix} = \sqrt{\frac{2}{3}} \begin{bmatrix} \frac{1}{\sqrt{2}} & \frac{1}{\sqrt{2}} & \frac{1}{\sqrt{2}} \\ 1 & -\frac{1}{2} & -\frac{1}{2} \\ 0 & \frac{\sqrt{3}}{2} & -\frac{\sqrt{3}}{2} \end{bmatrix} \begin{bmatrix} u_1 \\ u_2 \\ u_3 \end{bmatrix}, \quad (9.24)$$

$$\begin{bmatrix} i_0 \\ i_\alpha \\ i_\beta \end{bmatrix} = \sqrt{\frac{2}{3}} \begin{bmatrix} \frac{1}{\sqrt{2}} & \frac{1}{\sqrt{2}} & \frac{1}{\sqrt{2}} \\ 1 & -\frac{1}{2} & -\frac{1}{2} \\ 0 & \frac{\sqrt{3}}{2} & -\frac{\sqrt{3}}{2} \end{bmatrix} \begin{bmatrix} i_1 \\ i_2 \\ i_3 \end{bmatrix}. \quad (9.25)$$

In a four-wire system, the neutral current is the sum of three line currents:

$$i_N = i_1 + i_2 + i_3 = i_0\sqrt{3}. \quad (9.26)$$

So, the instantaneous real power is:

$$p(t) = e_a i_a + e_b i_b + e_c i_c = e_0 i_0 + e_\alpha i_\alpha + e_\beta i_\beta. \quad (9.27)$$

The p–q theory defines two instantaneous real powers, p_0 and $p_{\alpha\beta}$, and one instantaneous imaginary power, $q_{\alpha\beta}$.

$$\begin{bmatrix} p_0 \\ p_{\alpha\beta} \\ q_{\alpha\beta} \end{bmatrix} = \begin{bmatrix} e_0 & 0 & 0 \\ 0 & e_\alpha & e_\beta \\ 0 & -e_\beta & e_\alpha \end{bmatrix} \begin{bmatrix} i_0 \\ i_\alpha \\ i_\beta \end{bmatrix} = [P_0] \begin{bmatrix} i_0 \\ i_\alpha \\ i_\beta \end{bmatrix}. \quad (9.28)$$

The inverse matrix $[P_0]^{-1}$ allows an expression for load currents be written as a function of defined power terms,

$$\begin{bmatrix} i_0 \\ i_\alpha \\ i_\beta \end{bmatrix} = \frac{1}{e_0 e_{\alpha\beta}^2} \begin{bmatrix} e_{\alpha\beta}^2 & 0 & 0 \\ 0 & e_0 e_\alpha & -e_0 e_\beta \\ 0 & e_0 e_\beta & e_0 e_\alpha \end{bmatrix} \begin{bmatrix} p_0 \\ p_{\alpha\beta} \\ q_{\alpha\beta} \end{bmatrix}, \text{where } e_{\alpha\beta}^2 = e_\alpha^2 + e_\beta^2. \quad (9.29)$$

The instantaneous currents in $0\alpha\beta$ coordinates are:

$$i_0 = \frac{1}{e_0} p_0,$$

$$i_\alpha = \frac{1}{e_{\alpha\beta}^2} e_\alpha p_{\alpha\beta} + \frac{1}{e_{\alpha\beta}^2} \left(-e_\beta q_{\alpha\beta}\right) = i_{\alpha p} + i_{\alpha q}, \quad (9.30)$$

$$i_\beta = \frac{1}{e_{\alpha\beta}^2} e_\beta p_{\alpha\beta} + \frac{1}{e_{\alpha\beta}^2} \left(e_\alpha q_{\alpha\beta}\right) = i_{\beta p} + i_{\beta q}.$$

where:

i_0 is the instantaneous zero-sequence current,
$i_{\alpha p}$ is the instantaneous active current of the α-axis,
$i_{\beta p}$ is the instantaneous active current of the β-axis,
$i_{\alpha q}$ is the instantaneous reactive current of the α-axis,
$i_{\beta q}$ is the instantaneous reactive current of the β-axis.

Now, it is possible to calculate the instantaneous power components of the three axes.

$$\begin{bmatrix} p_0 \\ p_\alpha \\ p_\beta \end{bmatrix} = \begin{bmatrix} e_0 i_0 \\ 0 \\ 0 \end{bmatrix} + \begin{bmatrix} 0 \\ e_\alpha i_\alpha \\ e_\beta i_\beta \end{bmatrix} = \begin{bmatrix} e_0 i_0 \\ 0 \\ 0 \end{bmatrix} + \begin{bmatrix} 0 \\ e_\alpha i_{\alpha p} \\ e_\beta i_{\beta p} \end{bmatrix} + \begin{bmatrix} 0 \\ e_\alpha i_{\alpha q} \\ e_\beta i_{\beta q} \end{bmatrix}$$
$$= \begin{bmatrix} p_0 \\ 0 \\ 0 \end{bmatrix} + \begin{bmatrix} 0 \\ p_{\alpha p} \\ p_{\beta p} \end{bmatrix} + \begin{bmatrix} 0 \\ p_{\alpha q} \\ p_{\beta q} \end{bmatrix}. \quad (9.31)$$

So, the instantaneous real power is:

$$p(t) = p_0(t) + p_\alpha(t) + p_\beta(t) = p_0(t) + p_{\alpha p}(t) + p_{\beta p}(t) + p_{\alpha q}(t) + p_{\beta q}(t)$$
$$= e_0 i_0 + \frac{e_\alpha^2}{e_\alpha^2 + e_\beta^2} p_{\alpha\beta} + \frac{e_\beta^2}{e_\alpha^2 + e_\beta^2} p_{\alpha\beta} + \frac{-e_\alpha e_\beta}{e_\alpha^2 + e_\beta^2} q_{\alpha\beta} + \frac{e_\alpha e_\beta}{e_\alpha^2 + e_\beta^2} q_{\alpha\beta}. \quad (9.32)$$

The currents and powers in 123 axes can be calculated from the $0\alpha\beta$ terms through the inverse transformation. If [T] is the transformation matrix from 123 to $0\alpha\beta$ axes, it is possible to write:

$$\begin{bmatrix} i_1 \\ i_2 \\ i_3 \end{bmatrix} = [T]^{-1} \begin{bmatrix} i_0 \\ 0 \\ 0 \end{bmatrix} + [T]^{-1} \begin{bmatrix} 0 \\ i_{\alpha p} \\ i_{\beta p} \end{bmatrix} + [T]^{-1} \begin{bmatrix} 0 \\ i_{\alpha q} \\ i_{\beta q} \end{bmatrix} = \begin{bmatrix} i_{10} \\ i_{20} \\ i_{30} \end{bmatrix} + \begin{bmatrix} i_{1p} \\ i_{2p} \\ i_{3p} \end{bmatrix} + \begin{bmatrix} i_{1q} \\ i_{2q} \\ i_{3q} \end{bmatrix}, X_1,\ldots,X_r \quad (9.33)$$

where:

$$[T]^{-1} = \sqrt{\frac{2}{3}} \begin{bmatrix} \frac{1}{\sqrt{2}} & 1 & 0 \\ \frac{1}{\sqrt{2}} & -\frac{1}{2} & \frac{\sqrt{3}}{2} \\ \frac{1}{\sqrt{2}} & -\frac{1}{2} & -\frac{\sqrt{3}}{2} \end{bmatrix}. \quad (9.34)$$

In Equation 9.33, the different instantaneous currents, in 123 coordinates are presented. They are, respectively, the: zero sequence, active and reactive current. Besides:

$$i_{10} = i_{20} = i_{30} = \frac{i_0}{\sqrt{3}}. \quad (9.35)$$

The respective instantaneous powers are:

$$\begin{bmatrix} p_1 \\ p_2 \\ p_3 \end{bmatrix} = \begin{bmatrix} e_1 i_{10} \\ e_2 i_{20} \\ e_3 i_{30} \end{bmatrix} + \begin{bmatrix} e_1 i_{1p} \\ e_2 i_{2p} \\ e_3 i_{3p} \end{bmatrix} + \begin{bmatrix} e_1 i_{1q} \\ e_2 i_{2q} \\ e_3 i_{3q} \end{bmatrix} = \begin{bmatrix} p_{10} \\ p_{20} \\ p_{30} \end{bmatrix} + \begin{bmatrix} p_{1p} \\ p_{2p} \\ p_{3p} \end{bmatrix} + \begin{bmatrix} p_{1q} \\ p_{2q} \\ p_{3q} \end{bmatrix}. \quad (9.36)$$

All defined power terms ($p_{\alpha\beta}(t)$, $p_0(t)$ and $q_{\alpha\beta}(t)$), can be expressed as the sum of a constant term, equal to its average value in a period, and a variable term. The first one is designated with a capital latter and the second one with a "~" symbol over the corresponding term.

$$p_{\alpha\beta}(t) = P_{\alpha\beta} + \tilde{p}_{\alpha\beta}(t), \quad (9.37)$$

$$p_0(t) = P_0 + \tilde{p}_0(t), \quad (9.38)$$

$$q_{\alpha\beta}(t) = Q_{\alpha\beta} + \tilde{q}_{\alpha\beta}(t). \quad (9.39)$$

When the compensation aim is to get a constant source power ($P_0+P_{\alpha\beta}$), the basic equations for the compensation are:

$$p_{C0}(t) = p_{L0}(t) - P_{L0} = \tilde{p}_{L0}(t), \tag{9.40}$$

$$p_{C\alpha\beta}(t) = p_{L\alpha\beta}(t) - P_{L\alpha\beta} = \tilde{p}_{L\alpha\beta}(t), \tag{9.41}$$

$$q_{C\alpha\beta}(t) = q_{L\alpha\beta}(t). \tag{9.42}$$

where the subscript "C" is referred to the compensator, and the subscript "L" is referred to the load. These equations define the power terms. So, the compensation currents can be calculated in $0\alpha\beta$ coordinates.

$$\begin{bmatrix} i_{C0} \\ i_{C\alpha} \\ i_{C\beta} \end{bmatrix} = \frac{1}{e_0 e_{\alpha\beta}^2} \begin{bmatrix} e_{\alpha\beta}^2 & 0 & 0 \\ 0 & e_0 e_\alpha & -e_0 e_\beta \\ 0 & e_0 e_\beta & e_0 e_\alpha \end{bmatrix} \begin{bmatrix} \tilde{p}_{L0}(t) \\ \tilde{p}_{L\alpha\beta}(t) \\ q_{L\alpha\beta}(t) \end{bmatrix} \text{ where } e_{\alpha\beta}^2 = e_\alpha^2 + e_\beta^2. \tag{9.43}$$

Finally, in 123 coordinates, the compensation currents are:

$$\begin{bmatrix} i_{C1} \\ i_{C2} \\ i_{C3} \end{bmatrix} = \sqrt{\frac{2}{3}} \begin{bmatrix} \frac{1}{\sqrt{2}} & 1 & 0 \\ \frac{1}{\sqrt{2}} & -\frac{1}{2} & \frac{\sqrt{3}}{2} \\ \frac{1}{\sqrt{2}} & -\frac{1}{2} & -\frac{\sqrt{3}}{2} \end{bmatrix} \begin{bmatrix} i_{C0} \\ i_{C\alpha} \\ i_{C\beta} \end{bmatrix}. \tag{9.44}$$

It is possible to modify the above strategy to eliminate the neutral current. The next equation gives the neutral compensation for any supply voltage and any load,

$$i_{c0}(t) = i_{L0}(t). \tag{9.45}$$

The load instantaneous real power is:

$$p(t) = p_L(t) = p_{L0}(t) + p_{L\alpha\beta}(t). \tag{9.46}$$

The source power average value will be equal to a period average of the load instantaneous real power.

$$p_S(t) = P_S = \int_T p_L(t)dt = P_L = P. \tag{9.47}$$

where the subscript "S" is referred to the source magnitude. Besides, the load average power can be divided into two terms.

$$P = P_L = P_{L0} + P_{L\alpha\beta}. \tag{9.48}$$

For this,

$$p_{S\alpha\beta}(t) = P_{L0} + P_{L\alpha\beta}. \quad (9.49)$$

And the power relationship is satisfied:

$$p_L(t) = p_S(t) + p_C(t). \quad (9.50)$$

From the above equations, it is possible to write:

$$P_{L0}(t) + P_{L\alpha\beta} + \tilde{p}_{L\alpha\beta}(t) = P_{L0} + P_{L\alpha\beta} + p_{C0}(t) + p_{C\alpha\beta}(t)$$
$$= P_{L0} + P_{L\alpha\beta} + p_{L0}(t) + p_{c\alpha\beta}(t), \quad (9.51)$$

$$p_{c\alpha\beta}(t) = \tilde{p}_{L\alpha\beta}(t) - P_{L0}. \quad (9.52)$$

So, the completed control strategy is synthesised in the following equations:

$$i_{c0}(t) = i_{L0}(t),$$
$$p_{c\alpha\beta}(t) = \tilde{p}_{L\alpha\beta}(t) - P_{L0}, \quad (9.53)$$
$$q_{c\alpha\beta}(t) = q_{L\alpha\beta}(t).$$

9.4.2 Instantaneous d-q Theory

In d–q theory, the voltages and currents are expressed in $dq0$ axes, where the 0-axis is the same as that in $0\alpha\beta$ axes, and the d- and q-axes are α- and β-axes rotated by an angle $\theta = \omega t$, where ω is a constant rotational speed. This theory is usually applied to electrical machines, but it can be applied to active power filter control too [57–58].

The Park or d-q transformation allows a general three-phase electrical system, including unbalance and distortion conditions to be studied. The transformation from 123 to $dq0$ axes, referred to currents, is the following,

$$\begin{bmatrix} i_d \\ i_q \\ i_0 \end{bmatrix} = \sqrt{\frac{2}{3}} \begin{bmatrix} \cos\theta & \cos(\theta-120) & \cos(\theta+120) \\ -\sin\theta & -\sin(\theta-120) & -\sin(\theta+120) \\ \frac{1}{\sqrt{2}} & \frac{1}{\sqrt{2}} & \frac{1}{\sqrt{2}} \end{bmatrix} \begin{bmatrix} i_1 \\ i_2 \\ i_3 \end{bmatrix}. \quad (9.54)$$

The mentioned $i_d(t)$ and $i_q(t)$ signals are time variable, and their average values are I_d and I_q, respectively. So, each one can be divided into two terms, a constant value and a variable component.

$$i_d(t) = I_d + \tilde{i}_d(t), \tag{9.55}$$

$$i_q(t) = I_q + \tilde{i}_q(t). \tag{9.56}$$

The constant I_d and I_q components are, respectively, the fundamental harmonic positive sequence-of $i_d(t)$ and $i_q(t)$. The other components are the fundamental harmonic negative sequence and the rest of the harmonics of $i_d(t)$ and $i_q(t)$. The average values of these components are null [57].

This current decomposition is used to generate the compensation current reference in shunt active power filter applications. So, the source current will include the constant components I_d and I_q, and the shunt active conditioner will inject the variable components. In the practical implementation, it is possible to separate both parts with a low-pass filter. In 123 coordinates, the source currents will be:

$$\begin{bmatrix} i_{S1} \\ i_{S2} \\ i_{S3} \end{bmatrix} = \sqrt{\frac{2}{3}} \begin{bmatrix} \cos\theta & -\sin\theta & \frac{1}{\sqrt{2}} \\ \cos(\theta-120) & -\sin(\theta-120) & \frac{1}{\sqrt{2}} \\ \cos(\theta+120) & -\sin(\theta+120) & \frac{1}{\sqrt{2}} \end{bmatrix} \begin{bmatrix} I_d \\ I_q \\ 0 \end{bmatrix}, \tag{9.57}$$

and the compensation currents will be:

$$\begin{bmatrix} i_{C1} \\ i_{C2} \\ i_{C3} \end{bmatrix} = \begin{bmatrix} i_{L1} \\ i_{L2} \\ i_{L3} \end{bmatrix} - \begin{bmatrix} i_{S1} \\ i_{S2} \\ i_{S3} \end{bmatrix} = \sqrt{\frac{2}{3}} \begin{bmatrix} \cos\theta & -\sin\theta & \frac{1}{\sqrt{2}} \\ \cos(\theta-120) & -\sin(\theta-120) & \frac{1}{\sqrt{2}} \\ \cos(\theta+120) & -\sin(\theta+120) & \frac{1}{\sqrt{2}} \end{bmatrix} \begin{bmatrix} \tilde{i}_d(t) \\ \tilde{i}_q(t) \\ i_0(t) \end{bmatrix}. \tag{9.58}$$

In conclusion, this theory allows sinusoidal and balanced currents to be obtained, but independent of a symmetrical or distorted voltage supply. Besides, the APLC will be null only if the voltages are sinusoidal and balanced [58].

9.5 Practical Design

The control block calculates the compensation current references according to the APLC control strategy. The power circuit follows these references using a PWM method. The switching frequency of converter electronic devices has to be high to follow any reference. The limits are fixed by the actual technology. These commutations force no desired frequencies into the compensation currents, and consequently into the source current. So, it is necessary to include an inductance and/or a LC passive filter in the output converter to eliminate (or to mitigate) these no desired high frequencies. The parameters of these passive elements will condition the APLC performance.

The APLC design performance can be simulated via a software platform [59]. This process will allow the different model values to be contrasted and the stationary and dynamical APLC performance to be adjusted. This optimises the final implementation. In this case, the simulation probes have been developed in MATLAB® and Simulink® software. The Power System Blockset includes electrical elements to configure the source, the load and the filter used in the active compensation.

In Section 9.6, the laboratory prototype of a shunt active conditioner will be presented. The control block has been implemented in a DSP platform (DS1103 PPC controller board). This hardware includes a 400–MHz microprocessor where the control program is running. The outputs are the trigger signals of converter IGBT gates.

9.5.1 Component-design Considerations

As was mentioned above, with any modulation method, the compensation current includes no desired high-frequency harmonic components. A passive filter in the APLC output in used to filter these components [60].

The APLC power circuit presents a quadrangular voltage in the output. These positive and negative voltages imply an increasing or decreasing output current when the load is inductive. It allows the reference current calculated by the control block to b followed. The output current noise depends on commutation frequency. Today, the IGBT commutation frequency is around 20–25 kHz.

On the one hand, an output inductance L reduces some high-frequency components of the output current. The value of L and the voltage value of the converter DC side capacitors fix the gradient of the compensation current. So, they determine the active conditioner dynamical answer. The scheme of Figure 9.39 shows the APLC network connection with an output inductance L. The DC output voltage of the APLC power circuit is V_{DC} when the IGBTs voltage drop is not considered. A sinusoidal network waveform has $\sqrt{2}\ V_{NET}$ volts as its amplitude.

In Figure 9.39, i_S, i_L and i_C are the source, the load and the compensation current, respectively. The compensation current gradient is:

$$\frac{d i_C}{dt} = \frac{1}{L}\left(V_{DC} - \sqrt{2}\ V_{NET}\right). \tag{9.59}$$

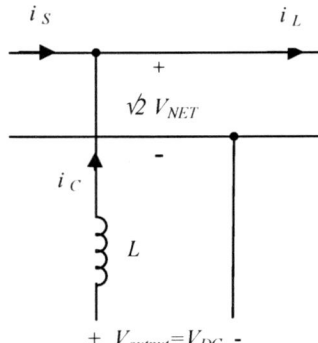

Figure 9.39. APLC network connection through an output inductance

The compensation current gradient increases when the converter DC voltage is high or the output inductance low. In this case, the dynamical answer is fast but the output current ripple is high. If the converter DC voltage is low or the inductance is high, the APLC answer will be poor but the current ripple will be minor. So, an equilibrated election of L and V_{DC} values is necessary [61–66] Two main design considerations are the following:

- To limit the harmonic amplitude corresponding to the commutation frequency.
- To get a compensation current gradient higher than the load current gradient.

The usual DC voltage and inductance values to limit the harmonic amplitude of commutation frequency satisfy:

$$\frac{I_{fc}}{I_1} = \frac{V_{fc}}{I_1 L 2\pi fc} < 5\%, \qquad (9.60)$$

$$V_{DC} > \left(L\frac{di_L}{dt} + \sqrt{2}V_{RED} \right), \qquad (9.61)$$

where f_c is the commutation frequency, and V_{fc} and I_{fc} are the rms values of voltage and current corresponding to this frequency.

On the other hand, the DC capacitors are designed to limit the ripple of the DC voltage, V_{output} in Figure 9.39, to around 2%. The capacitor voltage increases (or decreases) when there is active power circulation from the network to the APLC (or from the APLC to the network). The energy increment is:

$$\frac{1}{2}C\Delta V_C^2 = \left(P_C - P_{loss} \right) \Delta t \quad \Rightarrow \quad r = \frac{\Delta V_C}{V_C} = 0.02, \qquad (9.62)$$

where P_C is the power incoming to the compensator and P_{loss} is the consumed APLC power. In Equation 9.62, P_C has been considered constant. With an adequate control strategy, the power source supplies the load active power and the compensator losses. So, the compensator average power will be null.

The output inductance does not eliminate all no desired high frequencies of the compensation current. Besides, it is necessary to include a high-pass passive filter, e.g., a LC branch. Figure 9.40 shows a nonlinear compensation current flowing to the network through a LC filter and a $n:1$ transformer. If the transformer relation is $n < 1$, it is possible to work with low voltages in the converter DC side.

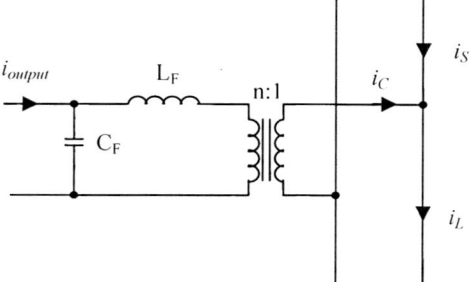

Figure 9.40. LC filter and network-connection transformer

For the high compensation current, the capacitor impedance, $1/C\omega$, has to be less than the inductance impedance, $L\omega$. So, these no desired components will flow to the capacitor and not to the network. The necessary L and C values to eliminate the commutation frequency component, f_C, will satisfy:

$$X_{C\,fc} << X_{L\,fc} . \tag{9.63}$$

If the source impedance is Z_{NET}, the equivalent impedance, referred to the APLC side, is $Z = n^2 Z_{NET}$. This network impedance will be bigger than the L_F and C_F impedances to ensure a correct filter performance.

$$X_{C\,fc} \text{ and } X_{L\,fc} << Z . \tag{9.64}$$

Figure 9.41 shows a general APLC output scheme, including the output inductance, the LC filter and the network impedance. The APLC output voltage is a quadrangular wave.

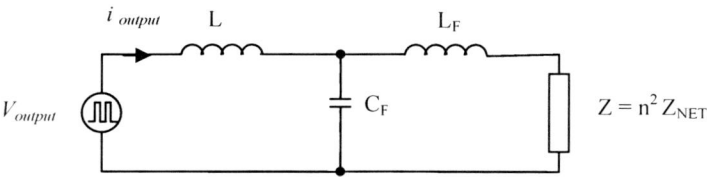

Figure 9.41. Equivalent scheme of a LC filter connected in the APLC output

The main waveforms are shown in Figure 9.42 to summarise the highpass filter performance.

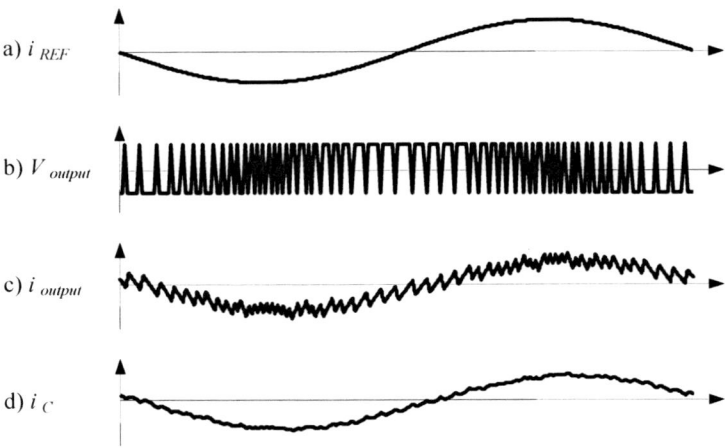

Figure 9.42. APLC output waveforms with an output inductance and a LC filter; **(a)** reference current, **(b)** output voltage, **(c)** AC output current and **(d)** compensation current

In this example, the current reference is sinusoidal, i_{REF}. The PWM control method allows the converter output voltage presented in Figure 9.42b to be obtained. The output current, i_{output}, and the final compensation current injected into the network, i_C, are presented in Figs. 9.42c, d. The quadrangular voltage is converted to a triangular current, and the LC passive filter eliminates the high frequencies.

In a practical design, it is usual to include a low resistor in series with the capacitor to limit other high-frequency currents. At these frequencies, the capacitor impedance is near to null. Besides, it is possible to place the passive LC filter in the APLC side or in the network side.

9.5.2 Simulation Analysis

In this section, a practical case is presented to check the shunt APLC design and their performance. The power system has been modelled in MATLAB® and Simulink® software with the Power System Blockset, including the power source, the nonlinear load and the shunt APLC, with the control and power blocks [65–66].

The aims of this design step are:

- To contrast the adequate performance of a shunt active power filter to compensate nonlinear loads with current harmonics.

- To analyse the source-current distortion, THD, before and after the active compensation.

- To determine the APLC dynamical performance.
- To check the efficiency of the passive filter after the selection of the passive element values.

In Figure 9.43, a general scheme of a simulated power system is presented. The nonlinear load is an unbalanced three-phase four-wire AC regulator. The APLC control inputs are the measurements of load voltages and currents (to calculate the compensation current references according to the control strategy) and the APLC compensation currents (to implement the closed-loop current control). The APLC control outputs are the trigger signals of the power-circuit IGBTs. The APLC output current is filtered by a passive step, and the compensation current is injected into the network through a transformer. A block in Figure 9.43 represents this conditioning step.

The main source and load parameters are:

$R_1 = 52\ \Omega$, $R_2 = 52 \times 1.2\ \Omega$, $R_3 = 52 \times 0.8\ \Omega$; $L_1 = L_2 = L_3 = 50\ 10^{-3}$ H;

$\alpha_1 = \alpha_2 = \alpha_3 = 60°$;

$V_{source} = 230$ V, $f = 50$ Hz.

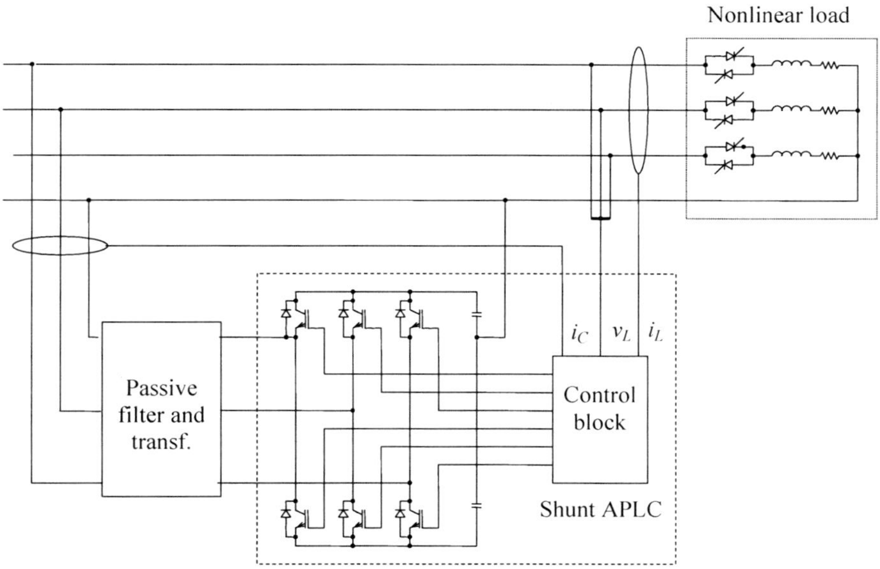

Figure 9.43. Three-phase, four-wire power system, with a nonlinear load compensated by a shunt APLC

Figure 9.44 shows the corresponding Simulink® block diagram. A power switch allows turning the shunt APLC ON/OFF. The values of passive filter elements and DC capacitors are obtained from Eqs. 9.60–9.64.

C_{dc} = 2200 µF; V_{dc} = 500 V;   *Transf. relation* 1:2;
L_{output} = 17 mH; C_F = 20 mF; L_F = 5 mH.

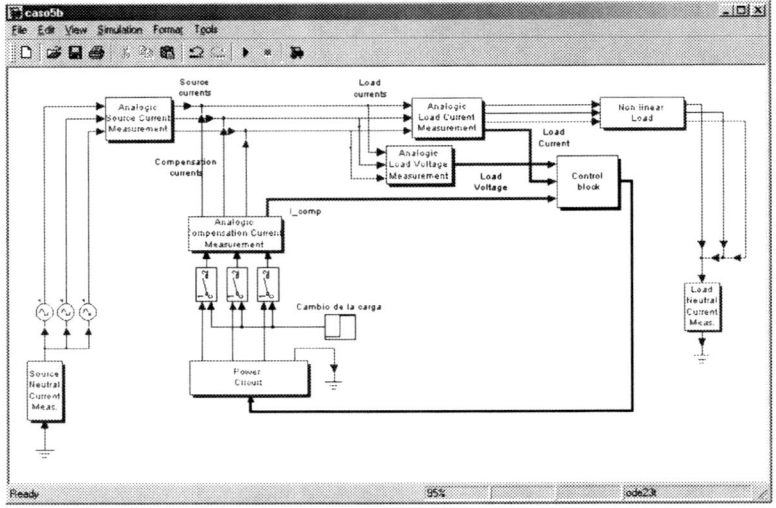

Figure 9.44. Simulink® block diagram of power system

The main simulation results are presented in the next figures. At time t=0.1 s, the shunt active filter is connected to the network, and at time t=0.2 s, an instantaneous change of load current is applied to study the dynamical performance of the compensator. Figure 9.45 shows the load voltage and current of one phase of the AC regulator.

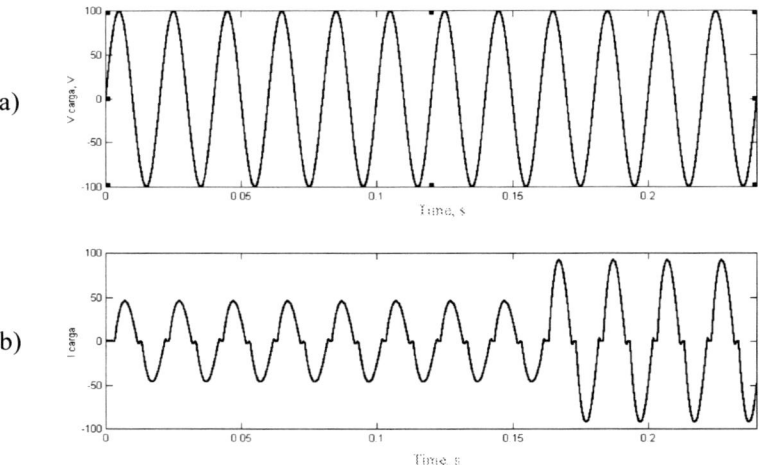

Figure 9.45. One phase of load; **(a)** load voltage, **(b)** load current.

Active Power-line Conditioners 275

The compensation current and the source current are presented in Figure 9.46. Before the active filter connection, the source current is equal to the load current because the compensation current is null.

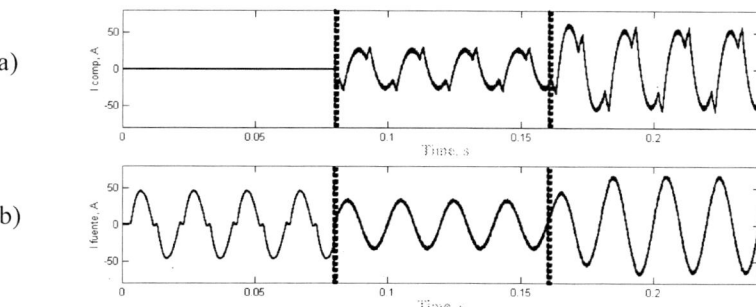

Figure 9.46 Main waveforms of the AC regulator compensation, referred to one phase; **(a)** compensation current, **(b)** source current

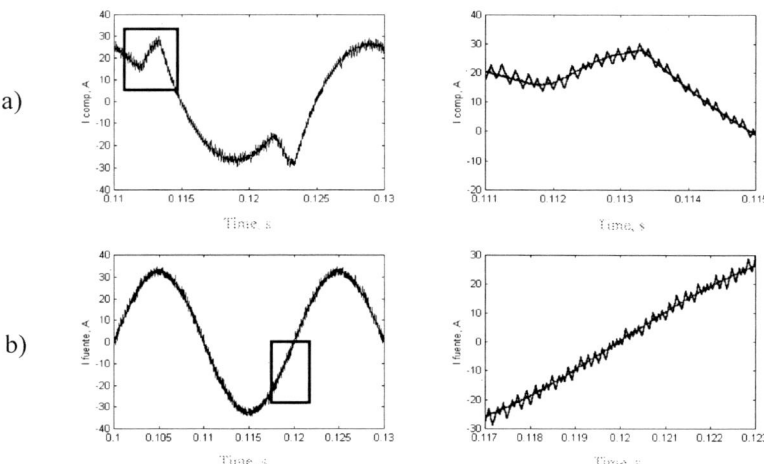

Figure 9.47. Some waveforms and detailed zooms; **(a)** compensation current and compensation current reference, **(b)** source current and control target

The control strategy allows the harmonic load currents, the reactive power and the inverse and zero sequence load currents to be compensated. So, after the compensation, the source currents have low distortion and a similar rms value, and the neutral source current is null. These results are illustrated in Figure 9.48.

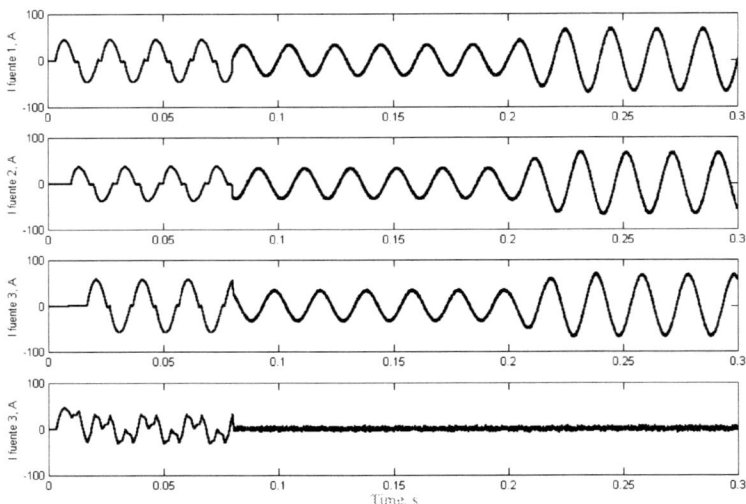

Figure 9.48. Three phases and neutral source currents

The initial and final source current distortions are similar in all the phases.

Before the compensation, THD of i_S = 17.67.

After the compensation, THD of i_S = 5.22.

9.6 APLC Prototyped Trough PC Acquisition Board

Nowadays, high–performance DSP-based shunt APLCs have been implemented, [67–71]. In this section, a dSPACE acquisition and control board installed in a PC have been used to control a shunt active power filter. This commercial hardware/software includes a microprocessor where the control program is running. The designed Simulink® control program is converted to C language and it is compiled to obtain a final executable program. This platform allows a real–time control. Figure 9.49 presents a scheme of the compensation system.

The real time interface, RTI, is a MATLAB® toolbox that makes the interface with dSPACE to obtain a real–time control. The dSPACE software includes some Mlib/Mtrace libraries components that allow access to dSPACE hardware from MATLAB®. The Simulink® control programs developed in a simulation platform can be completed with input and output components to make a real–time control block. The SystemBuild toolbox included in dSPACE converts the designed simulation control to C language and makes the executable program. This program is executed in the microprocessor of the acquisition and control boar.

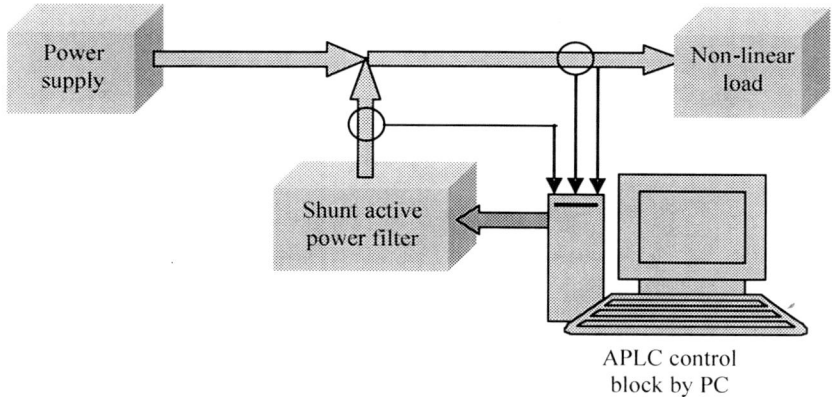

Figure 9.49. General scheme of the experimental system

Finally, the Controldesk software allows a graphical virtual desk to be configured, where the experiment can be supervised and controlled. It is possible to register the experimental results or to modify the desired control parameters during the experiment.

9.6.1 Experimental System

The experimental system is described now [72–73]. The source and load parameters are:

- Power supply: electrical network with line voltage $V_L = 400$ V and a Dyn11 distribution transformer, with accessible neutral wire.

- Issued load: three-phase AC regulator with two antiparallel SCRs and a RL branch. The references are:

    Semikron SCRs model MSS 40-800, 800 V, 55 A.

    Snubber of $C_S = 1$ µF, 250 V, and $R_S = 1$ kΩ, 15 W.

    Trigger modules of United Automation Ltd, model FC11AL

The load is that studied in the above section. Figure 9.50 shows a graphical scheme of the three-phase regultator and Figure 9.51 presents some details of the load configuration. In the next section the load parameter values will be presented.

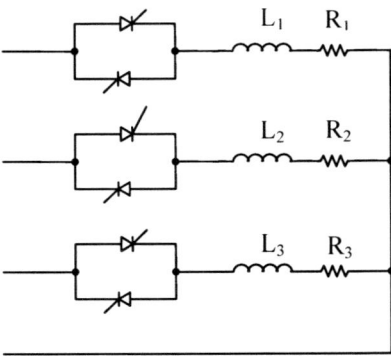

Figure 9.50. Scheme of a three-phase AC regulator

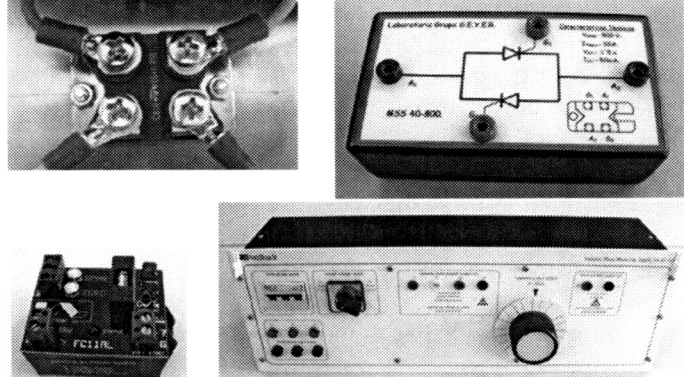

Figure 9.51. Some details of compensated load. From top to bottom, and from left to right: Connection of SCR module, SCRs block, SCR trigger block and trigger power supply

The main APLC control block inputs are the measures of voltage and current load to calculate the compensation current references. The measures of the compensation current allow the PWM control method to be executed to follow the reference ones. Besides, the DC voltage, the source current and the APLC output currents are measured to check the APLC design performance. The Hall-effect sensors used to measure the voltages and current are the following:

- Voltage sensor AC/DC LEM LV 25–600, with galvanic isolation, voltage supply –15 V/0 (GND)/+15 V, and configured to measure 0–7 A (maximum 14 A), with 35 mA output current maximum.

- Current sensor AD/DC LEM LA-35 NP, with voltage supply –15 V/0 (GND)/+ 15 V, configured to measure 0–600 V (maximum 900 V), with 25 mA output current maximum.

The voltage and current sensors used are shown in Figure 9.52. In the proposed experimental system, 9 voltage sensors and 12 current sensors have been used. The BNC connections simplify the developed cases.

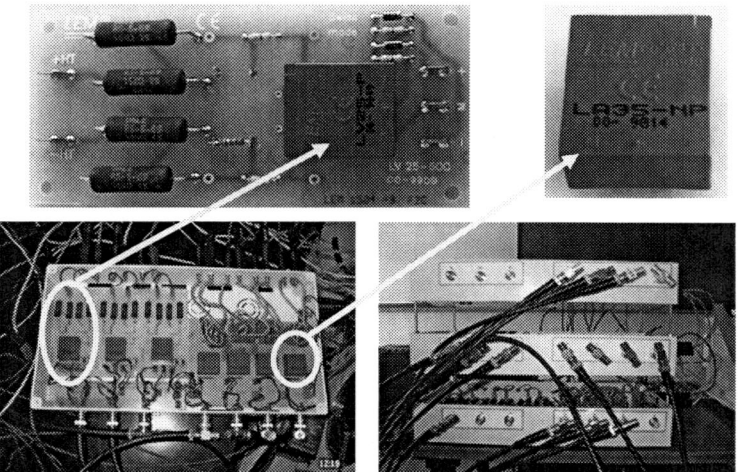

Figure 9.52. Upper left, current sensor; upper right, voltage sensor; bottom, sensor block with input and output BNC connections

The shunt APLC includes control and power circuit blocks. On the one hand, the control has been implemented in a DS1103 PCC Controller Board, which includes a microprocessor PowerPC 604e of 400 MHz. Figure 9.53 shows the acquisition and control board and their installation.

The main parameters of the DS1103 board are:

- 4 analog inputs of 16 bits, with a sample time of 4 µs.
- 4 analog inputs of 12 bits, with a sample time of 800 ns.
- Input voltage range ±10 V.
- 8 analog outputs of 14 bits with actualisation time of 5 µs.
- Output voltage range ±10 V.
- 6 input digital channels.
- Digital noise filters with maximum input frequency of 1.65 MHz.
- 4 input digital channel of 8 bits.

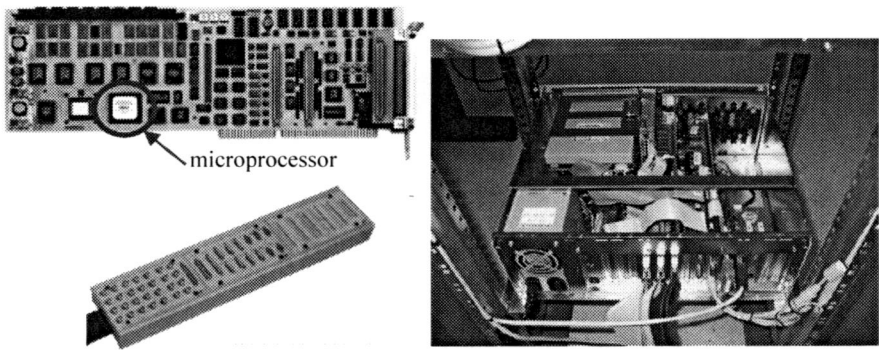

Figure 9.53. DS1103 PPC Controller Board with input/output interface, and PC installation details

The control program execution time is about 40 μs, including the input analog/digital conversion time, the execution time of the control block and the output digital/analog conversion time. So, the maximum real–time control frequency is 25 kHz in the presented case. This frequency is enough to get an adequate performance of converter IGBT devices.

As was mentioned above, it is possible to build a virtual desk to supervise and to control the APLC compensation, helped by ControlDesk software. Figure 9.54 shows the designed desk.

The supervision and control desk includes:

- Control to connect/disconnect the shunt APLC.
- Manual adjusts of SCR trigger angles.
- Figures to supervise the main experiment waveforms.
- Alarms to stop the experiment when the compensation current or the DC voltage is over a fixed limit.

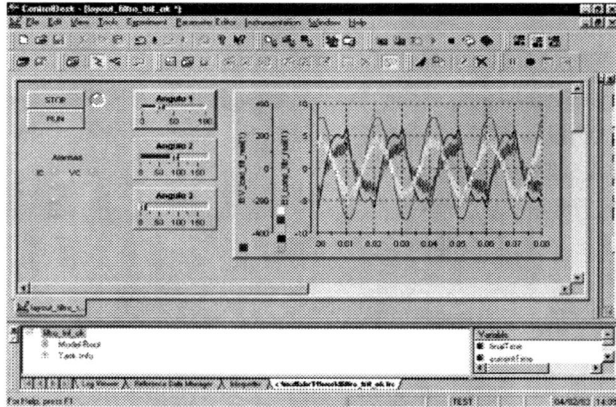

Figure 9.54. Instrumentation virtual desk developed by ControlDesk software to supervise and to control the APLC experiment

The APLC power circuit is a three-phase full-bridge inverter with 6 IGBTs and a split capacitor in the DC side. In this case, the power circuit includes an input board to adapt the IGBT trigger signals and some protection relays. Figure 9.55 shows some views of the inverter. A frontal view can be seen with the BNC connections and the trigger input board, the DC side capacitors and a detail of the IGBT branch connection.

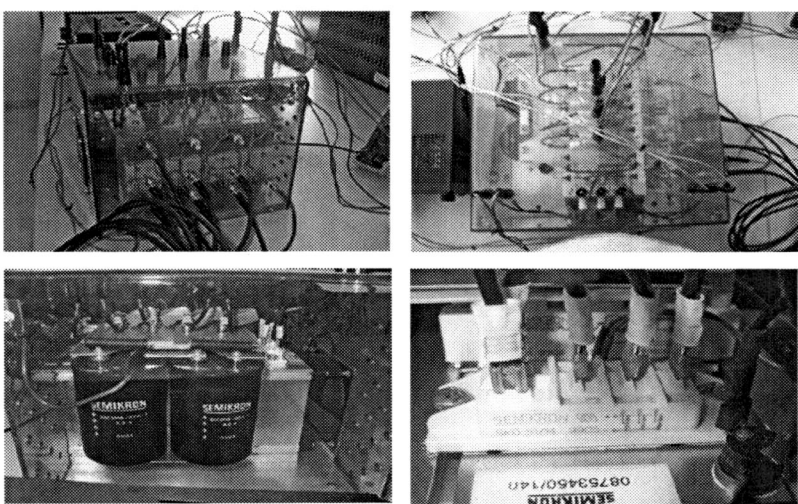

Figure 9.55. Semikron three-phase inverter and details

The power inverter is the Semikron model SKM50GB123. Its main technical characteristics are:

- IGBTs power modules SKM 50 GB 123 D; ON/OFF commutation signals, +15 V/–15 V; nominal commutation frequency, 20 kHz; nominal voltage and currents, V_N = 480 V p, I_N = 30 A p; maximum working temperature, T_{max} = 125°C.
- Trigger drivers SKHI 22 of 160 mA; drivers power supply 0/+15 V DC; CMOS logic (0 V–off, +15 V-on) through BNC connectors; inverse CMOS logic for error signals.
- Snubbers, 0.22 µF and 1600 V AC.
- Two DC electrolytic capacitors of 2200 µF, 400 V DC, 19.1 A at 40°C and 8.5 A at 85°C; three-phase rectifier SKD 51/14 for the capacitor supply; discharge resistors of 22 kΩ, 13 W, V_{max} = 534 V AC.
- Hysteresis dissipation temperature relay, ON at 50°C, OFF at 70°C; IGBTs temperature sensor, 10 mV/°C, 2.947 V at 20°C; forced ventilation, 220 V AC.

The APLC network coupling is made through an output inductance, three single-phase transformers and a passive LC filter with a resistor to reduce the high-frequency currents. An adequate transformer relation allows working at low voltage in the APLC DC side. Figure 9.56 shows a global scheme of the experimental platform and Figure 9.57 some of the installed elements. The control-block inputs are sensors 1–18.

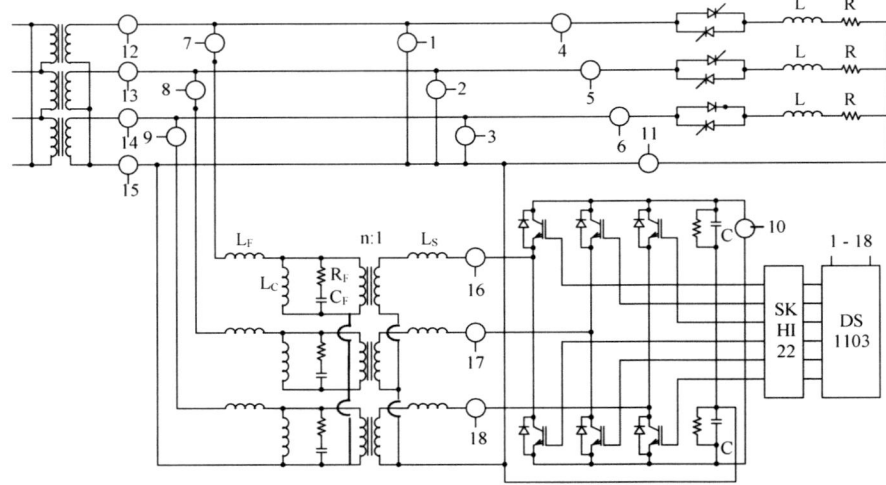

Figure 9.56. Three-phase APLC connection scheme

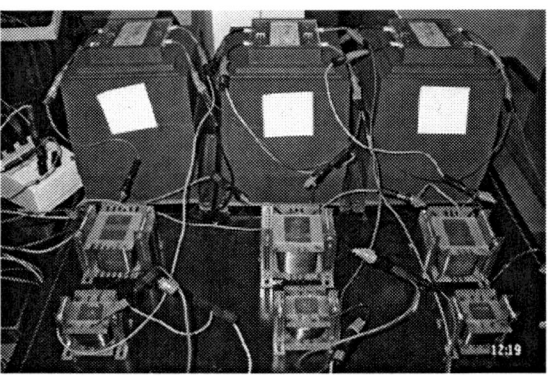

Figure 9.57. Some elements of APLC output, including the output inductance and the coupling transformers

The passive filter and the transformer characteristic of each phase are:

- Inverter output reactance, $L_S = 17$ mH, 3.7 Ω.
- Transformer, 4.5 kVA, 230/460 V (relation 1.9644:1), 50 Hz, $R_0 = 3052$ Ω, $L_0 = 7.35$ H, $R_{CC} = 1$ Ω, $L_{CC} = 3$ mH.
- Harmonic tramp capacitor, $C_T = 20$ μF, 480 V.
- Harmonic tramp resistor, $R_T = 80$ Ω (with regulation)

- Harmonic tramp reactance, $L_T = 5$ mH, 2 Ω.
- Reactive compensation inductance, $L_C = 500$ mH.

A global image of the experimental system used is shown in Figure 9.58. Some blocks are the three-phase power inverter (block 1), output inductance and coupling transformers (block 2), voltage and current sensors (block 3) and the PC interface (block 4).

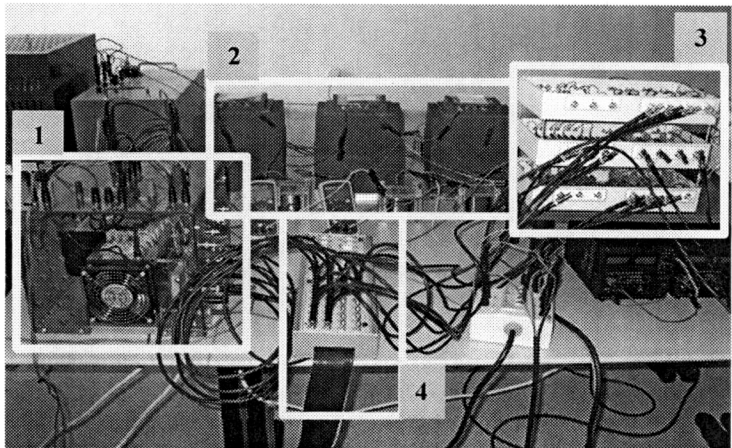

Figure 9.58. Global image of experimental APLC

9.6.2 Results of a Practical Case

The above-described experimental platform has been used to compensate a nonlinear load, a three-phase unbalanced AC regulator [74].

The system parameters are now summarized:

- Power supply: $V_{rms} = 230$ V.
- Load: a three-phase AC regulator; three branches with two SCRs models MSS 40-800 (800 V, 55 A) back-to-back, with snubbers $C_S = 1$ μF and $R_S = 1$ kΩ, and RL branch (phase 1, 52.2 Ω and 150 mH; phase 2, 52.2 Ω and 150 mH; phase 3, 52.2 Ω and 0 mH); the SCRs trigger angles are 90°, 90° and 0° respectively.
- APLC: four voltage sensors (for load voltages and DC capacitor voltage) models AD/DC LEM LA-35 NP; nine current sensors (for load, compensation and source currents) models AC/DC LEM LV 25–600; one power inverter SKM50GB123 of Semikron, with a three-phase configuration, three IGBTs branches and two capacitor in the DC side of 2200 μF; three single-phase coupling transformers, 230/460 V; passive LC filter, $L_S = 17$ mH, $C_T = 20$ μF, $R_T = 80$ Ω, $L_T = 5$ mH, $L_C = 500$ mH.

On the one hand, the stationary results of the shunt active compensation have been registered with a digital oscilloscope. Figure 9.59 presents the three-phase load voltages and currents. The voltages are balanced but the currents are unbalanced.

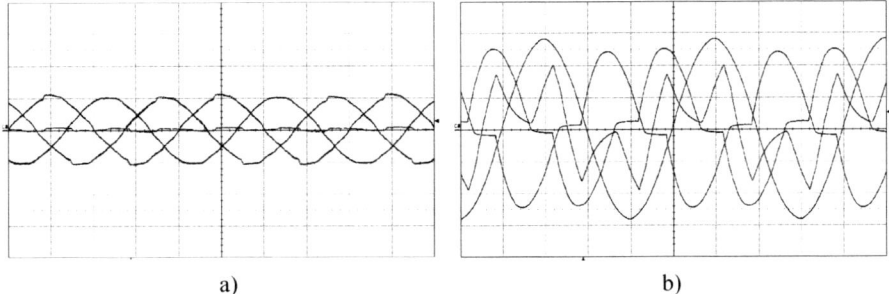

Figure 9.59. (a) Load voltages (300 V/div), (b) load currents (2.5 A/div)

When the APLC is connected, it injects the compensation currents into the system according the control strategy used. Figure 9.60a compares load voltage and current of phase 1, and Figure 9.60b compares the source current before and after the compensation. The source current became sinusoidal and inphase with the load voltage. So, the load-current harmonics and its reactive power are compensated.

Figure 9.61 shows a detail of the PWM control-method performance. The compensation current follows their reference through the application of variable pulses to upper and lower IGBT gates of the corresponding inverter branch. So, the compensation current increases or decreases around the reference signal.

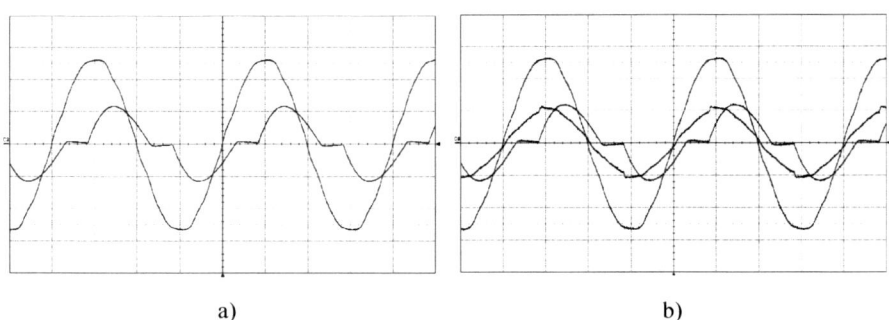

Figure 9.60. (a) Phase-1 load voltage and current (100 V/div, 5 A/div), (b) phase-1 source current before and alter the compensation (5 A/div)

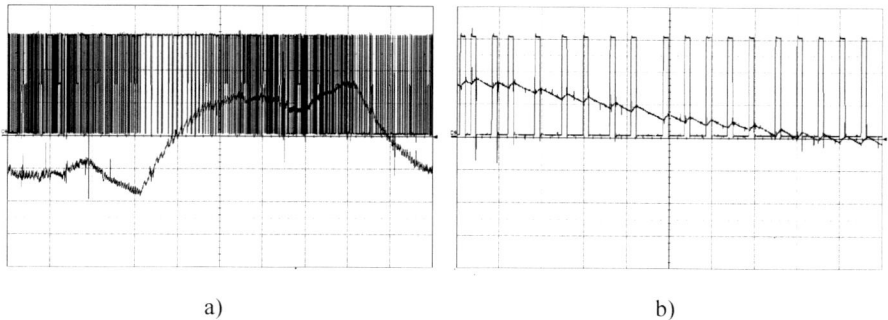

Figure 9.61. (a) Compensation current (2.5 A/div) and the corresponding upper IGBT pulses of the inverter branch; **(b)** a zoom

The compensation results of all phases are similar. Figure 9.62 shows the load currents to be compensated, and Figure 9.63 presents the compensation and source currents after the compensation. The source currents are sinusoidal and balanced. So, the neutral current is eliminated.

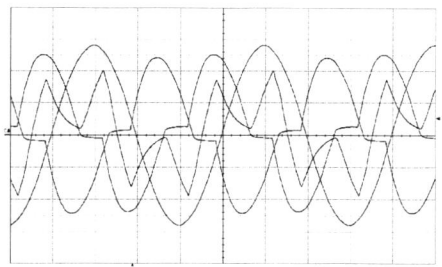

Figure 9.62. Three-phase and neutral wire load currents

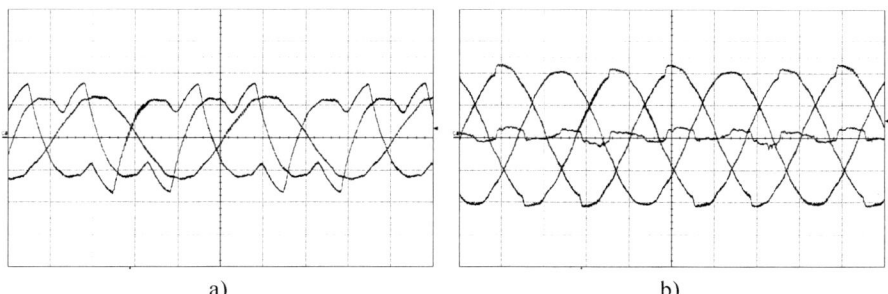

Figure 9.63. (a) Compensation currents injected by the APLC (2.5 A/div), and **(b)** source currents in the compensated system (2.5 A/div)

Table 9.5 hits the compensation results. The THDs of the source currents have been reduced.

Table 9.5. Compensation results of experimental case

	Phase	THD, i_S	cos φ	Power factor
Phase 1	Before	28.5	0.54	0.5
	After	2.5	1	0.99
Phase 2	Before	29.7	0.54	0.5
	After	2.6	1	0.99
Phase 3	Before	1.4	1	0.99
	After	1.9	1	0.99

The above results are referred to APLC stationary performance. Now, the APLC transient behavior is studied. For this, a load change is forced during the compensation. The evolution of the main waveforms is analyzed. The Controldesk software facilitates the capture of these signals using dSPACE board. Figure 9.64 shows the load change. It is a commutation of the resistor value of phase 3.

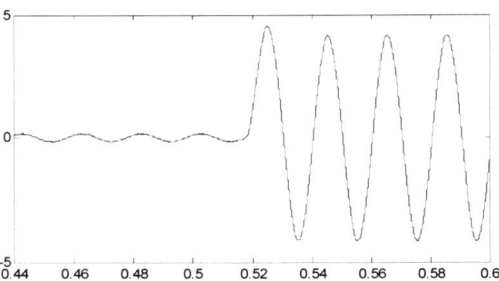

Figure 9.64. Phase-3 current evolution with a forced change

From the load voltage and current measurements, the control block calculates the compensation current references to obtain sinusoidal and balanced source currents. The evolution of a proportional quantity of calculated rms value of source currents is shown in Figure 9.65. The evolution time is around two periods. So, the active filter presents a fast dynamical answer.

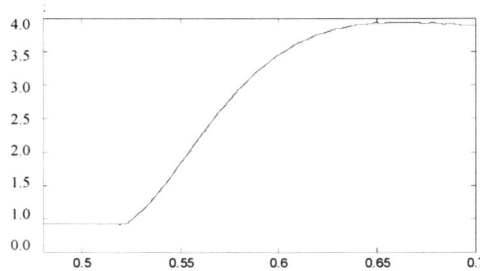

Figure 9.65. Evolution of calculated rms value of source current with a load change

Figure 9.66 shows the compensation and source current evolutions of phase 1. The transient period is not drastic.

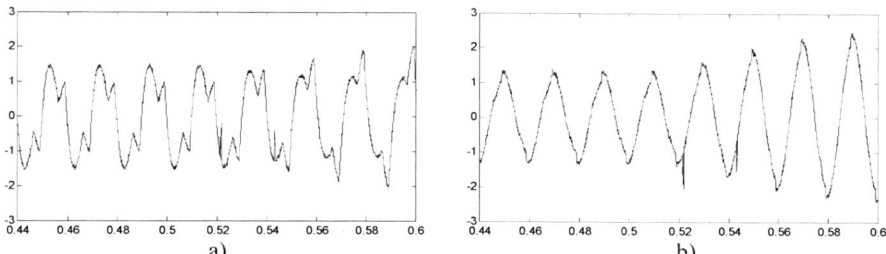

Figure 9.66. (a) Compensation current and (b) source current evolution, with a sudden load change

Finally, the inverter DC voltage evolution is presented in Figure 9.67. When the power is flowing from the network to the APLC (or *vice versa*), the DC voltage is increasing (or decreasing). With the implemented control strategy, the APLC absorbed power will be null in stationary conditions.

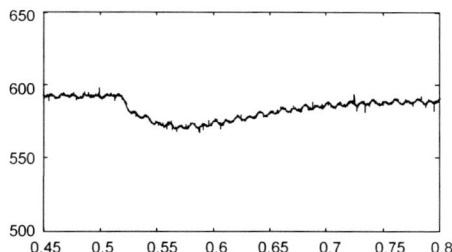

Figure 9.67. Inverter DC voltage evolution with a load change

The results allow affirming that the shunt active power-line conditioner makes an efficacy compensation of nonlinear three-phase loads, in particular when its loads present current harmonics. If the loads present voltage harmonics principally (as diode rectifiers with a capacitive branch), it will be necessary to use series or combined filters.

References

[1] Akagi. H. (1992). Trends in Active Power Line Conditioners. Proceedings of the IEEE Industrial Electronics, IECON'92, San Diego, pp. 19–24.
[2] Akagi. H. (1994). Trends in Active Power Line Conditioners. IEEE Transactions on Power Electronics, Vol. 9, No. 3, pp. 263–268.
[3] Akagi. H. (1996). New Trends in Active Filters for Power Conditioning. IEEE Transactions on Industry Applications, Vol. 32, No. 6, pp. 1312–1322.
[4] Akagi, H. (2000). Active and Hybrid Filters for Power Conditioning, ISIE'2000, Cholula, Puebla, Mexico.

[5] Peng, F. Z., Akagi, H., Nabae, A. (1988). A Novel Harmonic Power Filter, Record of PESC'88, pp: 1151–1159.
[6] Quinn, C. A., Mohan, N., Mehta, H. (1993). A Four-Wire, Current-Controlled Converter Provides Harmonic Neutralization in Three-Phase, Four-Wire Systems. Proceedings of Applied Power Electronics Conference and Exposition, APEC '93, pp. 841–846.
[7] Yanchao, J., Fei, W. (1998). 100 kVAr Generalized Active Power Filter. Conference Records of IEEE Industry Applications Conference. Thirty-Third IAS Annual Meeting, 1998, Vol. 3, pp. 2354–2359.
[8] Zhang, W., Asplund, G., Aberg, A., Jonsson, U., Loof, O. (1993). Active DC Filter for HVDC System-a Test Installation in The Konti-Skan DC Link at Lindome Converter Station", IEEE Trans. on Power Delivery, Vol. 8, No. 3, pp. 1599–1606.
[9] Montaño, J. C., Salmerón, P., (1999). Identification of Instantaneous Current Components in Three-Phase Systems. IEE Proc.-Sci. Meas. Technol., Vol 146, No. 5, pp: 227–233.
[10] Montaño, J. C., Salmerón, P., Prieto, J. (2005). Analysis of Power Losses for Instantaneous Compensation of Three-Phase Four-Wire Systems. IEEE Transactions on Power Electronics, Vol. 20, No. 4, pp. 901–907.
[11] Dugan, R. C., McGranaghan, M. F., Beaty, H. W. (1996). Electrical Power Systems Quality. McGraw-Hill. New York.
[12] Henderson, R. D., Rose, P. J. (1994). Harmonics: The Effects on Power Quality and Transformers. IEEE Transactions on Industry Applications, Vol. 30, No. 3, pp. 528–532.
[13] Merhej, S. J., Nichols, W. H. (1994). Harmonic Filtering for the Offshore Industry. IEEE Transactions on Industry Applications, Vol. 30, No. 3, pp. 533–542.
[14] Phipps, J. K., Nelson, J. P., Sen, P. K. (1994). Power Quality and Harmonic Distortion on Distribution Systems. IEEE Trans. on Industry Applications, Vol. 30, No. 2, pp. 476–484.
[15] Redl, R., Tenti, P., Van Wyk, J. D. (1997). Power electronics' polluting effects. IEEE Spectrum, Vol. 34, Issue 5, pp. 32–39.
[16] Habrouk, M. E., Darwish, M. K., Mehta, P. (2000). Active Power Filters: A Review. IEE Proc. Electr. Power Appl., Vol. 147, No. 5 pp. 403-413.
[17] Kriegler, U. (2000). Active filters-basic principles. PCIM 2000, 6th Power Quality Conference, Vol. 1, Nuremberg, Germany.
[18] Singh, B., Al-Haddad, K., Chandra, A. (1999). A Review of Active Filters for Power Quality Improvement. IEEE Transactions on Industrial Electronics, Vol. 46, No. 5, pp. 960–971.
[19] Tolbert, L.M., Hollis, H.D., Hale, P.S., (1996). Evaluation of Harmonic Suppression Devices. Conference Record of the IEEE Thirty-First Industry Applications Conference Annual Meeting, IAS '96, Vol. 4, pp.2340–2347.
[20] Ledwich G. Ghosh A. (2002). A Flexible DSTATCOM Operating in Voltage or Current Control Mode. IEE Proc. Generation, Transm. and Distribution, Vol. 149, No. 2, pp. 215–224.
[21] Aredes, M., Watanabe, E. H. (1995). New Control Algorithms for Series and Shunt Three-Phase Four-Wire Active Power Filters. IEEE Transactions on Power Delivery, Vol. 10, No. 3, pp: 1649–1656.
[22] Dixon, J. W., Venegas, G., Morán, L. A. (1997). A Series Active Power Filter Based on a Sinusoidal Current-Controlled Voltage-Source Inverter", IEEE Transactions on Industrial Electronics, Vol. 44, No. 5, pp. 612–620.
[23] Morán, L., Pastorini, I., Dixon, J., Wallace, R. (2000). Series Active Power Filter Compensates Current Harmonics and Voltage Unbalance Simultaneously. IEE Proc. Generation, Transm. and Distribution. Vol. 147, No. 1, pp. 31–36.

[24] Wang, Z., Wang, Q., Yao, W. (2001). A Series Active Power Filter Adopting Hybrid Control Approach, IEEE Transactions on Power Electronics, Vol. 16, No. 3, pp. 31–39.
[25] Akagi, H., Fujita, H. (1995). A New Power Line Conditioner for Harmonic Compensation in Power Systems. IEEE Transactions on Power Delivery, Vol. 10, No. 3, pp. 1570–1575.
[26] Fujita, H., Akagi, H. (1990). A Practical Approach to Harmonic Compensation in Power Systems -Series Connections of Passive and Active Filters-. Conference Record of Annual Meeting IEEE Industry Applications Society, Vol. 2, pp. 1107–1112.
[27] Fujita, H., Yamasaki, T., Akagi, H. (2000). A Hybrid Active Filter for Damping of Harmonic Resonance in Industrial Power Systems. IEEE Transactions on Power Electronics, Vol. 15, No. 2, pp. 215–222.
[28] Peng, F. Z. (1998). Application Issues of Active Power Filters, IEEE Industry Applications Magazine, pp: 21–30.
[29] Peng, F. Z. (2001). Harmonic Sources and Filtering Approaches. IEEE Industry Applications Magazine, Vol. 7, No. 4 pp. 18–25.
[30] Aredes, M., Heumann, K., Watanabe, E. H. (1998). An Universal Active Power Line Conditioner. IEEE Trans. on Power Delivery, Vol. 13, No. 2, pp. 545–551.
[31] Fujita, H., Watanabe, Y., Akagi, H. (1999). Control and Analysis of a Unified Power Flow Controller. IEEE Trans. on Power Electronics, Vol. 14, No 6, pp. 1021–1027.
[32] Kamran, F., Habetler, T. G. (1998). Combined Deadbeat Control of a Series-Parallel Converter Combination Used as a Universal Power Filter, IEEE Transactions on Power Electronics, Vol. 13, No 1, pp. 160–168.
[33] Prieto, J., Salmerón, P., Vázquez, J. R., Pérez, A. (2002). A Series-Parallel Configuration of Active Power Filters for VAR and Harmonic Compensation, Proceedings of International Conference on Industrial Electronics IECON'02, Seville, Spain, Vol. 4, pp. 2945–2950.
[34] Ghosh A. Ledwich G. (2002). Power Quality Enhancement using Custom Power Devices. Kluwer Academic Publishers. Boston.
[35] Kazmierkowski, M. P. Malesani, L. (1998). Current Control Techniques for Three-Phase Voltage-Source PWM Converters: A Survey. IEEE Trans. on Industrial Electronics, Vol. 45, No. 5, pp. 691–703.
[36] Yunus, H. I., Bass, R. M. (1996). Comparison of VSI and CSI topologies for single-phase Active Power Filters", Proccedings of IEEE Power Electronics Specialists Conference, pp. 1892–1896.
[37] Rahman, M. A., Radwan, T. S., Osheiba, A. M., Lashine. A. E. (1997). Analysis of Current Controllers for Voltage-Source Inverter. IEEE Transactions on Industrial Electronics, Vol. 44, No. 4, pp. 477–485.
[38] Akagi, H. (2005). Active harmonic filters. Proceedings of IEEE, Vol. 93, No. 12, pp. 2128–2141.
[39] Aredes, M., Häfner, J., Heumann, K. (1997). Three-Phase four-wire shunt active filter control strategies. IEEE Transactions on Power Electronics, Vol. 12, No. 2, pp. 311–318.
[40] Carrara G. Gardella S. Marchesoni M. Salutari R. Sciutto G. (1992). A New Multilevel PWM method: A Theoretical Analysis. IEEE Transactions on Power Electronics, Vol. 7. No. 3, pp. 497–505.
[41] Lai J. S., Peng F. Z. (1996). Multilevel converters- A New Breed of Power Converters. IEEE Transactions on Industry Applications, Vol. 32 No. 3. pp. 509–517.
[42] Kazmierkowski, M. P. Krishnan R. Blaabjerg F. (Editors). (2002). Control in Power Electronics. Selected Problems. Academic Press, Elsevier Science. San Diego.
[43] Rahmani, S., Al-Haddad, K., Fnaiech, F. (2002). A New PWM Control Technique Applied to Three-Phase Shunt Hybrid Power Filter. Proceedings of International Conference on Industrial Electronics IECON'02, Seville Spain, Vol. 1, pp. 727–732.

[44] Van der Broeck H. W., Skudelny H. C., Stanke G. V. (1988). Analysis and Realization of a Pulsewidht Modulator Based on Space Vector. IEEE Transactions on Industry Applications, Vol. 24. No. 1, pp. 142–150.
[45] Buso S. Malesani L. Mattavelli P. (1998). Comparison of Current Control Techniques for Active Filter Applications, IEEE Transactions on Industrial Electronics, Vol. 45, No. 5, pp. 722–729.
[46] Dixon, J., Tepper, S. Moran, L. (1996). Practical Evaluation of Different Modulation Techniques for Current-Controlled Voltage Source Inverters. IEE Proceedings Electric Power Applications, Vol, 143, Issue 4, pp. 301–306.
[47] Grady, W. M., Samotyj, M. J., Noyola, A. H. (1990). Survey of Active Power Line Conditioning Methodologies. IEEE Transactions on Power Delivery, Vol. 5, No 3, pp: 1536–1542.
[48] Montaño, J. C., Salmerón, P. (1998). Instantaneous and full compensation in three-phase systems. IEEE Transaction on Power Delivery, Vol. 13, No. 4. pp. 1342–1347.
[49] Montaño, J. C., Salmerón, P. (2002). Strategies of instantaneous compensation for three-phase four-wire circuits. IEEE Transactions on Power Delivery, Vol. 17, No. 4, pp: 1079-1084.
[50] Sonnenschein, M., Weinhold. M. (1999). Comparison of Time-Domain and Frequency-Domain Control Schemes for Shunt Active Filters. ETEP Vol. 9, No. 1, pp. 5–16.
[51] Akagi, H., Kanazawa, Y., Nabae. A. (1983). Generalized Theory of the Instantaneous Reactive Power in Three-Phase Circuits. Proceedings IPEC83, Tokio, Japan, pp. 1375–1386.
[52] Akagi, H., Kanazawa, Y., Nabae. A. (1984). Instantaneous Reactive Power Compensators Comprising Switching Devices without Energy Storage Components. IEEE Transactions on Industry Applications, Vol. IA-20, No. 3, pp. 625–630.
[53] Akagi, H., Ogasawara, S., Kim, H. (1999). The Theory of Instantaneous in Three-Phase Four-Wire Systems: A Comprehensive Approach. IEEE Industry Applications Conference, Vol. 1, pp. 431–439.
[54] Cavallini, A., Montanari, G. C. (1994). Compensation strategies for shunt active-filter control. IEEE Trans. on Power Electronics, Vol. 9, No. 6, pp. 587–593.
[55] Salmerón, P., Montaño, J. C., (1996). Instantaneous Power Components in Polyphase Systems under Nonsinusoidal Conditions. IEE Proc.-Sci. Meas. Technol., Vol 143, No. 2, pp. 151–155.
[56] Salmerón, P.; Herrera, R. S. (2006). Distorted and Unbalanced Systems Compensation within Instantaneous Reactive Power Framework. IEEE Transaction on Power Delivery, Vol. 21, No. 3, pp. 1655–1662.
[57] Soares V., Verdelho P., Marques G. D. (2000). An Instantaneous Active and Reactive Current Component Method for Active Filters, IEEE Transactions on Power Electronics, Vol. 15, No. 4, pp. 660–669.
[58] Herrera, R. S., Salmerón, P. (2007) Instantaneous Reactive Power Theory: A Comparative Evaluation of Different Formulations. IEEE Transactions on Power Delivery, Vol. 22, No. 1, pp. 595–604.
[59] Litrán, S. P., Montaño, J. C., Salmerón, P., Alcántara, F. J., Vázquez, J. R. (1999). Control de un filtro activo de potencia para compensación en sistemas trifásicos de cuatro conductores. 6as Jornadas Luso-Espanhola de Enghenharia Electrotécnica, Lisboa, Portugal, Proceedings, Vol IV, Cap II, pp. 203–209.
[60] Duke, R. M., Round., S. D. (1993). The steady-state performance of a controlled current active filter, IEEE Trans. on Power Electronics, Vol. 8, No. 3, pp. 140–146.
[61] Jou, H. L., Wu, J. C., Chu, H. Y. (1994). New single-phase active power filter. IEE Proceedings Electric Power Applications, Vol. 141, No.3, pp. 129–134.

[62] Superti Furga, G., Tironi, E., Ubezio, G. (1997). Shunt active filter for four wire low-voltage systems: theoretical operating limits and measures for performance improvement. ETEP, Vol. 7, No. 1, pp. 41–48.
[63] Thomas, T., Haddad, K., Joos, G., Jaafari, A. (1998). Design and Performance of Active Power Filters. IEEE Industry Applications Magazine, Vol. 4, No. 5, pp. 38–46.
[64] Wu, J. C., Jou, H. L. (1996). Simplified Control Method for the Single-Phase Active Power Filter. IEE Proceedings Electric Power Applications, Vol. 143, No. 3, pp. 219–224.
[65] Vázquez, J. R., Salmerón, P. (2003). New active filter control using neural network technologies. IEE Proceedings Electric Power Applications, Vol. 150, No. 2, pp. 139–145.
[66] Alcántara, F. J., Salmerón, P. (2005). A New Technique for Unbalanced Current and Voltage Measurement with Neural Networks. IEEE Trans. On Power Delivery, Vol. 20 No. 2, Mayo 2005. pp. 852–858.
[67] Bonifacio, G., Schiano, A. L., Marino, P., Testa, A. (2000). A New High Performance Shunt Active Filter Based on Digital Control. IEEE Power Engineering Society Winter Meeting, 2000, Vol. IV, pp. 2961–2966.
[68] Singh, B., Singh, B. N., Chandra, N. A., Al-Haddad, K. (2000). DSP-Based Implementation of an Improved Control Algorithm of a Three-Phase Active Filter for Compensation of Unbalanced Non-Linear Loads. ETEP Vol. 10, No. 1, pp. 29–35.
[69] Jacobs J., Detjen D., Karipidis C. U., De Doncker R. W. (2004). Rapid Prototyping Tools for Power Electronic Systems: Demonstration with shunt Active Power Filters. IEEE Transactions on Power Electronics, Vol. 19, No. 2, pp. 500–507.
[70] Verdelho, P., Marques, G. D. (1997). An active power filter and unbalanced current compensator. IEEE Transactions on Industrial Electronics, Vol. 44, No. 3, pp. 321–328.
[71] Tey L. H., So P. L., Chu Y. C. (2005). Improvement of Power Quality using Adaptive Shunt Active Filter. IEEE Transactions on Power Delivery, Vol. 20 No. 2, pp. 1558–1568.
[72] Vázquez, J. R., Salmerón, P., Prieto, J., Pérez, A. (2002). Practical Implementation of a Three-Phase Active Power Line Conditioner with ANNs Technology. Proceedings of International Conference on Industrial Electronics IECON'02, Seville, Spain, Vol. 1, pp. 739–744.
[73] Salmerón, P., Vázquez, J. R. (2005). Practical design of a Three-Phase Active Power-Line Conditioner Controlled by Artificial Neural Networks. IEEE Transaction on Power Delivery, Vol. 20, No. 2, pp: 1037-1044.
[74] Salmerón, P., Montaño, J. C., Vázquez, J. R., Prieto, J., Vallés, A. P. (2004). Compensation in Nonsinusoidal, Unbalanced Three-Phase Four-Wire Systems with Active Power Line Conditioner. IEEE Transactions on Power Delivery, Vol. 19, No. 4, pp. 1968–1974.

10

Distributed Generation

Juan Carlos Gomez Targarona and Medhat M. Morcos*

Instituto de Protecciones de Sistemas Eléctricos de Potencia,
Departamento de Electricidad y Electrónica,
Universidad Nacional de Río Cuarto,
Ruta 8, Km. 603,
(5800) Río Cuarto, Córdoba, Argentine.
Email: jcgomez@ing.unrc.edu.ar

Department of Electrical and Computing Engineering,*
Kansas State University,
289 Rathbone Hall,
Manhattan, KS 66506, USA.
Email: morcos@ksu.edu

10.1 Introduction

The growing concern on climate changes and the ever-present oil crisis have recently increased the interest in distributed generation (DG) from renewable and traditional (but highly efficient) sources. Deregulation and growth of competitive supply markets also magnify this interest. In general, distributed resources are defined as sources of electrical power that are not directly connected to a bulk power-transmission system. They include both generators and energy-storage technologies, with a power rating of 10 MW or less [1].

Currently, distribution-system engineers are divided between DG advocates and adversaries, each having their own valid reasons. One of the main reasons for this conflict is just "fear of what could happen" due to the lack of practical knowledge on traditional power systems having a high level of DG penetration. Still fresh in the minds of many engineers is the wrong approach that has been taken in the past by large computer manufacturers when they underestimated the growing market of personal computers; an analogy with the present DG situation that is easy to make.

The impact of DG on distribution systems depends on the penetration level, a presence of approximately 20% on energy resources for the years 2010–2020 is predicted. During the first part of the current decade fewer industrial and commercial businesses have expressed an interest in adopting onsite DG equipment [2]. After the widespread blackouts that took place in the USA and several European countries by the end of 2003 and their related huge costs, industries and utilities started to change their philosophy.

The widespread use of DG or distributed resources has created a series of new problems, some of them closely related with power quality such as selective coordination of overcurrent protection and its effects on continuity (service quality), in addition to control of the voltage-sag magnitude and duration by using overcurrent protective devices (PD) considered as included in product quality. The most feared effect is the islanding that can be highly risky for user equipment, utility elements and personnel, but can be desirable from the power-quality point of view. Interaction and impact of DG on network operation, fault detection, fault clearing, and reclosing operations are also of great concern [3].

It has been suggested that the widespread use of DG defines old and new scenarios, calling for highly sophisticated equipment (using communications, electronics, and microprocessors) as the only way to cope with the current distribution-system scheme. Several protection concepts have been presented recently, many are so contradictory that protection studies become even harder. The doubts and contradictions have delayed the approval of the IEEE P1547 Draft, "Standard for Interconnecting Distributed Resources with Electric Power Systems," which has been under study for several years and has been recently issued [4].

Most of the utilities have established new connection rules for DG, also changing their own distribution systems. One of the changes is the avoidance of any single-phase protective device positioned between the utility transformer and the three-phase DG. Also, it has been realized that DG protection needs a longer time (up to 1 s) to detect faults and operate, thus longer reclosing intervals are used in utilities.

The ANSI/IEEE Standard 1001 presents the following out-of-range trip settings [5]:

voltage higher than 137%, maximum trip time of 2 cycles;
voltage higher than 106%, and lower than 88%, maximum trip time of 6 cycles;
voltage lower than 50%, maximum trip time of 6 cycles;
frequency is considered normal between 59.3–60.5 Hz (60 Hz base), with 15 cycles of trip delay when out of tolerance.

It has been reported that the use of novel fast-clearance protection systems to protect DG would improve transient stability and also improve power quality (by reducing voltage-sag duration). Several issues are currently being considered, for example:

1. Whether to disconnect the DG very rapidly or keep it connected as long as possible.

2. Impairment or improvement of the traditional overcurrent coordination with DG addition.
3. Islanding advantages or disadvantages from the voltage-sag ride-through point of view.

The answers to these issues are different, depending on what portion of the power system is considered. In the authors' opinion more research is needed with regard to classical protection coordination, as fuses are still widely used in distribution systems. One possible reason is based on the fuse simplicity and reliability, being able to operate even in extremely hard environments. New protection systems need to be studied or restudied for this application, until all the traditional protection possibilities have been exhausted.

The application of DG is very useful for deferring investments, improving the operation of the existent distribution networks, and increasing the customer's reliability (perhaps only for the DG owner and not for the other customers connected to the same system) but it creates new challenges for the engineers involved. When the original idea of DG application came out, utilities were very sceptical about their capability, especially resistant to introduce this technology in their own systems, and refused to allow DG installation on customer circuits.

Besides, distribution systems have been mainly designed for energy flowing in just one direction. If the DG equipment does not belong to the utility, the customer decides about its connection to or disconnection from the system, based on its own needs. If the DG belongs to a cogeneration scheme, in the case of a blackout the customer could keep the equipment feeding the critical loads operating as an isolated circuit or island, and needing resynchronisation when the external supply is restored. This scheme created more than one problem to the utility when it is applied during voltage sags caused by nearby faults [6].

The effect of DG greatly depends on the technology used and on the network characteristics. The main aspect is that now the energy source is not unique and then the energy flow could be changed or reversed under normal or abnormal conditions.

There are several types of power generators that can be used as DG that, in order of importance, are:

- wind turbines;
- fuel cells;
- photovoltaic cells;
- small and micro-turbines;
- internal combustion engines.

Each of these power generators has its own advantages and disadvantages, such as easy installation based on its small size, fast start, generates voltage distortion, and uncertain power availability. In spite of the distributed-generation promotion carried out by utilities, they have been expressing concerns related with potential distribution-system problems.

10.2 General Impact of Distributed Generation on Power Quality to Strong or Weak Electrical Systems

The strength of the existent distribution system where DG is connected is very important from the point of view of the impact on power quality. Some power-quality phenomena are very noticeable when they take place in weak systems and pass practically unnoticed when they happen in strong distribution networks. In other words, the importance of the events is closely related with the system regulation.

10.3 Dissimilar Effect of the Different Distributed Resources Technologies

Depending on the available energy source, DG can be of dissimilar technologies, such as: wind turbines, photovoltaic cells, fuel engines, gas turbines, vapour turbines, fuel cells, *etc.*, the first mentioned being the currently more developed, reaching today a unit power of 2 to 3 MW. Each technology has its own advantages and drawbacks. The DG deployment depends among other factors on geographic situations, weather conditions, electric-energy cost and availability, utility tariff policy, existence of waste having usable calorific power, *etc.*

10.4 Coordination between Overcurrent Protection and Sensitive Equipment Voltage-sag Immunity

Distribution-system protection interaction with power quality, especially related to "product quality", is one of the aspects where the DG has higher impact. Besides, most of the voltage sags are generated by faults whose duration is ruled by the protective-device operation speed. This speed is modified when the DG is supplying part of the fault current and is also supporting the voltage system. Thus, voltage-sag studies are widely affected by any DG connected to the network.

One of the important issues that the new scenario will introduce is overcurrent protection, where there will be changes in system behaviour and flow of power under short-circuit conditions. Normally, there is no concern with the DG when its power is less than 10% of the minimum load demand in the feeder [6].

The most immediate consequence is the need for verification of the protective-device breaking capacity, which might not be enough due to the increase of the available short-circuit power. The short-circuit current and transient behaviour of the generator that provides power through inverters – such as photovoltaic cells, fuel cells, wind turbines, and induction generators – are completely different from the synchronous generator response. Induction generators directly connected to the supply will show a special behaviour when a short-circuit takes place. Due to the fact that this type of generator gets its excitation from the mains, it is not able to maintain the short-circuit currents for a relatively long time.

The situation when the mains supply is intentionally or unintentionally disconnected from the power system having at least one DG, which continues operating with this single source, is called *islanded-mode operation*.

Due to the increase of power-electronics applications in industry, residences and elsewhere, users have become very sensitive to voltage perturbations, among which voltage sag is the event of most concern. An aspect that requires special treatment is the coordination between overcurrent protection and the voltage-sag ride-through capability of sensitive equipment.

Voltage sag is almost universally considered as "a nonpermanent voltage reduction with values between 10% and 90% of the rated voltage and duration between ½ cycle and a few minutes". The ability of sensitive equipment (SE) to withstand voltage sags without dropout is called the *ride-through capability*. Reductions in voltage between 0% and 10% are considered as there is no perturbation since the voltage is included in the normal zone of voltage variation in distribution systems. Voltage reduction of more than 90% or applied voltage less than 10% are considered as an interruption. The zero voltages with durations shorter than three minutes correspond to the phenomenon known as microinterruption [7].

Many typical ride-through capability curves have been proposed as guidelines, presented by institutions such as the U.S. Department of Commerce-Federal Information Processing Standards Publication No. 94, the Computer Business Equipment Manufacturing Association (CBEMA), SEMI F47 presented by semiconductor manufacturers together with EPRI, and the Information Technology Industry Council (ITIC). The current version of the ITIC curve, revised in 1996, has reached prominent widespread use. These guidelines are normally given in graphical form, having two well-bounded parts corresponding to the overvoltage and undervoltage withstand values.

These curves represent an important tool for manufacturers, users, and system designers. The present work is confined to only the lower part of the voltage-sag ride-through capability curves, adopting the Computer Business Equipment Manufacturing Association (CBEMA) curve as a guideline. Fig. 10.1 shows the lower legs of the CBEMA and ITIC curves [7].

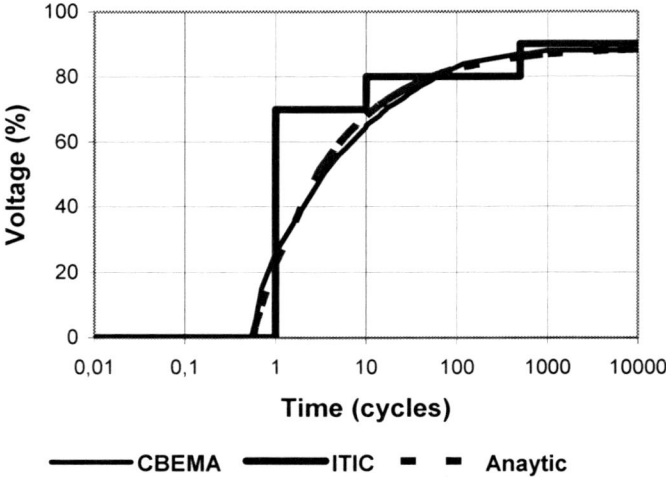

Figure 10.1. CBEMA and ITIC sensitive-equipment-immunity curves

10.4.1 Application of Specific Energy Concept

The sensitive-equipment ride-through capability can be considered as directly related to the energy stored in kinetic, electrostatic, or electromagnetic form that is the source used to keep the apparatus working while the supply voltage is below the device needs. Due to cost and competitive reasons, the manufacturer supplies as low a ride-through capability as possible, which causes users to need to add auxiliary equipment in order to withstand the frequent voltage sags. Many technical articles and papers on the auxiliary equipment are readily available, however, only a few are related to the sensitive equipment incorporations tending to improve its immunity.

The main exception is related to one of the most sensitive applications, the adjustable-speed drives (ASD), where the consequences of equipment dropout (or shutdown) are extremely expensive. Besides, an extensive study on personal computers has been published where it is estimated that 80 to 90% of voltage-sag problems have durations shorter than one second. The increase of the ride-through capability to this time duration will produce a considerable improvement in the computer behaviour [8].

The methodology of increasing the energy-storage ability was adopted by a few manufacturers of household appliances, like digital clocks, VCRs, phone answering machines, TVs, *etc*. A few sensitive-equipment customers are starting to specify the ride-through capability when ordering new equipment, for instance whether the equipment follows the CBEMA-ITIC curve, then doing the corresponding ride-through coordination study. The CBEMA curve in Fig. 10.1, shows the existence of a nearly smooth voltage – duration correlation. The ITIC curve is given in a stepwise shape; however its average really follows the CBEMA curve [9]. It would be of great interest if the manufacturer provided information about the ride-through capability in a way that is compatible with CBEMA-ITIC, such as a voltage – time curve instead of giving one point only (actual methodology) related to this curve.

IEEE Standard 1346 gives three values of voltage tolerance for six sensitive equipments, which correspond to the typical upper, average, and lower ranges available in the market. The cited sensitive equipment are: programmable logic controller (PLC), PLC card input, adjustable-speed drive (ASD), control relay, motor starter, and personal computer [10].

The SE ride-through capability is affected by the pre-sag voltage, which could confirm the concept that a voltage over 87% is used for storage, where this energy is to be released in the case of voltage-sag occurrence. The value 87% of the rated voltage is the minimum voltage specified by the CBEMA curve, for which the equipment should work indefinitely without any performance limitations [9].

The stored-energy concept leads to the presence of a variation law related to the product of the squared voltage and time. The third curve in Fig. 10.1, identified as analytic, represents the relation:

$$V^2 t = \text{constant}, \tag{10.1}$$

which shows a very good agreement with the CBEMA curve. In this case, the voltage is expressed in per cent and the time in cycles, resulting in a constant value of 4400.

The previously defined methodology (CBEMA-ITIC curves) agreed on considering the value of maximum voltage drop as the magnitude of nominal voltage sag. The difficulty in determining the magnitude of nominal voltage sag is in assigning the duration, since in the real world the voltage variation is usually not so sharp (square sag); its waveform depends on the system and equipment characteristics.

Another difficulty in the voltage-sag characterization is the presence of a progressive or evolved fault, which starts as single phase and after a short time becomes a multiple-phase fault, or modifies the current without changing the fault type [11].

The application of the "energy concept" will lead to the change of Eq. 10.1 to:

$$\int v^2 \, dt = k, \tag{10.2}$$

where $\int v^2 \, dt$ will be called the specific energy in analogy to $I^2 t$ which is used extensively in overcurrent-protection studies. The introduction of the integral for each sensitive equipment would solve the ambiguity issue, giving a full voltage-sag characterization. The proposed methodology is able to consider the value of the initial voltage sag (or small step), which is normally present due to the voltage developed across an arcing fault. This effect is also noticeable on double-supply systems, which is precisely the situation when a DG exists, after the first operation of the circuit breaker takes place.

The ride-through capability of three-phase equipment is not the same for single-phase voltage sags as for three-phase voltage sags. In the case of single-phase faults, the nonfaulted phases usually present an overvoltage, which overfeeds two phases of the three-phase equipment. The application of the specific-energy concept provides an understanding of why a three-phase equipment suffering 18% voltage sag during 6 cycles remains undisturbed. During this event the other two phases were increased by 16%, which gives a specific energy fall of only 5%, a value that can be withstood by the equipment without any difficulty during the circuit-breaker operating time [9].

It has been reported that one phase-to-phase voltage must sag to below 75% of nominal and stay below 95% for six or more cycles to disrupt operations. Also, three-phase faults are more devastating to the plant because the phase-to-phase voltage does not have to drop as much on all three phases as one event affecting just one phase-to-phase voltage. The application of the specific-energy concept, as expressed by Eq. 10.2, will clarify and simplify the analysis.

It is not difficult to understand the concept of specific energy for the SE ride-through capability, however, this value should be correlated with something similar to the probability of occurrence in the system under analysis, and the system-performance indices. Besides, the three-phase SE will have a unique value of ride-through specific energy, which should be compared with the single-phase value or three-phase composite value, depending on the characteristics of the event under study (voltage sag). Furthermore, it should be noted that the probability of three-phase voltage sags is much less than the presence of single-phase dips. Also, the type of transformer connection modifies the voltage-sag effect. For instance, a single-phase fault in the utility system is seen as a two-phase voltage sag on the customer side of a delta-wye utility transformer.

The specific-energy concept has been proposed for the analysis of SE immunity, showing its application not only to individual SE but also to sensitive industrial plants, production lines, and parts of distribution systems [6]. The effect of voltage sag on some specific sensitive equipment is not only related to the magnitude and duration, but also to the phase-angle jump (unusual voltage zero-crossing position), such as ASDs. The phase-angle jump effect is not considered in this work.

10.4.2 Transformation of Protective Device Time–Current Curves into Time–Voltage Curves, for Voltage-Sag Coordination Studies

The present study is initially done on a simple distribution system where the effect of DG presence can be easily analysed. The system is typical of rural, industrial, or low-power distribution schemes. Most of the distribution systems are operated radially due to the simple and cheap overcurrent protection. The change of the whole distribution protection in order to allow the unlimited DG penetration is a very expensive, and perhaps unrealistic, idea. The operation would be optimised under the present situation in order to do only a certain corrective work.

It should be taken into account that the type of fault is strongly influenced by the generator and isolation transformer connections. Transformers with an ungrounded primary would not contribute significantly to single line-to-ground faults, thus the utility protection would delay its trip signal. In this case, a temporary fault can become permanent with the disadvantage that a momentary interruption turns into a sustained one and the voltage sag is now followed by an outage [6]. The present study is carried out from two points of view, voltage sags and overcurrent coordination.

A. Radial system from the voltage-sag coordination point of view

The classical study of overcurrent protection–voltage sag coordination is given using the single-phase equivalent circuit shown in Fig. 10.2, where PCC indicates the point of common coupling and SE specifies the sensitive equipment (equipment having low voltage-sag ride-through capability). Basically, the PCC will be defined as the point of the circuit where the SE current is generally separated from the distorted or too high current path. In this case it is the short-circuit main flow path as can be seen in Fig. 10.2 [11].

Steps of the coordination methodology have been reported in [9].

The voltage applied to SE during the three-phase fault is:

$$V_{SE} = 1 - I_f Z_s \text{ [pu]}, \tag{10.3}$$

which can be rewritten as,

$$V_{SE} = Z_f / (Z_s + Z_f) \text{ [pu]}. \tag{10.4}$$

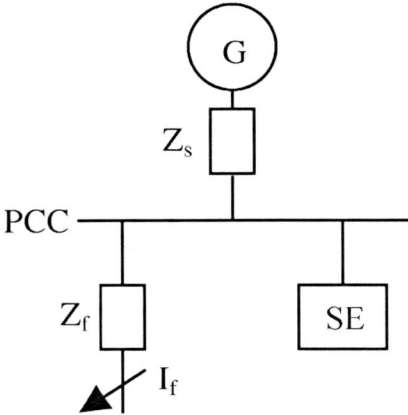

Figure 10.2. Simple radial circuit for voltage-sag studies

When DG is introduced into the system, the circuit of Fig. 10.2 is transformed into the complex form of Fig. 10.3, where the classical concept of PCC, which considers that the energy-flow direction is unique, is no longer valid. The circuit now includes the DG impedance (Z_d) and the interconnection impedance (Z_i), where Z_d is a nonconstant value, usually represented by the DG transient impedance. Also, each of the two branches (where SE and the faulted branch are connected) could be located anywhere between the DG and the power-system supply points.

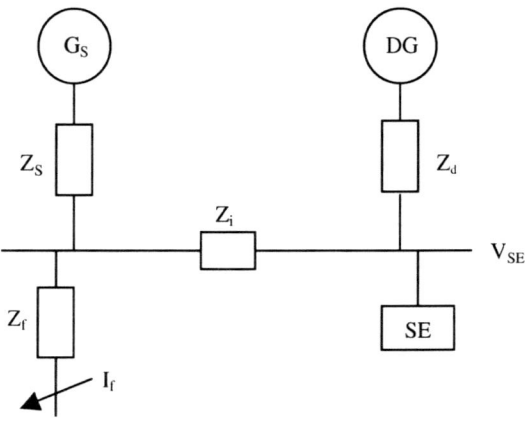

Figure 10.3. Circuit with DG for voltage-sag studies

Depending on the locations, four main cases can be analysed.

Case I – with the faulted branch connected at the power system side and the SE at the DG side. Case II – both branches are at the system side. Case III – both branches at the DG side. Case IV – the SE connected at the power-system side and the faulted branch at the DG side.

For each of the four extreme cases the circuit behaviour and equations would be different. From a quick analysis of the cited cases, it can be deduced that:

1. Cases I and IV: DG power-supply combination determines a minimum voltage-sag magnitude at the SE connection point. These minimum values, expressed in pu, are $Z_i / (Z_d + Z_i)$ for case I, and $Z_i / (Z_s + Z_i)$ for case IV.

2. Cases II and III: the power supply is reinforced by the DG collaboration, increasing the short-circuit current at the PCC, with Z_i in series with Z_d in the first case, and with Z_s alone in the second case. Depending on the power system and DG relative sizes, impedance Z_i can represent a substantial short-circuit current limitation. For these two cases there are no minimum voltage-sag values.

In general, DG can be considered as mitigating voltage sags in two ways: increasing the fault level (reducing system impedance Z_s to Z_s in parallel with Z_d), and keeping up the voltage at the neighbouring load (V_{SE}) by feeding the fault. For the circuit shown in Fig. 10.2, the coordination study (carried out by converting the protective device time–current characteristic curve into a time–voltage curve) is relatively easy.

When both types of data, protective device and SE-immunity characteristics are overlapped in the voltage–time graph, the interception (if any) between the ride-through-capability curve and the protective device curve indicates the SE dropout possibility. The comparison is not direct, because the fuse TCC shall be transformed in the voltage–time curve, which can be easily done by subtracting the voltage drop caused by the fault current in the PCC impedance (Z_s) from the system voltage, shown in Fig. 10.2. In order to fill the graph, as the work is done in percentile values, the only necessary information from the circuit is the short-circuit power at the PCC. In this way the voltage-sag depth is determined for a few selected time durations, with eight values being enough due to the regularity of the fuse TCC curve. For very high short-circuit currents (or faults near the PCC), the voltage sag is practically zero, needing a very fast protection that interrupts the current in less than half a cycle in order to avoid sensitive-equipment disruption. Due to device characteristics, the only protectors with sufficient speed are current-limiting fuses and in some extension CL circuit breakers. If for some reason (for instance the existence of limiting inductances) the current does not reach this high value, any device of the non current-limiting type can be used, directly comparing the TCC and voltage–time curves, as shown in Fig. 10.4. In the figure, the motor start and load-pickup characteristics have also been drawn, in order to determine the mutual effect of the protective device, sensitive equipment and overcurrents due to faults, motor starts or load pickups [12].

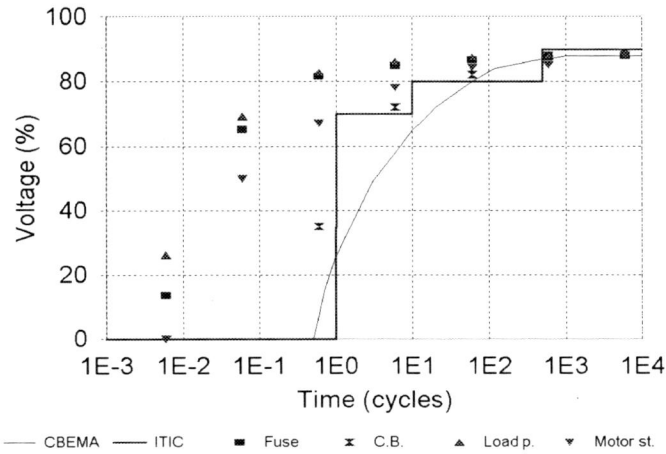

Figure 10.4. TV characteristics of fuses, circuit breakers, sensitive equipment, motor start and load pickup

Now, besides the circuit complexities, an additional uncertainty is provided at the instant when the DG circuit breaker opens, changing the circuit behaviour and altering both fault-current and voltage-sag magnitudes. The study of this situation is given below, using the specific-energy concept.

B. Radial system from the overcurrent coordination point of view

Due to its simplicity and uniformity from the time–current characteristic point of view, the study will be carried out for fuses, explaining the main differences if circuit breakers are used instead. The DG presence in the circuit can cause fuse blowing, since the fuse current now has two components, one coming from the supply (already considered during the distribution system design), and the other from the DG, which creates an acceleration of the fuse operation. Figure 10.5, which is derived from Fig. 10.3, shows a typical radial circuit with five possible fault locations and subsequent fuse operations that require a special analysis.

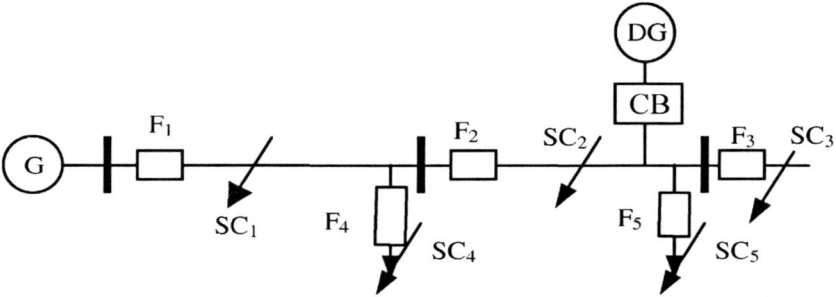

Figure 10.5. Typical rural radial system with distributed generation

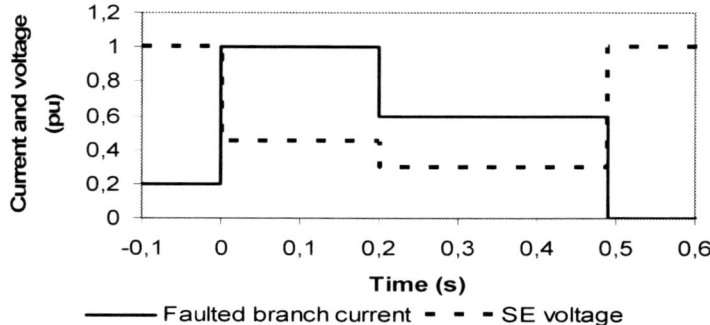

Figure 10.6. Faulted-branch current and SE voltage for a circuit having a double supply

1. Fault SC_1: current mainly supplied by the power system (generator G) would cause fuse F_1 to operate, and if the DG circuit breaker is not fast enough it also would cause fuse F_2 to operate. The operation of fuse F_2 would cause some confusion on the task aimed at locating the fault, but, would allow that part of the circuit to continue to be supplied by the DG (islanding). It must be remembered that islanding is not always allowed by the local authorities but sometimes is desirable from the users' point of view.

2. Fault SC_2: current supplied by the power system would cause fuse F_2 operation. The DG circuit breaker would also trip (in a shorter or longer time).

3. Fault SC_3: current supplied by the power system as well as by the DG would cause fuse F_3 operation. If the fuses F_3 and F_2 have been properly coordinated without considering the DG, the new situation would enhance the coordination by accelerating fuse F_3 fault interruption. The DG circuit breaker can also open depending on its coordination with F_3. When the power system can not supply the loads without the DG or if this generation is for some reason necessary, it would be desirable for the DG breaker not to open.

4. Fault SC_4: the only fuse that needs to be operated is F_4. Fuse F_1, if properly coordinated, would not operate. The inconvenience comes from fuse F_2 operation caused by the DG supplied current (when fuse F_2 rated current is similar to or lower than that of F_4), leaving part of the system operating as an island. This situation is possible when an important load is supplied from the feeder but its frequency of occurrence is very low. The possible existence of this situation needs to be analysed further.

5. Fault SC_5: the only fuse that needs to operate is F_5. Fuse F_2, if properly coordinated would not operate. Also, it is desirable that CB remains closed.

From the cases mentioned above, the most conflictive situation is presented by fault SC_4, which requires special consideration.

The coordination analysis is done using the graph shown in Fig. 10.6. Now, the specific energy of current, $I^2t(t)$, needs to be applied instead of the classical time–current characteristic $I(t)$, due mainly to the large variation of the DG current supplied during the early part of the short-circuit process. In this conflictive situation, the specific energy of G and DG currents through F_4 (prearcing + arcing) needs to be lower than the F_1 prearcing energy due to G, and also lower than F_2 prearcing energy due to DG. The new problematic application, when DG is connected in the system, is analysed below.

C. SE Voltage-sag Ride-through Capability
The study is carried out considering the faulted branch being protected by fuses and that the DG possesses a circuit breaker commanded by a definite-time relay. The analysis is directed to determine if the voltage sag caused by the fault current, with the consequent DG disconnection, would cause SE dropout. The specific-energy concept can be applied to this study based upon its capability to analyse PD and SE behaviour as time functions [9]. Figure 10.6 shows the voltage-sag and current magnitudes due to a change in source impedance when the DG circuit breaker opens. The fault impedance was assumed to initially produce a fault

current (1 pu) five times the prefault load, and a voltage sag to 0.45 pu. It was also assumed that the DG breaker opens after 200 ms, reducing the current to 0.6 pu and the voltage sag to 0.3 pu. The PD operation — 490 ms after the fault started — has restored the voltage to 1 pu. The comparison between the voltage-sag specific energy and the SE ride-through-capability curve would indicate the SE withstand or dropout conditions. In other words, the SE would drop out if the energy deficit (due to 0.45 pu voltage for 200 ms and 0.30 pu voltage for 290 ms) was higher than the SE withstand-energy deficit.

A generic circuit — derived from that shown in Fig. 10.3 is presented in Fig. 10.7, where the interconnection impedance has been divided into three components in order to allow the branch locations to be considered. The components are derived from the original impedance Z_i multiplied by three factors, whose sum is always equal to one. The three factors transforming Z_{id}, Z_{im} and Z_{is} allow the representation of any particular case. Table 10.1 shows the values for the previously described cases, I through IV.

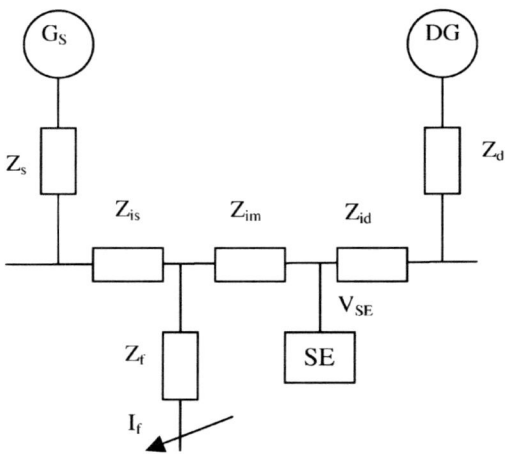

Figure 10.7. Generic circuit with DG for voltage-sag studies

Table 10.1. Impedance factors for the four extreme cases

Impedances	Cases			
	I	II	III	IV*
Z_{id}	0	1	0	0
Z_{im}	1	0	0	1
Z_{is}	0	0	1	0

*Note: for this case Z_s needs to be replaced by Z_d

The formula below, developed from the circuit in Fig. 10.7, is the general expression for the SE voltage calculation:

$$V_{SE} \text{ [pu]} = 1-(Z_s+Z_{is})*(Z_d+Z_{id})/(Z_f*Z+(Z_s+Z_{is})*(Z_d+Z_{id}+Z_{im})) \tag{10.5}$$

where, $Z = Z_s + Z_i + Z_d$.

The situation after the DG breaker opens can be easily calculated using the above expression and considering Z_d as infinite. In conclusion, the coordination task to be carried out consists of the estimation of the specific energy of the fuse current as a function of time, together with the calculation of the specific energy of the SE voltage, also as a function of time. The new situation needs to be considered after the DG breaker opening, if it happens. If the SE specific-energy value is reached before fuse opening, the SE would suffer a dropout. The aim is to have the fuse open before the SE stored specific energy is exhausted. The SE lack of specific energy is greatly influenced by the operating time of the DG breaker due to the fault backup given by the DG, as can be concluded from Fig. 10.6.

The procedure can be applied also to determine the improvement in the SE ride-through capability when a DG of a given power is connected somewhere to a simple radial circuit. Figure 10.8 shows an example of the procedure (the values correspond to Fig. 10.6); the thick curve represents the increase in the specific energy (of current) as a function of time, whereas the thin curves, dotted and solid, represent the defective specific energy (of voltage), respectively. The fuse would operate when its curve reaches the prearcing + arcing specific energy, for our case 0.066 pu^2 s (of current, pointed by the upper arrow), and the SE would drop out for a value of 0.205 pu^2 s (of voltage shown by two collinear arrows). It can be seen that the DG breaker tripping at 200 ms (point from which dotted and solid lines depart) causes the SE dropout at 420 ms (left, lower arrow), before the fuse operation, which takes place at 490 ms (shown by upper arrow). The dropout could be avoided by delaying the breaker operation to 380 ms (indicated by a slope change of the thin solid line). The acceptability of the circuit-breaker delay needs to be carefully considered due to the possible damage to equipment and hazard to personnel, in spite of being a weak distribution system.

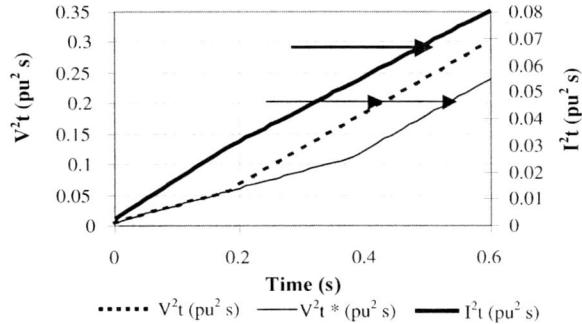

Figure 10.8. Comparison of fuse and SE specific energies

D. Voltage-sag and overcurrent protection coordination using time — voltage characteristics

The electric fuse, or simply "fuse", is the most commonly used protective device in distribution systems, being manufactured in a series called a homogeneous series, defined by ANSI/IEEE C37.41 and IEC 60282. The homogeneity signifies that any fuse is similar to the one having higher or lower rated current, with the only difference based on the load-current capability. The characteristic curves of dissimilar fuse ratings are parallel to each other due to the use of logarithmic scales on both axes [13]. The same homogeneity can be found in circuit breakers, having much more tailoring possibilities. As a result of this homogeneity the time–current characteristic curves for all PD (of the same series) can be approximated by one curve only on the graph, expressing the horizontal axis in multiples of the current rating instead of A or kA. Applying the selected PD individual curve can later solve this approximation.

In order to clarify the idea, the study would begin with the analysis of the simple radial circuit shown in Fig. 10.2, from which the expression for the voltage applied to faulted SE was obtained. When a fuse is used as a protective device connected on the faulted branch, the fault current can be expressed as a multiple of the fuse current rating, with I_{rf} being the fuse current rating and k_f the multiplier. For example, in Fig. 10.9 the fuse under study (rated current 200 A) having a given current value will operate in a time of 1 cycle, and after doing the corresponding conversion from the time–current characteristic into the time–voltage characteristic, the voltage drop due to this current will create a voltage sag to 80%. Sensitive equipment would withstand during this time period a voltage sag as deep as 20%, thus the equipment will withstand this event. As the entire fuse characteristic is to the upper left of the SE curve, the equipment will be protected by fuse operation.

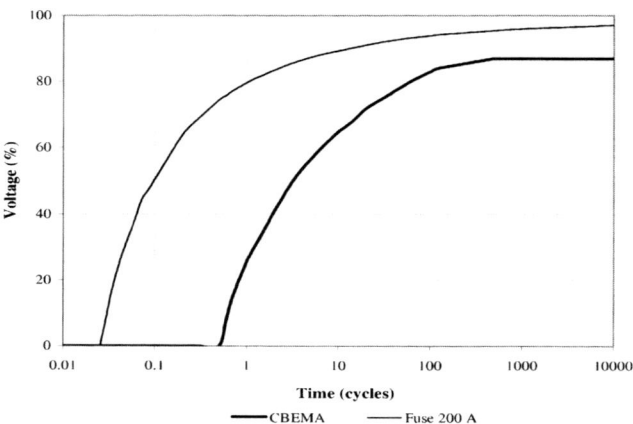

Figure 10.9. Example of fuse time–voltage characteristic and the CBEMA curve

A similar procedure can be followed when circuit breakers are applied, using the methodology for the $I(t)$ characteristic transformation into the $V(t)$ characteristic using Z_s as the conversion factor [9].

Equation 10.3 can be rewritten as,

$$V_{SE} = 1 - k_f\, I_{rf}\, Z_s \text{ [pu]} \tag{10.6}$$

where the product $I_{rf}Z_s$ represents the system impedance voltage drop, considering the faulted-branch as the only circuit load. If there are other loads connected, the faulted-branch load can be represented as a pu value of the total load, expressed by k_1, which is always less than one.

Equation 10.6 now becomes:

$$V_{SE} = 1 - k_f\, k_1\, I_{rf}\, Z_s \text{ [pu]}. \tag{10.7}$$

This expression can be represented on a time–voltage graph, together with the SE immunity data given either by CBEMA-ITIC or IEEE standards, as shown in Fig. 10.9. For this work, the CBEMA curve was selected based upon its worldwide application and also due to its regular form. The fuse curve has been determined for both Z_s and I_{rf} values.

A generalization can be obtained by changing the vertical axis scale into logarithmic, then the curves representing fuse ratings, source impedances, and load percentage would shift horizontally, taking the point defined by a 100% voltage and 0.01 cycle as the new origin. In other words, once the study has been done under basic conditions, it is very easy to improve the results by changing any of the constants in Eq. 10.7, which leads to the shifting of the fuse–time curve. The changeable constants are k_f, k_1, I_{rf}, and Z_s, which implies modifications on the short-circuit current, load sharing, fuse current rating, and source impedance, respectively. Any of the four constants has the same weight on the curve shifting, besides the final curve location can be obtained by a combination of changes in some, or all, of the four constants. Figure 10.10 shows the change obtained by the modification of the source impedance.

When the conditions are changed, for instance the new fuse characteristic has moved to the right and down as shown in Fig. 10.10, the point corresponding to 1 cycle will be withstood by the SE, but as some crossing points are shown, some voltage-sag events will cause the SE dropout.

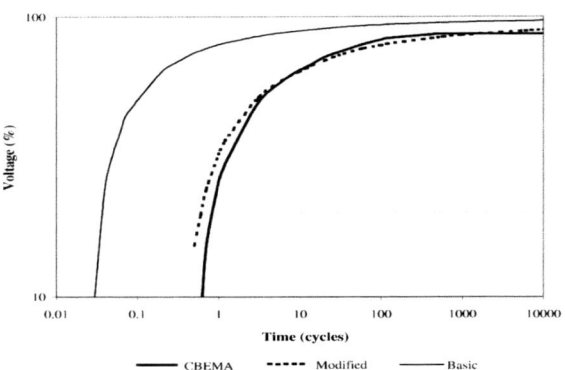

Figure 10.10. Similar to Fig. 10.9 with source impedance modified

Following the same analysis, when DG is used, the SE and the faulted branch have two energy sources, the mains and the DG. When the fault starts, PD of both sources would detect the overcurrent and one of the devices would be open at first producing a sharp change in the source impedance. The proposed methodology would allow the SE ride-through analysis to be carried out, as the device opening produces a change in the source impedance, as described above. For instance, if a short-circuit takes place, the voltage sag detected by the SE would be 80%. If after 1 cycle the source impedance is changed, the voltage would fall to 40%. As the fuse curve is still to the left of the immunity curve, the protective device would clear the fault avoiding SE dropout. If the circuit modification produces a slightly lower curve, the SE would not be able to withstand the voltage sag due to the intersection of the two curves, Fig. 10.10.

The methodology can be applied, based on the specific-energy concept, which signifies that the stored energy is slowly exhausted at first during one cycle (voltage sag 80%), then quickly consumed, since the voltage sag is now 40%. As the total specific-energy deficit is still lower than the dropout curve, the SE would ride through the event.

In the case of using more sophisticated PD, data management is not as easy as with fuses, due to the possible time–current characteristic changes. Modern devices allow the modification to be done with both instantaneous and delayed trips, as well as with multiples of the pickup current. Figure 10.11 shows both the CBEMA immunity curve and one of the available characteristics after being transformed into a time–voltage curve. The characteristic curve is formed by several horizontal and vertical segments that can be moved either vertically or horizontally, so each location would have its own constants.

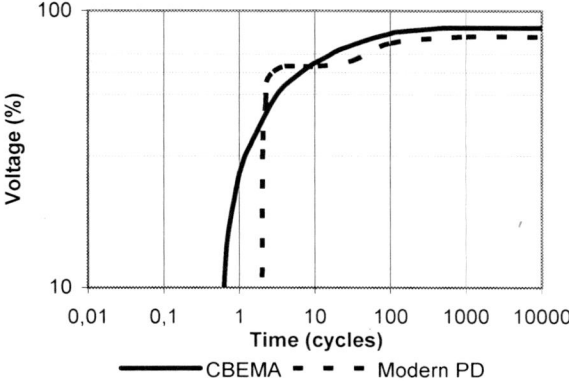

Figure 10.11. Coordination of CBEMA and modern PD curves

It would be very helpful if the PD manufacturers provided the basic characteristics and constants for allowing any modifications – related to source impedance, branch load, and short-circuit current – to be done. The basic curve can be completely defined by two figures, the specific energy and the minimum voltage for steady-state operation.

The different coordination methodologies described above present a short guide for the improvement of the protection–voltage-sag coordination. By using the general expression represented by Eq. 10.5, it is possible to consider the faulted branch and DG connection point together with the supply/DG impedance ratio. The temporal behaviour of the applied protection scheme having dissimilar operation characteristics can be easily calculated using the specific energy instead of the time–current characteristic approach. In the case of having several SE connected to the same branch, the envelope of the combined immunity curves would be used instead of the individual SE $V(t)$ characteristic. In this way, both the instant at which the protective device operates and the SE dropout time can be calculated, and by comparing the two values it can be determined whether the SE would ride through the event or would drop out. The methodology provides the distribution-system engineer with the necessary tools in order to do the protection coordination/power quality tradeoff.

E. Coordination study with variable-supply scheme

As was previously explained, the coordination study between the overcurrent protection device (PD) and the SE voltage immunity located in parallel branches or feeders, is done in graphical form, comparing the adapted time–current characteristic (TCC) of the protective device with the previously cited CBEMA curve. An adapted protective device TCC is a curve transformed into a time–voltage characteristic (TVC), that represents the voltage sag that the protective device allows to be applied to the SE under study. Voltage sag applied to SE is directly the voltage at the PCC or the supply voltage minus the voltage drop across Z_1 due to the short-circuit current [12].

Figure 10.12 shows the new scheme of the system with utility and distributed generation, connected by an islanding circuit breaker (ICB), where DG may be representing more than one device. When the islanding circuit breaker is closed the source impedance is approximately the parallel combination of the utility (area EPS) and DG impedances. When ICB is opened the source impedance jumps to a larger value, thus in the presence of a fault, the short-circuit fault current will be greatly attenuated and the operating time of the protective device will be increased subsequently. Thus the coordination study needs to be done only for the two branches on the DG side of the ICB, with maximum and minimum rated currents of the protective device.

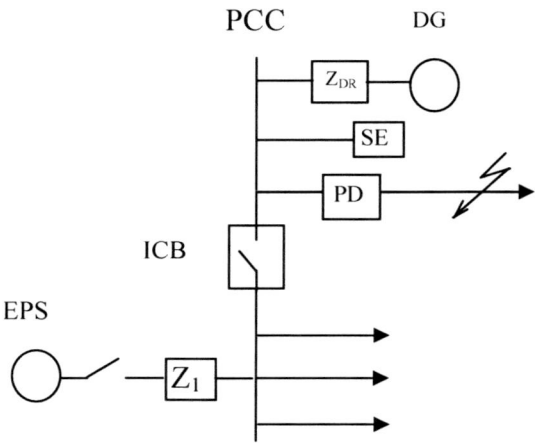

Figure 10.12. System with utility and distributed resource

Considering that the utility and DG have nearly the same *voltage-regulation* value, the voltage-sag magnitude at the SE connection point (or PCC) is given by,

$$V_s = V_{EPS} - (Z_1 \,//\, Z_{DR})\, I_{sc} \text{ [pu]}, \tag{10.8}$$

where,
- V_s = voltage-sag value
- V_{EPS} = electric power system voltage
- Z_1 = utility impedance
- Z_{DR} = distributed resource impedance
- I_{sc} = short-circuit fault current

The product ZI_{sc} in Eq. 10.8 can be identified. During the protective-device TVC calculation, once the rated current has been adopted, each selected time value will correspond to one value of the voltage-sag magnitude. I_{sc} is expressed as a function of the PD rated current, thus the fault current will be a number of times the rated current.

Normally, fuses are built in a homogeneous way with the same structure and with characteristic curves approximately parallel, thus, increasing the number of fuse elements will increase the rated current. Hence, the I_{sc} value expressed as a multiple of the fuse rated current has a multiplier that increases with the decrease in the fuse rated current and *vice versa*. A short-circuit current of 1000 A will trigger the operation of a fuse of 100-A rated current in nearly 1.4 cycles and the same current will melt the 200-A rated fuse in approximately 20 cycles. In conclusion, the product $Z\, I_{sc}$ can be considered as a function of the source impedance and PD rated current. Thus, a reduction in the short-circuit power (islanding mode) is equivalent to an increase in the rated current of the protective device and *vice versa*.

The previous analysis, although applied to fuses, is applicable also to circuit breakers and to any other type of low-voltage protective device.

The effect of a protective-device rated current can be seen in Fig. 10.2, where the TVCs of the two fuses mentioned above where plotted. To clarify the procedure, the methodology for obtaining a point on each curve is as follows. Considering an application with a load rated current of 1000 A and a source impedance of 4% (source impedance voltage drop due to rated current is of the same value as the voltage regulation), with two feeders coming out from the busbar. They are protected by fuses with rated currents of 100 A and 200 A, respectively.

An arbitrary time of 10 cycles (200 ms) was selected for the explanation, for which the fuses will need currents (to melt) of 600 A and 1200 A, respectively. The fuse-rated currents expressed in pu of the circuit-rated current are 0.1 and 0.2, respectively, then the base current for 10 cycles is 6000 A. Voltage sags that the fuses allow to be applied to SE connected to the parallel feeders are:

- For the 100-A fuse; $V_s\,(\%) = 100 - (0.04\;\; 0.1\;\; 6000) = 97.6\%$;
- For the 200-A fuse, $V_s\,(\%) = 100 - (0.04\;\; 0.2\;\; 6000) = 95.2\%$.

The same procedure allows the calculation of the complete fuse TVC curves.

The protecting-device rated current is crucial for the coordination between the protecting device and the SE immunity curve. This is due to the fact that, as the rating current increases the curves approach each other, and when they intersect there are points of miscoordination.

If, during parallel operation of utility and DG, due to any cause, the IB opens, the circuit suffers a source-impedance increase, modifying the situation. Figure 10.13 was drawn for a similar situation as before, but changing the source impedance from 0.04 to 0.06 pu and maintaining similar fuse-rated currents. It can be seen that the protection given by the 100-A fuse is still satisfactory, but the curve representing the 200-A fuse now intersects the immunity curve of the sensitive equipment, which might signify SE dropout.

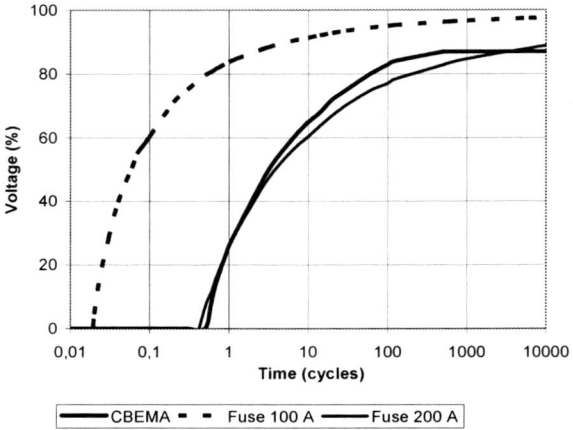

Figure 10.13. Fuse time–voltage characteristics and CBEMA curve for DG alone

It may be deduced that the protective-device TVC has been transformed into a zone bordered by the curve locations when the supply is the utility plus DG and the new curve location having DG as only the supply. In the present case, it is the zone enclosed by the positions of the 200-A fuse curves for both source impedances.

10.4.3 Mitigation of Distant Voltage-sag Penetration into Industrial Premises by Using Semirigid Connection

As was mentioned above, the ability of sensitive equipment to withstand or ridethrough these voltage sags without dropout is a function of the magnitude of voltage reduction and its duration. The curve proposed initially by the Computer Business Equipment Manufacturer Association (CBEMA) has been adopted as the graph of immunity. This means SE will withstand voltage sags represented by points given by their depth and duration as coordinates, which are to the left and above the graph line. The points below the curve and to the right will represent voltage sags that produce SE dropouts.

The concept of having points that represent perturbation depth, duration, and location on the CBEMA graph is only applicable if the voltage-sag magnitude is constant during the event duration. Voltage sag caused by a short-circuit, possesses a depth that depends on the fault current and on the event location, and the time taken for the protective device to disconnect the faulted area. In this case there is no control on the voltage-sag depth but on the duration that depends on the delay in the protective-device response.

For the many cases where the voltage magnitude is far from being constant, the analysis related to SE dropout can be done by using the energy integral during the perturbation and comparing it with the energy necessary for SE energy. If the available energy is higher than that required by the SE, the equipment will withstand the event without dropout [9].

The concept is based on the energy deficit and operating voltage; the ride-through (or dropout) occurrence is determined by comparing the SE prefault stored energy with the energy deficit during the voltage-sag event. Based on the above explanation, the possibility of SE ride-through will be increased if the depth of the voltage sag can be reduced during part of the perturbation duration. For the case of distribution systems with DG, the extra generators can be kept connected in order to support SE, thus reducing the energy supplied to the fault.

Using a connection between DG equipment and the supply system, whose rigidity is variable depending of the working conditions, can carry out this task. If the system is under steady-state conditions, the connection is "rigid" (or of very low impedance). When the presence of an external fault is detected, the interconnection must become nonrigid – having a significant impedance – in order to support the SE and reduce the fault-current contribution.

Figure 10.14 shows a simple distribution system having a DG with a variable interconnection. Z_v&CB represent the combination of limiting impedance and bypass circuit breaker, supplying a rigid connection when the circuit breaker is closed, switching to nonrigid connection when open.

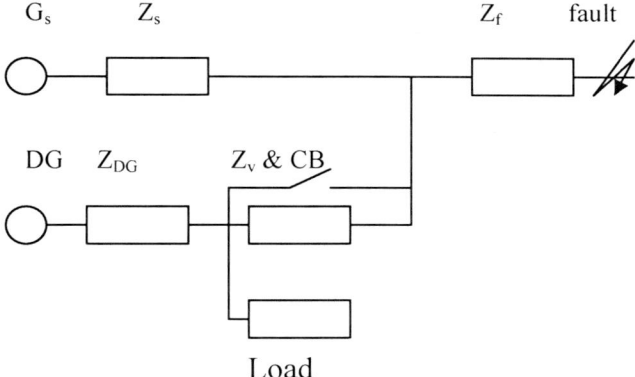

Figure 10.14. Simplified distribution system with embedded DG

Figure 10.15 shows the voltage applied to the SE for a situation similar to that presented in Fig. 10.14 when a fault takes place, assuming impedances and faults that result in voltage sags to 30% (0.3 pu) and 80% (0.8 pu), followed by a voltage swell to 110% (1.1 pu). The duration was assumed by considering the circuit breaker needs 40 ms to detect and transfer the fault current and that the fault is cleared in 160 ms after the fault start. As a portion of the system was severed from it, the assumption was that the voltage would be increased to 110%. It was also assumed, as in the normal situation, that the system power is at least 20 times higher than the DG power in order to neglect the change in fault duration due to the current supplied by the DG. As this hypothetical voltage sag is not of constant depth, the CBEMA curve cannot be directly applied; the use of the specific-energy concept was necessary.

The energy deficit suffered by the customer's SE, without using the nonrigid connection, would be:
$ED_r = (87-30\%)^2 \, 0.16 \text{ s} = 520\%^2 \text{ s}$.

Using the nonrigid connection:
$ED_{nr} = (87-30\%)^2 \, 0.04 \text{ s} + (87-80\%)^2 \, 0.12 \text{ s} = 136\%^2 \text{ s}$.

In spite of the unconventional unit, the energy deficit was reduced from 520 to 136. In other words the voltage-sag effect is reduced to 26% by using a nonrigid connection, thus the possibility that SE would withstand the perturbation is greatly increased.

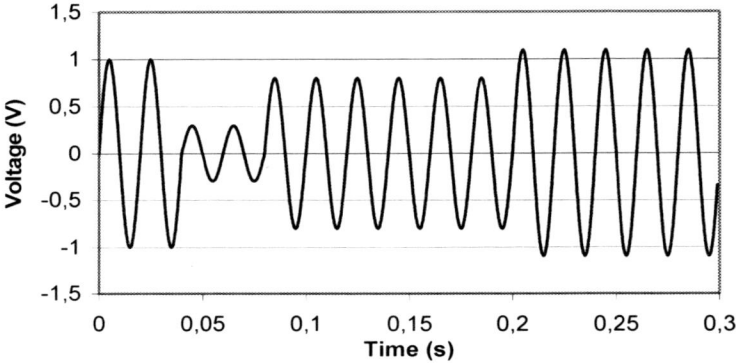

Figure 10.15. Voltage applied to SE during a hypothetical fault

– *Numerical example:*

A typical distribution system with a customer having his own DG was modelled using MATLAB. The system characteristics are: rated voltage 13.2 kV, rated power 30 MVA, DG rated power 3 MVA. A three-phase fault was assumed 10 km downstream of the DG location. Voltage calculations for different fault-currents and limiting-impedance values on two circuit locations were done. The circuit locations were assumed where the DG owner has the SE and where another customer is connected to the main 13.2 kV feeder. For all the calculations the CB detection plus commutation time was kept constant at 40 ms. The comparison is made between the specific energy of SE that belongs to the customer with and without DG and Z_v&CB.

Table 10.2 shows the analytical results where the comparison can be made for different Z_v and fault-current values. The current values are expressed as a percentage of the three-phase short-circuit current located in the part of the distribution system mentioned above.

Table 10.2. Specific-energy deficit depending on Z_v and short-circuit currents

Impedance	I_{fault}	Specific-energy lack with DG and Z_v&CB	Specific-energy lack without DG and Z_v&CB
Ohm	% s-c	%² s	%² s
j 15.7	3	19.22	438.9
j 15.7	6	30.60	276.4
j 15.7	6	26.04	388.7
j 15.7	10	33.31	757.1
j 6.28	50	47.87	1062.6
j 3.14	100	60.83	955.5
j 31.4	100	64.80	952.2

The table shows the remarkable effect on the reduction of the specific-energy deficit caused by the application of a nonrigid connection.

10.4.4. New Overcurrent-protection Schemes using Intelligence

There is a strong trend aimed to add flexibility to the traditional distribution fuse. The cited flexibility can also be called intelligence, where an intelligent fuse is one that is able to make decisions, for instance doing opening and reclosing operations, discriminating direction and detecting phase currents from earth currents.

The current widespread application of distribution fuses is based upon their low cost, high reliability that arises mainly from their simplicity, and from the application of well-known physical principles. Since the earliest fuse, the first scientific reference was by Sir Edward Nairne during 1773 and at first official US fuse patent granted to Thomas Edison by 1880, the main improvements have been aimed at the use of better materials, to extend the current and voltage application ranges, and towards the development of faster and cheaper construction techniques [14, 15].

After these early improvements many changes of the original fuse design have been presented, in order to extend the low current interruption capability having different acceptation or success degrees.

During more than 120 years of fuse application many improvements have been published at conferences and magazines, but just a few have been readily available and of stable market position. The phrase "New fuse design or type for specific applications …" has been groundlessly expressed many times. No critical changes towards adding intelligence to the traditional fuse have been produced lately.

A literature survey — of the innovative additions to the early fuse designs — was conducted to analyse the presented DG problem, just two original ideas have arisen as the application of chemical charges and the use of saturable transformers [15]. The two ideas can be summarised as follows.

a) Application of chemical charges:

The idea was originally presented by Muth and Zimmermann in 1938, based upon the fuse blowing due to the ignition of a chemical charge caused by any externally controlled heater, the device described as the combination of the accuracy of a relay with the cheapness of the fuse. Later, the same idea was pursued and developed, especially on the ignition control system, introducing a device called "limiter" in 1963, which has been on the market ever since. The same concept was further developed and applied to the Automatic Seccionalizer, coming out in the 1980s. In the same decade many other designs using this principle were presented, developed especially for medium-voltage distribution systems, having dissimilar market success. One of them, called the electronic power fuse loaded with up-to-date or last-generation electronics has a relatively good foothold in the market, in spite of its high cost. By 1990 a technical paper was published that presented a new design applying the concept to low-voltage DC systems, called a Smart Fuse. Five years ago, limitations on the application of this chemically charged device in distribution systems were emphasised giving recommendations about how to avoid frequent misapplications [16, 17].

b) Use of saturable transformers:

During the 1970s an interesting idea was proposed, related to the availability in a single fuse cutout of a double fuse time–current characteristic. This double TCC was obtained by using a current transformer whose working zone included the saturated and nonsaturated areas, changing the two paths current sharing depending on the overcurrent level. The design was proposed for the solution of coordination problems between the main feeder and branch protections. Initially, the current transformer was located by the detachable fuse link, lately being attached to the fixed cutout part [18].

One of the most important needs is the availability of medium-voltage fuses designed especially for distribution systems having embedded generation, where particular requirements shall be considered. Figure 10.6 shows a typical system having embedded distributed generation. In Fig. 10.6 the sensitive-equipment and faulted-branch locations are shown in distributed generation and main supply sides respectively. Currently, there is a big movement in the field of distribution generation having DG mainly due to the present strict power-quality requirement, voltage-sag presence is one of the biggest issues [6].

From the voltage-sag viewpoint, it is very important to keep the DG connected and feeding the system during fault events, due to its capability to back up the system voltage, increasing the possibility that the sensitive equipment would remain operative. On the other hand, there is a serious risk of keeping part of the system working under "islanding" conditions. In addition, there are difficulties in getting selective coordination for all possible operating conditions, having the fault energy coming from either the main supply or from the DG, the DG presence in the system is the decision of the DG owner and not of the main utility.

The solution to the listed drawbacks can be done by using complex and expensive protection schemes, as many of the needed tasks are beyond the reach of a traditional fuse. Among the intelligence factors that a current fuse needs to have, are directionality and discrimination if ground is included in the fault path. In addition, the solution must be of low cost so as to keep the complex fuse as an attractive possibility if compared with other choices.

Both ideas have been combined in a design that analytical studies have shown to be very auspicious. As the design is entering the patenting process, no further details can be provided here [15].

10.5 Impact of DG on Recloser–Fuse Coordination

The recloser is an invaluable tool for the reliability of rural and low-power distribution systems. The success of its application is based on the fact that most faults are of a temporary nature or start as temporary (approximately 85 to 90% of faults), and if momentarily interrupted they are self-eliminated during the reclosing interval. In the case of a recurrent fault (or a permanent one), another fast (or delayed) operation is provided in order to allow the operation of the down-stream fuse, isolating the faulted-system sector in this way. At each reclosing operation the adjacent customers are subject to a new voltage sag. Some utilities especially those with a high percentage of urban and suburban loads, have changed their reclosing-protection practices, eliminating the second instantaneous trip (fuse-saving operation), as can be seen in Fig. 10.16 [19].

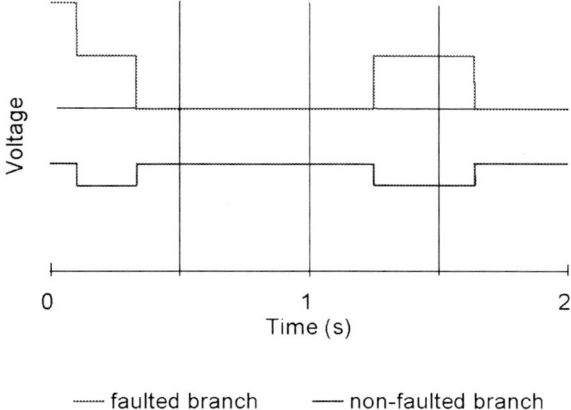

----- faulted branch   —— non-faulted branch

Figure 10.16. Voltage sags generated by a recloser without the second instantaneous operation

Generally, the first fast operation eliminates 85% of the faults, 4% are cleared during the second reclosing interval, and only 1% are eliminated in the third interval. Ninety percent of faults are removed without fuse operation or maintenance calls [20]. Unfortunately recloser and DG are irreconcilable; if DG is left connected during the recloser fault-clearing operation, fault clearing can be incorrect and there is a high risk of DG damage when reconnected.

One of the main problems is the possible lack of coordination during the fast operation of the first recloser when DG is connected at any point between the recloser and the fuse nearest, and upstream, to the fault. The current flowing through this fuse is supplied by the power system as well as by the DG, both can cause fuse operation during the first fast recloser operation, making the first reclosing interval useless for fault self-clearing. During the reclosing time the DG, if it remains connected, will supply current to the transient fault avoiding its deionisation. Experimental studies have indicated that a current of 5 A supplied by the DG during the reclosing interval would be large enough to sustain the fault, preventing the needed self-clearing from taking place [21]. Thus, the recloser's main advantage is eliminated, and the protected equipment will suffer unnecessary "through faults", shortening the equipment life span and increasing repair and maintenance cost. Also, customers will experience several long interruptions, losing the benefit supplied by recloser applications. The temporary fault is then considered permanent and the recloser would move to delayed operation and possibly to a lockout position. In this way the advantage of clearing nearly 85% of the faults at the first operation is eliminated.

The recloser user can normally adjust the reclosing interval in order to meet its own requirements. A reclosing interval of 2 s is normally used; a value that traditionally was able to solve most of the existing problems. At present there is a strong tendency to reduce this value to 0.5 s in order to improve the power quality of customers having digital clocks, reducing customers' complaints due to the nuisance of digital-clock blinking. The use of instantaneous reclosing when DG remains connected worsened the previously described problem.

However, if DG is connected to the circuit it is necessary to increase this interval in order to be sure that the DG breaker has separated the generation before the recloser reconnection. Some utilities are requiring a 5-s reclosing interval on feeders having DG [6].

The problem of out-of-phase connection could cause physical damage to the DG. This is an old issue that has been studied for induction motors — when suffering short interruptions or voltage sags — that run as induction generators during the event. The study normally concludes with the resolution of keeping the motor connected for at least 500 ms in order to keep the production line in operation as long as possible [22]. The main difference from DG is that the motor slows down and perhaps stalls, subsequently reducing its emf. In the case of DG, the generator excitation maintains the emf and the machine is kept running by the prime mover, causing a speed increase [3]. It can be concluded that there is a need for an engineering tradeoff between keeping DG connected and then improving the continuity of part of the system (islanding and its own risks), or quickly disconnecting it in order to increase the reliability of the whole distribution system, which is detrimental to the users' quality of power.

10.6 Harmonics Generated by Distributed Generators

Synchronous generators, when well designed and built, do not generate any harmonic voltage. On the other hand, generators using modern switched inverters (such as induction, photovoltaic, fuel cell, etc.) should theoretically have a low level of low-order harmonics that the distribution engineer is normally concerned with. By controlling the switching, a relatively clean current waveform can be achieved, which does not mean that there are no harmonics problems. Underground cables seem to create natural resonance on the utility distribution system in the 25th to 35th harmonic range. Some discussions have been started in the USA with respect to the agreement with the stringent IEEE Standards 519 limitations on generators of some popular inverters. It is argued that this equipment can be considered as a negative load, in which case the limits change from 5% to 20%. Besides, there is no problem to introduce in the inverter the filtering function in order to reach the specified limits, but the cost would increase [23].

10.7 Flicker Due to Wind Gusts and Tower Shadow

According to the IEEE Standard Dictionary of Electrical and Electronics Terms, flicker is the "impression of fluctuating brightness or colour, occurring when the frequency of an observed variation lies between a few hertz and the fusion frequency of images". If the voltage variation is large enough or in a certain critical frequency range, the equipment performance can be affected. The main problem of voltage fluctuations is usually the effect on lighting loads, mainly when the illumination of the lamp varies with frequencies between 1 and 10 Hz. Additionally, the phenomenon can slightly affect other types of loads such as motors, electronic devices and process controllers. Voltage fluctuations may originate in the power system, but most frequently in the equipment or load connected to it. The main voltage-fluctuation generators are arc furnaces, welders, alternators, and motors [24, 25].

The application of wind turbines in isolated form or in a wind farm is becoming widespread, thus the quality of the generated voltages and their impact on the power grid, especially the high-power fluctuations resulting in flicker, are of great concern to utilities and customers.

The main cause of flicker is the frequent switching on and off when the wind powers the generator to a speed around the cut-in speed. The utility to which the generators are connected usually limits the switching operation number to no more than 3 to 4 times per hour. One of the possible solutions is to keep the rotor at standstill until the wind reaches a stable speed beyond the cut-in speed, not letting it free until the speed is exceeded by at least 1 m/s, and similarly to the switching-off criterion. In this way, a certain hysteresis in the switching is introduced, which reduces the flicker issue. This problem is very noticeable when many individual generators form the wind farm, and obviously the wind energy is not uniformly distributed among them.

Another utility compulsory limits is related with the network minimum short-circuit level, which should be at least 20 times the wind-generator rating. This limit is not easy to fulfill due to the fact that strong winds are available in the non-highly urbanised areas, where the supply systems are usually very weak.

The wind turbine generator is also characterised by a low-frequency mode of oscillation. Random wind fluctuations, wind shear and tower shadow effects (blade passage along the tower), may excite this mode producing large ripple on the drive-train torque as well as on the generated electric power, being noticeable as a voltage flicker at approximately 0.5% and a frequency between 2 and 4 Hz.

A flicker coefficient for the wind turbine, a function of the specific wind turbine and average wind speed, has been proposed in Denmark (one of the most developed countries with respect to wind turbines), which should be between certain limits in order to get installation approval. This coefficient is closely related to the short-term flicker emission P_{st} previously defined and allows its calculation taking into account the system short-circuit power [24].

$$P_{st} = [\cos \varphi_{sc} + dQ/dP \sin \varphi_{sc}] S_{wn} / S_{sc} I_w f_w(W) \qquad (10.9)$$

where:
φ_{sc} = short-circuit angle
Q = reactive power
P = active power
S_{wn} = wind turbine rated power
S_{sc} = supply system short-circuit power
I_w = average value of wind turbulence intensity
$f_w(W)$ = flicker coefficient (wind speed function)

The joint effect of several wind turbines on the P_{st} value can be assessed by the application of a methodology analogous to the procedure designed for the flicker generated by arc furnaces.

One of the most popular wind-energy conversion systems uses slip power recovery. These systems employ a double-output induction generator, which is connected directly to the grid by stator and through a static converter by a rotor. The existence of ripple in the generated torque may damage the drive-train components; besides fluctuations of the generated electric power may cause a considerable flicker in weak grids. Several control strategies of the static converter have been introduced in order to provide damping to the oscillation mode. One of the proposed control systems follows the wind-speed variations resulting in a large reduction of torque ripple and consequently control the voltage fluctuations.

10.8 Ferroresonant Overvoltages

There are some reports available that indicate that the application of DG through a service transformer could lead to the presence of ferroresonant voltages. This ferroresonance phenomenon, not exclusive to DG application, is presented when a grounded delta-wye transformer is used as DG service equipment, having the DG on the wye side and the delta side is supplied by an underground cable. If one of the phase voltages from the supply is missing, the DG protection disconnects the equipment, then a new series circuit through the transformer magnetising inductance and the cable capacitance is formed that leads to high voltages and currents. If the phenomenon is not detected for a relatively long time, the overvoltages can damage the transformer and its arresters [26].

10.9 Conclusions

The new scenario including distributed generation, changes the well-known protection concepts. Sensitive-equipment protection against voltage sags can be given for overcurrent protective devices, provided that careful coordination has been carried out. The readily available protective device TVC now moves into a zone that shall be above the SE immunity curve if effective coordination is desired. A simple methodology has been presented, and its fundamentals and difficulties are explained. The analysis is simple when fuses are used as protective devices when using more sophisticated equipment having variable TCC. The methodology has been specifically explained when applied to circuits having distributed generation, which produces a sharp impedance change. The application of a nonrigid connection between the customer's DG and the system supply can be used as an able tool to reduce the effect of voltage sags caused by faults located outside the customer circuit. Coordination studies could be facilitated if PD manufacturers would provide a new set of data for their products. Distribution systems is still a field where fuses are strongly suited, thus any effort towards sustaining and increasing their presence would be worthwhile. Medium-voltage distribution fuse technology is urgently in need of new and fresh ideas. The information needed to carry out the proper study under the new situation is available. By using the traditional PD and wise judgment of the distribution engineer, a reasonably good coordination can be reached at a reasonable cost.

References

[1] Borbely, A. M., Kreide, J. F. (2001), Distributed Generation, The Power Paradigm for the New Millennium, CRC Press, Boca Raton, FL
[2] Gómez, J. C., Morcos, M. M. (2005), Coordination Analysis of Voltage Sag and Overcurrent Protection in Electrical Systems with Distributed Generation, IEEE Transactions on Power Delivery, 20: 214–218
[3] Gómez, J. C., Morcos, M. M. (2002), Coordination analysis of voltage sag - fuse characteristics in electrical systems having embedded generation, Second International Symposium on Distributed Generation: Power System and Market Aspects, October, Stockholm, Sweden.
[4] IEEE Standard 1547 (2003), IEEE Standard for Interconnecting Distributed Resources with Electric Power Systems
[5] ANSI/IEEE Standard 1001 (1988), IEEE Guide for Interfacing Dispersed Storage and Generation Facilities with Electric Utility Systems
[6] Dugan, R. C., (2002) Impact of DG on Reliability, Transmission & Distribution World, October: 50–55
[7] IEEE Standard 446 (1995), IEEE Recommended Practice for Emergency and Standby Power Systems for Industrial and Commercial Applications

[8] Djokic, S., Desmet, J., Vanalme, G., Milanovic, J., Stockman, K., (2005) Sensitivity of personal computers to voltage sags and short interruptions, IEEE Transactions on Power Delivery, 20: 375–383
[9] Gómez, J. C., Morcos, M. M. (2002), Voltage sag and recovery time in repetitive events, IEEE Transactions on Power Delivery, 17: 1037–1043
[10] IEEE Standard 1346 (1998), IEEE Recommended Practice for Evaluating Electric Power System Compatibility with Electronic Process Equipment
[11] Gómez, J.C., Morcos, M.M.(2000); Effect of distribution system protection on voltage sags; IEEE Power Engineering Review, 20, 66–68
[12] Gomez, J. C., Campetelli, G. N. (2000), Voltage sag mitigation by current limiting fuses; IEEE IAS Annual Meeting, October, Rome, 8–12, 5: 3202–3207
[13] International Standard IEC 60282 (2005), High-Voltage Fuses, 6th edn
[14] Wright, A., Newberry, P.G. (1982), Electric Fuses, Peter Peregrinus Ltd., IEE, London
[15] Gómez, J. C. (2003), Intelligent fuse for M.V. distribution systems: a current need, 7th International Conference on Electric Fuses and their Applications, Gdansk, Poland, 50–54
[16] Lapple, H. (1952), Electric Fuses, Butterworths Scientific Publications, London
[17] S&C (1991), S&C Fault Fiter: Electronic Power Fuse, Descriptive Bulletin 441–30
[18] Aubrey, D. (1974), New 11 kV expulsion fuses for overhead lines, Electrical Times, August 1
[19] Dugan, R. C., McGranaghan, M., Beaty, H. W. (1996), Electrical Power Systems Quality, McGraw-Hill, New York
[20] Electrical distribution system protection (1990), Cooper Power Systems Bulletin 90020
[21] Gómez, J. C., Tourn, D. H., Amatti, J. C. (2003), Experimental determination of the reclosing time self-extinguish current for its application to distributed generation - reclosers coordination studies, CIRED, May, Barcelona, Spain
[22] Gómez, J. C. (2006), Efecto del Aumento de la Velocidad de la Protección contra Sobrecorriente en la Calidad de Potencia, VIII Simposio Iberoamericano sobre Protección de Sistemas Eléctricos de Potencia, May, Monterrey, Mexico
[23] Dugan, R. C., McDermott, T. E. (2000), Distributed Generation and Power Quality, PQA 2000 North America, May, Memphis, USA
[24] Morcos, M. M., Gómez, J. C. (2002), Flicker Sources and Mitigation, IEEE Power Engineering Review, 22: 5-10
[25] Sorensen, P., Tande, J. O. Sondergaard, L. M., Kledal, J. D. (1996), "Flicker emission levels from wind turbines," Wind Engineering, 20: 39–46
[26] Dugan, R. C., McDermott, T. E. (2002), Distributed Generation, IEEE Industry Applications, March / April, 19–25

11

Electronic Loads and Power-quality

Antonio Moreno-Muñoz and J. J. G. de la Rosa*

Área de Electrónica,
Departamento de A. C., Electrónica y Tecnología Electrónica,
Universidad de Córdoba,
Campus de Rabanales,
E-14071 Córdoba, Spain
Email: amoreno@uco.es

Área de Electrónica. Departamento de ISA, TE y Electrónica*,
Universidad de Cádiz,
Avda. Ramón Puyol, S/N,
E-11202-Algeciras-Cádiz, Spain

11.1 Introduction

Today's businesses depend heavily on electrical services for lighting, general power, computer hardware and communications hardware. With the generalised use of computers, adjustable-speed drives (ASDs) and other microelectronics loads, the subjects related to power-quality and its relationship to vulnerability of commercial and industrial plants are becoming an increasing concern not only to the utility companies but, what is more, to the end-customer.

Although it is common that public opinion considers utilities as the source of power-disturbance problems, they frequently argue that there are circumstances beyond their control. Things like lightning, large switching loads, nonlineal load stresses, inadequate or incorrect wiring and grounding or accidents involving electric lines. These can create problems to sensitive equipment if it is designed to operate within narrow voltage limits, or it does not have adequate ride-through capabilities to filter out fluctuations in the electrical supply [1]. Modern scientific instruments, computing and communications equipment are highly reliable but they are peculiarly sensitive to certain kinds of electrical perturbations. Some of the larger computing platforms are sensitive to earth-bound noise, and these require special treatment.

In the 21st century, instrumentation and control operations require high quality and ultrareliable power in the quantities and time frames that have not been experienced before. It has been estimated that more than 30% of the power currently being drawn from the utility companies is now heading for sensitive equipment, and this is increasing [2]. As of today, no standard exists that clearly defines the roles and responsibilities of the energy provider, the high-tech equipment manufacturer, or the high-tech facilities themselves in mitigation of PQ-event-caused losses [3]. In the new and dynamic deregulated electricity environment, this is opening business opportunities for acute energy providers.

Reliability is the ability of the power system to supply energy within accepted standards and in the amount desired. For high-value critical-process or computing programs, the availability in "nines", as shown in Table 11.1, is often stated as being the adopted reliability.

Table 11.1. Comparative levels of availability

Nines	Availability	Unavailability	Lost time per year		System type
1	90%	10%	876	h	Unmanaged
2	99%	1%	87.6	h	Managed
3	99.9%	1.0E-3	8.76	h	Well managed
4	99.99%	1.0E-4	0.876	h	Fault tolerant
5	99.999%	1.0E-5	5.3	min	Highly available
6	99.9999%	1.0E-6	31.8	s	Very H. A.
7	99.99999%	1.0E-7	318	ms	Ultra H. A.
8	99.999999%	1.0E-8	31.8	ms	Ultra H. A.

From a service-interruption service perspective, an urban power supply is delivered with a reliability in the range of 5 to 6 nines, while rural electric customers typically experience a low reliable power, in the range of 2 to 3 nines. Reliability is measured using various indices characterising frequency, duration, and magnitude of adverse effects on the electric supply. It has been recognized that measures of reliability should include some other power-quality issues (such as voltage-dip and swell disturbances) that are becoming increasingly significant in the digital age [4]. For high-value critical-process information services contemporary thoughts on this topic are quite differently focused; modern power-system reliability relates more to[5]:

- Security. Reducing the vulnerability of the information technology equipment and electricity infrastructures.
- Availability and Reliability. Supporting extreme bus voltage reliability, for example "eigth nines" (i.e. 99.999999 availability), or nine nines or even higher.
- Quality. Assuring power-quality for very large numbers of digital devices.
- Self-healing and redundant. Using distributed energy sources (DERs) to improve reliability.

Electronic Loads and Power-quality 327

Table 11.2. Disturbances on computer-related equipment

Electromagnetic disturbance	Equipment problem	Cause of disturbance	Threshold level of disturbance	Duration of disturbance
Impulsive and oscillatory transient	Lock equipment up, processing glitches of an unpredictable nature, miscellaneous soft errors, hard-disk crash, power-supply failure, circuit-board failure	Lightning, power network switching (large capacitors or inductors), SCR-controlled loads, variable-speed drives, photocopiers and operation of other loads	0 to 4 pu rated rms voltage	50 ns to 1 ms and oscillatory 0.3 ms to 5 μs
Voltage dip (sag)	Miscellaneous, soft errors, reset-reboot	Power-system fault, large load start up, faulty circuit breakers and loose wiring	0.9 pu to 0.1 pu rated rms voltage	Short duration: 0.5 to 30 cycles; 30 cycles to 3s; 3 s to 1 min Long duration >1min
Voltage swell	Miscellaneous soft errors, fuses or circuit breakers may be tripped, power supply shut down, circuit board damage	Power-system fault and large load disconnect	1.1 to 1.8 pu rated rms voltage	Short duration: 0.5 to 30 cycles; 30 cycles to 3 s; 3 s to 1 min Long duration >1 min
Interruption	Service disrupted, equipment shut down, hard disk crash, power-supply failure	Power-system faults, local circuit breaker trip, loose wiring and equipment failure	< 0.1 pu rated rms voltage	Short duration: 0.5 to 3s; 3 s to 1 min Long duration:>1min
Earth current (caused by earth fault or other cause)	Disruption of communications systems, destruction of computer and communications hardware	Earth currents to a building's earthing system may return to the main switchboard via other paths		

11.2 Electromagnetic Disturbances

Allen and Segall [6] conducted one of the most respected studies and, although based in the USA in 1974, their report is still quoted today. They monitored AC power to IBM equipment at 200 locations in 25 cities across the USA, and recorded the various AC power anomalies that disrupted the equipment operation during a two-year time span. The Allen–Segall study concluded that 88.5% of AC power problems were transient related. Allen and Segall found that the most disruptive (49%) of power disturbances stemmed from oscillatory, decaying transients. Other studies [7] recorded nearly double the frequency of disturbances at Bell Telephone sites. Today, most electronic equipment uses switch-mode power supplies whose susceptibility to transients and common-mode noise is far greater than traditional linear power supplies.

Clean, uninterrupted power is critical for communication and computer systems. If power fluctuates for just a few moments, data can become corrupted or lost. Internal system communications can lock up and require a reboot, damaging sensitive components and interrupting crucial procedures. A number of power events can affect computing and mainframes, servers, personal computers, these are summarised in Table 11.2 [8, 9, 10].

The well-known ITIC (Information Technology Industry Council) curve [11], formerly named CBEMA (Computer Business Equipment Manufacturer's Association) curve, is generally used to evaluate operational voltage limits for electronic-equipment power supplies. The curve is considered to be a typical design objective for computer-hardware designers. The curve establishes the magnitude and duration limits within which input-voltage variations do not affect the reliability of the electronic equipment. As can be seen, the steady-state tolerance envelope is in the range of ±10% from the nominal voltage. Within this range, the equipment will behave properly. For shorter-time events the tolerance is expanded. For example, voltage-dips down to 70% of nominal are permitted for up to 0.5 s, while on the other hand, voltage swells can be permitted rises of up to 120%. This sensitivity curve only applies to IT equipment; other equipment generally does have an entirely different sensitivity characteristic, unfortunately unknown for the majority of equipment [2]. The ITIC curve was specifically derived for use in the 60-Hz, 120-V distribution voltage systems. The guideline expects the European user to exercise their own judicial decision when translating that curve on to equipment operating under 50-Hz, 230-V distribution voltage systems [12].

11.3 The Rabanales Campus Case Study

The Rabanales Biotechnology Campus [13] is supplied through a 20-kV feeder emanating from the near substation of "La Lancha". This feeder serves 17 buildings that includes the data centre, the departmental R&D laboratories, the veterinary hospital, the main library, the lecture hall building, the "Lucano" residence hall, the sports facilities and the train station.

Each building houses a 1000-kVA delta-wye transformer. This is used to step down 20 kV to 230/400 V for the panel boards distributed throughout the building floors.

While the campus server room is served via an uninterruptible power supply (UPS), the rest of the campus sometimes experiences generalised power perturbations. This situation has led to subtle problems such as computer lockups and electronics component wear, as well as more devastating damage including data loss, disk crashes and burnt electronics (Table 11.1). As a result, an analysis of power disturbances at the Rabanales campus has been carried out.

To identify the most likely causes of the problems, onsite inspections of equipment and installations were conducted. These included a walk-down of the facility's electrical system to inspect the condition of equipment as well as interviews with facility electrical personnel and the end-users of the failed equipment. Electronic equipment that had been particularly sensitive to power disturbances was identified and collected. At the same time, electromagnetic compatibility and general equipment literature was reviewed.

Each building has an air-conditioning system and elevator, both powered via a separate branch circuit. Meanwhile, in each floor, a subpanel serves both the fluorescent lights and the electronic equipment separately. These power circuits are likely to contain more than 120 personal computers and terminals, several printers, copiers, facsimile machines, in addition to chromatography equipment, spectrometers, ultraviolet spectrophotometers and numerous microprocessor-based control and instrumentation devices.

After taking measurements over a 13-day period on a typical building entrance, it was evident that the voltage harmonics level was relatively high. Campus starts activity at 7.00 am and is characterised by a rise in the harmonic level.

As expected, the dominant harmonic was the fifth harmonic, which follows the typical daily load pattern. This peaked at 2% with total harmonic distortion (THD) being 2.41%. However, the 6% compatibility limit for that harmonic order – as dictated by the EN 50160 standard – was not violated.

Electronic-equipment power supplies were the most likely cause of this 5th harmonic level. In general, the harmonic pollution levels from work activity peaked around mid-afternoon and were lower during the morning. These measurements were taken during the summer, so the likely cause of the higher afternoon levels would have been air-conditioner actuation between midday and 7.00 pm. Lighting loads would also be lower in the mornings.

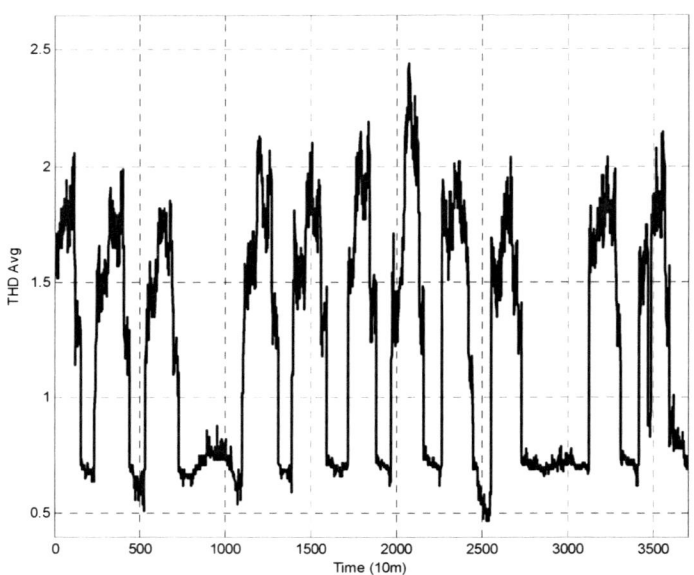

Figure 11.1. Voltage THD week pattern at LV typical building entrance

The current 3rd, 5th and 7th harmonic levels were found to be below EN-61000-3-4 standard limits. However, the total harmonic distortion – coming in at 27.54% – violated the 20% limit reccomended for the entire installation.

To trace the harmonic source, measurements were taken across the system as the closer you get to a source, the higher the current-distortion level. In this instance, current distortion was found in the organic-chemical laboratory. Here, there had been continuous disconnection of the residual-current circuit breakers caused by the impact of spectrometer and chromatography equipment currents.

The current THD maximum for the entire laboratory was 58.98%, while the maximum value for the neutral current was 83.51%. The phase current 3rd, 5th and 7th harmonic levels had maxima of 20%, 11% and 5% respectively, while the neutral current 3rd, 5th, 7th and 9th harmonic levels peaked at 35%, 30%, 12% and 23% respectively.

The highest degree of current distortion was found in the campus server room, with the THD actually reaching 100%. The phase current 3rd, 5th and 7th harmonic levels had a maximum of 61%, 50% and 38% respectively.

After monitoring the campus for a year, results showed that disturbances were within the voltage-tolerance envelope of the ITIC curve. This indicated that voltages were stable with only a daily fluctuation and the occasional minor dip and swell.

But although voltage outages were not recorded, findings over the relatively short power audit are not conclusive indications of either the long-term facility or the utility's continuity of service. Indeed, during the monitoring time, and after some recent power disturbances, computer power supplies were damaged – most likely due to transients – and had to be replaced.

11.4 The Infrico Case Study

The Infrico facility [14] is supplied through a 25-kV feeder emanating from a distant substation. The facility has its own 630-kVA delta-wye transformer that step down the 25-kV to 230/400 V for the panel boards distributed through the building. The Infrico plant is an up-to-date semi automatic factory devoted to the manufacturing of refrigerators and cold systems for the hostelry and pastry industry. Recently it has incorporated a flexible manufacturing system (FSM) from a world-leading supplier; this is a modular sheet metal FSM for the assembly of the stainless steel cabinets manufactured for the different refrigerators included in the Infrico catalogue.

The cells integrated in the FMS feature either integrated bending, right-angle shearing or laser cutting, enhanced by including a hydraulic or servo electric turret punch and its automatic loading-unloading function. In addition, an automatic material-handling module has been incorporated that can operate as an automatic raw-sheet storage and buffer storage between two fabrication processes or skeleton handling system (including even a possibility of sorting according to material type).

The Infrico maintenance department required the collaboration of the local utility, but after numerous and neglected complaints the analysis of power disturbances was proposed to the authors. As before, in order to identify the most likely causes of problems detected, onsite inspections of equipment and installations were conducted over the first week of this study.

Aiming to reduce the voltage-dip effects on Infrico operations a measurement campaign was conducted, in which a monitor device was connected to the low-voltage building entrance and to three more locations inside the industrial facility.

The monitoring device selected was a portable, standalone, three-phase power-quality analyser. Some of the key monitor requirements were: ability to transfer the surveyed data to an inhouse computer program, appropriate numerical storage and low cost and ease to use. The duration of this monitoring campaign was from May 2004 through April 2005. Although the device installed can monitor various power-quality disturbances, only rms variations (dips and swells), interruptions, and outages are considered.

A voltage-dip survey is a very common tool to obtain statistical information on the power-quality performance of a site after the period of observation. The combined information obtained from these surveys gives a good impression of the quality of the supply in the facility.

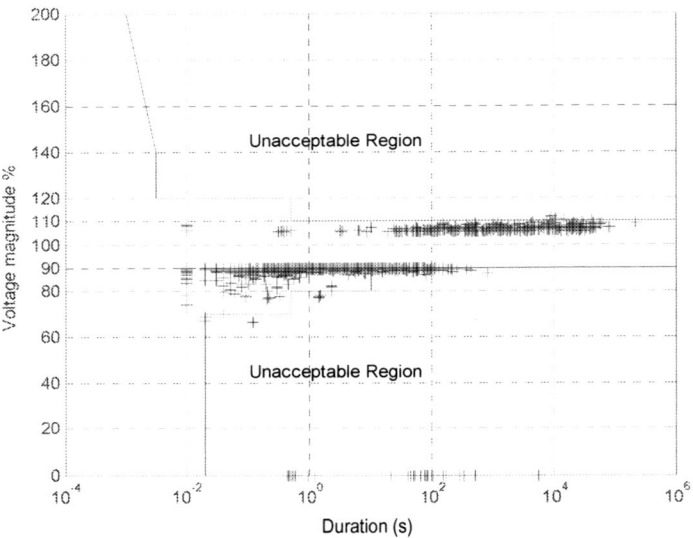

Figure 11.2. Anomalies detected at LV locations over the year on an ITIC curve

Figure 11.2 illustrates all the rms variations that were recorded at the facility (all four locations, all phases) during the monitoring period. This kind of diagram is known as a magnitude–duration scatter plot. It also translates information from the well-known ITIC (Information Technology Industry Council) curve [11]. This curve is generally used to evaluate operational voltage limits for electronic-equipment power supplies. We consider this curve only as a reference. The study indicated that an average site will experience about 45 voltage-dips and five momentary interruptions every month.

A common way of presenting voltage-dip survey results is from a density table. Following the method recommended in IEEE Standard 493 [15] and IEEE Standard 1346 [16], it breaks the dips and interruptions down into count versus magnitude versus duration. Each bin of the table gives the (average) number of voltage-dips within the given range of magnitude and the given range of duration. One step further of presenting the results is through a cumulative table. It shows the number of events per site per year that are more severe than the given magnitude and duration.

Electronic Loads and Power-quality 333

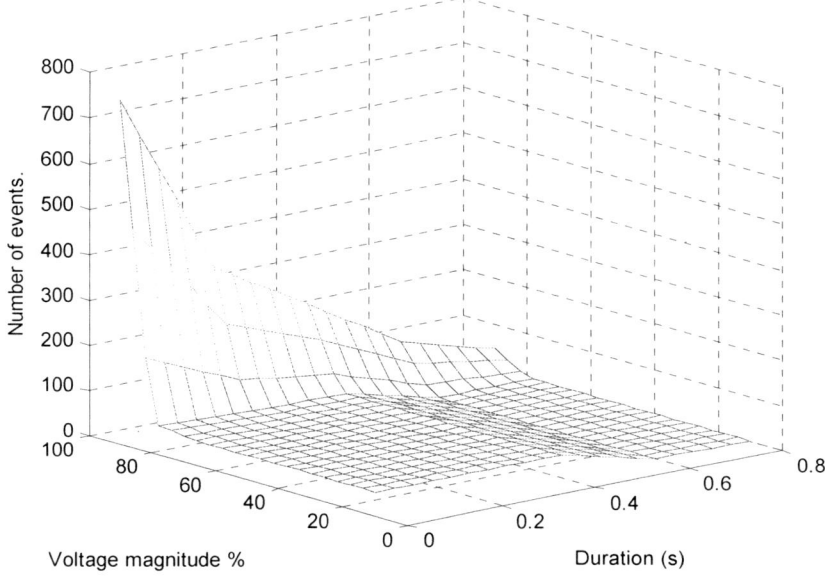

Figure 11.3. Cumulative dip frequency

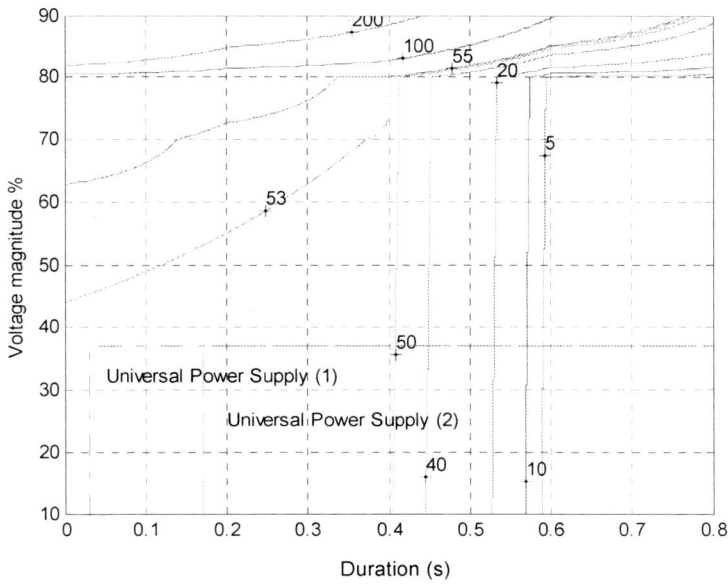

Figure 11.4. Voltage-dip contour chart

The values obtained in the cumulative voltage-dip table can be interpreted as values of a two-dimensional function that gives the cumulative dip frequency as a function of magnitude and duration. For the figure intermediate function values have been calculated by using an interpolation algorithm. The voltage-dip contour chart corresponding to the Figure 11.3 is shown on the Figure 11.4, with contours indicated for dip frequencies equal to 5, 10, 50, 100 and 200 events per year.

The sensitivity of equipment (power supply in the figure) to voltage-dips is usually expressed only in terms of the magnitude and duration of the voltage-dip. For this purpose, a "rectangular voltage-tolerance curve" is used. This curve indicates that voltage-dips longer than the specified duration and deeper than the specified voltage magnitude will lead to a malfunction. The information obtained from a device can be directly related to the voltage-dip contour chart corresponding to the facility. Thus, in this case, if a power supply can tolerate voltage-dips down to 37% for durations up to 20 ms, the equipment will trip 53 times per year.

The main conclusion of this study is that the occurrences of voltage-dips at the facility were, indeed, greater than normal. Each event results in 30 min of downtime. Following the financial evaluation method proposed in annex A of the IEEE 1346 Standard [13], each disruption costs about 1300 €. The results from the audit show that from the total disruptive dip per month, seven occur when the plant is in production.

Table 11.3. Operating and maintenance comparative cost of mitigating equipments

Mitigating equipment	Typical cost (€)	% of initial costs per year
Transient suppressor	15–100	5
UPS	500–1000/kVA	25
Standby UPS	100–1000/kVA	25
Shielded isolation transformer	20–60/kVA	15
Line conditioner	≤ 300/kVA	10
Written pole MG set (15 s)	1100/kVA	5
Dynamic dip corrector	184/kVA	5
CVTs	1000/kVA	10
Low-speed flywheel (15 s)	265–400/kVA	5
High-speed flywheel (15 s)	750/kVA	7
Static transfer switches	150,000 (10 MVA)	5
Ultracapacitors (10 s)	~1000/kVA	5
Fuel cells	~1500/kVA	10

11.5 Mitigations Technologies

Table 11.3 provides an example of cost for some general mitigating technologies used for end-user equipment protection [17, 18, 19, 20, 21]. Achieving "high-nines" reliability it is not possible using the existing bulk supply system without considerable enhancement. Such ultrahigh reliability protection of communication and computing equipment can be achieved by the use of power-electronics.

Recently, new technologies, like custom power devices based on power-electronic concepts, have been developed to provide protection against power-quality problems. Generally, custom power devices [22] are divided into three categories such as static series compensator like the DVR, static shunt compensator like distribution static compensator (DSTATCOM), and static series and shunt compensator like the unified power-quality conditioner (UPQC).

The essence of DVR is that a voltage is injected via a series-connected boost transformer to achieve the load voltage previous to the voltage-dip. This power-electronic converter comprises the following major components, namely: a capacitor bank, pulse-width-modulated (PWM) inverters, AC harmonic filters, injection transformers and an optional energy-storage system. Within the family of DVRs there exist various compensation techniques.

It is possible to correct voltage-dips using conventional technologies, such as uninterruptible power supply (UPS) systems. This implies that the amount of energy that a UPS is required to store is based upon the long duration of a typical voltage outage, this may not be necessary in the case of a voltage-dip. In addition, a UPS would need to be able to withstand not only the load current, but also the full load voltage.

However, from the perspective of most end-users, nowadays the most cost-effective technology continues to be a double-conversion online uninterruptible power supply (UPS). As opposed to offline or line-interactive UPS (only recommended for outages), the double conversion topology is the only recognised as true "online" topology for use in high-reliability applications. This is an universal correction device, because it has typically been designed to protect against all types of power contamination and create clean, perfect sine-wave power for downstream systems. Critical equipment and ICT systems are completely isolated from raw utility power and all its irregularities, using the AC–DC dual conversion scheme: incoming utility AC power is conditioned and converted to DC power through the rectifier, a small portion of which is used to charge the UPS battery; the remaining DC power travels to the inverter, which produces new, perfect sine-wave AC power to deliver to the instruments.

In addition, in the selection of a UPS it is important to consider the harmonic distortion that the rectifier can reflect back onto the facility power for the rest of the facility. Some of the best UPS are equipped with input power-factor-correction (PFC), which uses a unique IGBT (insulated gate bipolar transistor) rectifier to maintain an input power factor of 0.98 lagging for practically distortion free (< 3% THD) input. With this active solution the use of the costly and bulky passive input filter is overcome, allowing a 20% reduction in UPS power consumption.

Even with a uninterruptible power supply (UPS), it takes careful configuration planning to achieve the "five to six nines" of reliability demanded by high-tech facilities. The main element that contributes to system reliability is redundancy. In its basic form, distributed redundancy involves creating two (redundant) UPS system buses and redundant power-distributed systems. This eliminates as many single points of failure as practical; all the way up to the load equipment's input terminals. In order to provide "fault tolerance", some method of allowing the load equipment to receive power from both UPS power buses must be provided.

The most advanced online UPS are scalable, with advanced options that are designed for maximum uptime, even those applications where the power simply cannot be interrupted under any circumstances, including routine maintenance. These designs include parallel redundancy, a microprocessor-controlled system, a power-factor correction rectifier, an internal bypass, and multiple monitoring and communication options.

To protect against fast power system failures, such as circuit-breaker trips or a power-system fault, you need a fast switching approach. Static transfer switches (STSs) have been applied to accomplish very fast break-before-make transfers between two AC power sources. It is important that the two AC power sources be designed as independent as practical to eliminate any common failures.

Overall, the factors that improve reliability are:

- module redundancy;
- increased UPS bypass source reliability;
- individual battery systems;
- simplified operator interfaces and procedural safeguards;
- IGBT inverter technology (due to lower parts count) ;
- use of recognized and agency-listed standard components;

On the contrary, factors that decrease UPS reliability are:

- complicated switchgear systems;
- shared controls or single points of failure;
- common battery systems;
- system complexity;
- poor environmental conditions;
- UPS topologies with narrow input voltage and frequency windows or topologies.

11.5.1 Leakage Current

Leakage current can be defined as all currents, including capacitively coupled currents, which can be conveyed between exposed conductive surfaces and earth or other exposed conductive surfaces. Specifically, earth leakage current is current flowing from the mains part through or across insulation into the protective earth supply connection. Enclosure leakage current is current flowing from the enclosure to the operator or user, with the operator or user referenced to earth or to another part of the enclosure. Leakage current should be kept within acceptable limits for protection against electric shock.

Today's electronic equipments, with pulse-width-modulation (PWM) switch-mode power supplies, generate HF harmonic currents that can disturb sensitive equipment. If this is the case, the solution could be to insert an isolation transformer with a screen or to use ferrite cores, which attenuate disturbances of several tens of MHz in magnetic materials. However, electromagnetic compatibility (EMC) standards state that these HF currents must be shunted to earth, resulting in the presence of EMI filters and thus capacitors between phases and the frame.

According to the number of loads, their contribution to the network's "leakage" capacity can be significant or even important. Measurements taken on a variety of electrical-power networks show that capacity varies considerably from network to network and covers a range of a few µF to a few dozen µF.

The use of filters does, however, present certain difficulties that it is important to be aware of when defining the protection of the electrical installation: standard filters often have a much lower voltage withstand than that of the electrical equipment; these filters are therefore more vulnerable to common-mode overvoltages and they may need closer protection by a voltage suppressor (varistor); what is more, these filters are at the origin of 50 Hz leakage currents that should be taken into account in the electrical distribution.

These leakage currents are limited by standard EN 60950 to less than 3.5 mA, and are usually much lower, from 0.5 to 1 mA per device supplied via a power socket, but can be higher for power equipment in a fixed installation. UPS or certain computer equipment conforming to EN 60950 (marked "high leakage current") can have values up to 5% of In. For instance, in a 15 KVA UPS the leakage current can reach the level of 1 A. These leakage currents add up if the devices are connected to the same phase, which explains the need to divide up their electrical distribution. And if these devices are connected to all three phases, these currents cancel each other out when they are balanced. As a consequence, leakage currents due to the filters, when there are large amounts of electronic equipment, can cause accidental operation of high sensitivity (30 mA), or even medium-sensitivity (0.3 A) RCCB.

In order to guard against nuisance tripping, the recommendation is that the permanent leakage current must not exceed 0.3 IΔn in the TT and TN systems, and 0.17 IΔn in the IT system. It is for this reason that in normal practice no more than three power sockets are protected by one 30-mA RCCB. Finally, if the installation is equipped with a surge arrester, the RCCB sensor should not be placed on the flow path of the current generated by the lightning.

However, protection equipment has been the subject of numerous improvements. For example, nowadays RCCBs: are unaffected by steep edge disturbances and transient currents, are immune to pulsed unidirectional currents, allow overvoltages due to lightning to flow to earth via the lightning arrester without tripping (RCCB with slight delay on tripping).

Table 11.4. Additional requirements for equipment having higher leakage current

Permanently wired or supplied by an industrial plug and socket outlet (EN 60309-2)	The equipment shall be permanently wired to the fixed installation, or be supplied by an industrial plug and socket to
Protective conductors (clause 5.2.5 of EN 60950)	The equipment should have internal protective conductors of not less than 1.0 mm^2 cross-sectional area.
Label (clause 5.1.7 of EN 60950)	The equipment should have a label bearing the following warning or similar wording fixed adjacent to the equipment primary power connection.
	WARNING
	HIGH LEAKAGE CURRENT
	Earth connection essential before connecting the supply

11.5.3 Installation Recommendations

The general recommendation is to follow the well-known ITE installation general guidelines [23, 24], from which the most generally forgotten are:

- Each electronic branch circuit should have individual phase, neutral, and ground conductors. The use of shared neutral conductors as with multi-wire branch circuits is not recommended.
- Connect laser printers and heavy-duty equipment on individual 20-A branch circuits. Laser printers and heavy-duty copiers produce high current surges that cause an increase in the neutral-to-ground voltage. This action can potentially damage other electronic equipment localised on the same branch circuit.

- Ensuring that the earthing system for computer devices has the lowest possible impedance can most readily avoid the problem of earth-fault damage.
- Install transient voltage-surge suppression (TVSS) devices at the service entrance to protect it from utility power problems and lightning strikes.
- At electrical distribution panels, where hard-wired devices help suppress internal voltage transients from spreading to other circuits in your facility.
- At telecommunications and cable local area network circuits, which are extremely vulnerable to voltage transients.
- Install individual TVSS devices at the point of use, where sensitive equipment connects to electrical outlets.
- Problems caused by earth currents flowing in communications cables can be overcome by single-point earthing in the communications cabling system, using communications circuits that are isolated from earth, or installing fibre-optic cable.
- Specify active harmonic conditioner or cancelling transformers downstream of the UPS when neccesary to guarantee a voltage distortion lower than 5% at the point of use, *i.e.* at the input of the computer equipment.
- Increase of the size of the neutral conductor.

11.6 The Improvement of Electronic Power Supplies

It has been discovered that the 85% of computer malfunctions attributed to poor power-quality are caused by voltage-dip or interruptions of under one second duration [1]. Voltage dips at the terminals of sensitive equipment are often due to faults occurring at a much higher voltage level. Even though the load current is small compared to the fault current, the changes in load current during and after the fault still strongly influence the voltage at the equipment terminals.

The typical ride-through capability of power supplies ranges from 10 ms to 30 ms, but this time interval is too short to be of much help. If the switching power supply was modified to have a ride-through capability of one second, a low-cost power-conditioning system could be used rather than an expensive, and often over-specified, UPS installation. It has been demonstrated that there is a large opportunity to embed solutions to voltage-dip problems into advanced electronic tools without requiring large-scale mitigation solutions at the utility- or facility-wide level.

A so-called embedded solution is the result of a product re-engineering effort, typically undertaken by the manufacturer, to reduce the sensitivity of the equipment to variations in the quality of the supplied power. These measures can range from fairly minor design substitutions for the most sensitive components in a system to major product revisions that incorporate new technologies and possibly require revamping the power scheme inside the equipment.

If you wish to improve the robustness of DC power supplies in a system, three options exist. The first involves upsizing the existing power supply. The voltage-dip ride-through time available from a typical linear or switch-mode power supply is directly related to the loading, so power supplies should not be running at or near their maximum capacity. Upsizing by at least twice the nominal load will help the power supply to ride through voltage-dips. This can also be accomplished by adding another identical supply and sharing load with the existing unit.

The second option is to change to three-phase input DC power supplies. Past studies have shown that these devices have robust responses, including both standard linear and switch-mode DC power supplies using a three-phase input scheme.

A final route could be to use universal input-switching power supplies in every possible location. Typically, the universal input-type power supply has a voltage range of 90 to 264 V AC [25].

11.6.1 Converter Topology for Universal Input Switching Power Supplies

The typical AC–DC converter has a two-stage scheme: a diode rectifier and a DC filter. Under this topology, the diode rectifier normally supplies the DC–DC converter in the power supply and then to the DC–AC converter in adjustable-speed drives (ASDs). The DC-link capacitor actuates as energy storage and directly influences the ride-through capability of the system. As a nonlinear load, this classical rectifier configuration induces current harmonics in the network phases, which are limited by international standards such as EN-61000-3-2, EN-61000-3-4 and IEEE-519. The most widely used technique to decrease the THD and consequently increase the power factor of the rectifier has consisted in substituting the diode bridge by an active rectifier, with the regenerative-capability benefits incorporated [26, 27, 28].

However, the cost-effective solution usually proposed has been to place a single-switch PFC boost follower preregulator at the front end of the diode bridge. This is because the boost converter topology has continuous input current that can be shaped through the use of a multiplier and average current-mode control to achieve a near-unity power factor [29]. The boost-converter is traditionally designed to have a fixed output voltage greater than the maximum peak-line voltage. However, the boost voltage does not have to be well regulated or fixed because in the next stage, the step-down converter or the inverter can be designed to handle the voltage variations. As long as the boost voltage is above the peak input voltage the converter will regulate properly.

As a consequence, it has been proposed [30, 31, 32, 33] considering that the regulation capability inherent to the above topology ensures that, during dip conditions, the DC-bus voltage operates within reasonable bounds from the nominal value. With this improvement, the power supply can continue to operate for voltage-dips as low as 41 per cent of nominal (even at full load). Thus, this type of supply should be specified for instrumentation and control equipments. What is more, using this type of power supply for workstations and PCs will lead to excellent voltage-dip ride-through. However, manufacturers of digital equipment may well under-invest in PQ, primarily because it would increase the cost of their products in highly competitive markets whose customers have not yet shown a willingness to pay extra for better power conditioning.

11.6.2 Behaviour Against Balanced Dips

Voltage dips can be either balanced or unbalanced, depending on the causes. If the individual phase voltages are equal, the dip is balanced. If the individual phase voltages are different or the phase relationship is other than 120°, the dip is unbalanced.

There are several methods for voltage-dip classification. The ABC classification is the oldest classification and the one most commonly used. This is likely due to its simplicity. This method distinguishes between seven types of three-phase unbalanced voltage-dips (Table 11.5).

Table 11.5. Propagation of dips through transformers

Fault type	Dip location		
	I	II	III
3-phase	A	A	A
3-phase-to-ground	A	A	A
2-phase-to-ground	E	F	G
2-phase	C	D	C
1-phase-to-ground	B	C	D

Three-phase unbalanced voltage-dips through transformers change dependent on the transformer type. The ABC classification also describes the propagation of dips through transformers. The origin and transformation of the seven types are given in Table 11.5 and Table 11.6.

Results from different power-quality surveys generally indicate that the most common voltage-dips in three-phase power systems are types A, C and D [34]. For type-A voltage-dips, the voltages in all the three phases drop in magnitude by the same amount. Voltages are affected differently in the three phases with a type-C or -D dip.

Table 11.6. ABC voltage-dip classification

Type A:

Three-phase fault

$\overline{U}_a = hU$

$\overline{U}_b = -\frac{1}{2}hU - j\frac{\sqrt{3}}{2}hU$

$\overline{U}_c = -\frac{1}{2}hU + j\frac{\sqrt{3}}{2}hU$

Type B:

Single-phase fault

$\overline{U}_a = hU$

$\overline{U}_b = -\frac{1}{2}U - j\frac{\sqrt{3}}{2}U$

$\overline{U}_c = -\frac{1}{2}U + j\frac{\sqrt{3}}{2}U$

Type C:

Two-phase fault (or secondary type voltage-dip, *e.g.*, single-phase fault as seen behind a Yd or yD transformer)

$\overline{U}_a = U$

$\overline{U}_b = -\frac{1}{2}U - j\frac{\sqrt{3}}{2}hU$

$\overline{U}_c = -\frac{1}{2}U + j\frac{\sqrt{3}}{2}hU$

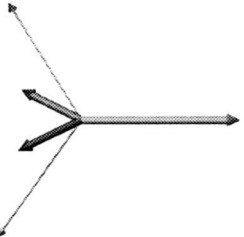

Type D:

Secondary-type voltage-dip, *e.g.*, two-phase fault as seen behind a Yd or yD transformer.

$\overline{U}_a = hU$

$\overline{U}_b = -\frac{1}{2}hU - j\frac{\sqrt{3}}{2}U$

$\overline{U}_c = -\frac{1}{2}hU + j\frac{\sqrt{3}}{2}U$

Table 11.6. (continued)

Type E:

Two-phase-to-ground fault

$\overline{U}_a = U$

$\overline{U}_b = -\frac{1}{2}hU - j\frac{\sqrt{3}}{2}hU$

$\overline{U}_c = -\frac{1}{2}hU + j\frac{\sqrt{3}}{2}hU$

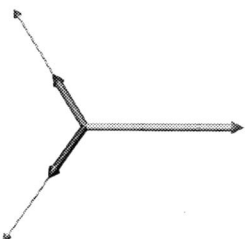

Type F:

Secondary-type voltage-dip, e.g. two-phase-to-ground fault as seen behind a Yd or yD transformer.

$\overline{U}_a = hU$

$\overline{U}_b = -\frac{1}{2}hU - j\frac{1}{\sqrt{12}}(2+h)U$

$\overline{U}_c = -\frac{1}{2}hU + j\frac{1}{\sqrt{12}}(2+h)U$

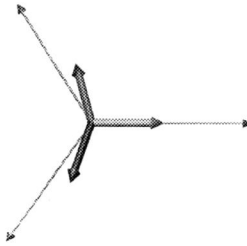

Type G:

Secondary-type voltage-dip, e.g. two-phase-to-ground fault as seen behind a YNy, Yyn, Yy or Dd transformer.

$\overline{U}_a = \frac{1}{3}(2+h)U$

$\overline{U}_b = -\frac{1}{6}(2+h)U - j\frac{\sqrt{3}}{2}hU$

$\overline{U}_c = -\frac{1}{6}(2+h)U + j\frac{\sqrt{3}}{2}hU$

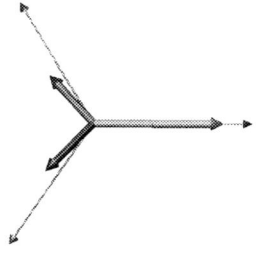

In this section we analyse the effects of type-A voltage-dips. The severity of a dip is defined through its "characteristic complex voltage" U. The three complex voltages for a type-A dip are the following:

$$\overline{U}_a = hU$$
$$\overline{U}_b = -\tfrac{1}{2}hU - \tfrac{1}{2}jhU\sqrt{3},$$ (11.1)
$$\overline{U}_c = -\tfrac{1}{2}hU + \tfrac{1}{2}jhU\sqrt{3}$$

where: $h=0\ldots 1$, dip depth.

The central component in the boost converter is the input inductor. The value of this inductor determines various critical operational aspects of the converter. If the value is too low, the input-current distortion will be high and will result in a low power factor and increased noise at the input. This will require more input filtering. If the inductor value is too high, then for a given operating current the required size of the inductor core will be large and/or the required number of turns will be high. So, a balance must be reached between distortion and core size [35]. In a boost converter the DC output voltage is calculated from

$$U_o = \frac{\sqrt{2}U_i}{1-D}. \tag{11.2}$$

As shown in Figure 18.6 of [36], the regulation of the DC-bus voltage output implies that the duty factor D decreases as the inductance current i_{Lref} increases, with D_{min} the smallest at the peak of i_{Lref}. Therefore to find the value of the inductor the maximum duty factor D_{max} for minimum line voltage is obtained from

$$D_{max} = 1 - \frac{\sqrt{2}U_{i\,min}}{V_o}. \tag{11.3}$$

The maximum value of ripple current occurs when the duty factor D is 50%. The peak current does not generally occur at this point because the inductor current follows the incoming mains voltage. In continuous conduction-mode operation (CCM), the typical peak-to-peak ripple current Δi_L range is between 15% and 25% of the peak current I_p. Thus, the inductance L is given by

$$L = \frac{\sqrt{2} \cdot U_{i\,min} \cdot D_{max}}{f_{PWM} \cdot \Delta i_L}. \tag{11.4}$$

Initially, we can admit that during a dip, the boost-regulator action sees the DC-bus voltage drop as an instantaneous low value of the line voltage sine wave. Consequently, D increases, as explained above, up to the maximum duty factor D_{max}. This action tries to ensure that the same active power is supplied by the rectifier during the dip as before the dip. This means that the input current will have to increase as the line voltage decreases. Commercially, the current is limited by rectifier-device current ratings through the short-time overload factor protection. As a consequence, the rectifier is usually prepared to withstand only 60% of the nominal line voltage [37, 25].

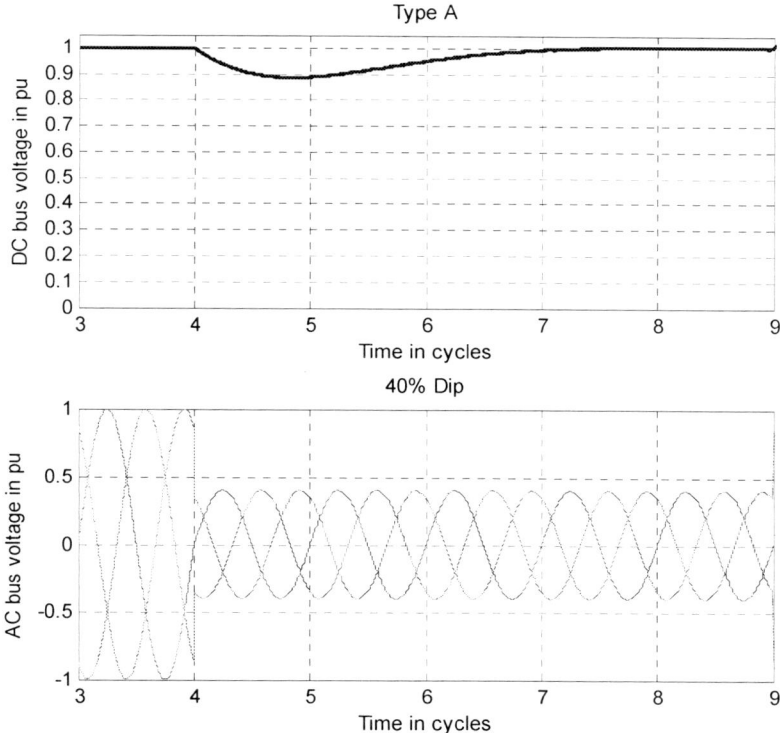

Figure 11.5. Voltage during a 40% three-phase unbalanced dip of type A: DC-side voltages (*top*) for a small capacitance and AC-side voltage (*bottom*).

Therefore, as is shown in Figure 11.5, to withstand a 40% of nominal line voltage it has been necessary to overdimension the rectifier. If this same rectifier is now supplied by 30% of the nominal line voltage, as can be seen in Figure 11.6, once the current has reached the maximum permitted (for a 40% dip), the DC voltage output will be below its nominal value in a new equilibrium point forever.

In this situation, if the dip increases, the output capacitor can discharge. The discharging of the capacitor will be determined by the load connected to the DC bus. The value of this capacitor is determined by several application-specific requirements: the switching frequency, the ripple voltage and current and the holdup time. The holdup time is the time that the output voltage stays within a specific range after the input supply has been removed. The capacitance required to provide a holdup time of t_{max} seconds is given by

$$C = \frac{2 \cdot P_{o\,max} \cdot t_{max}}{V_{o\,max}^2 - (V_{o\,max} - V_{drop})^2}. \qquad (11.5)$$

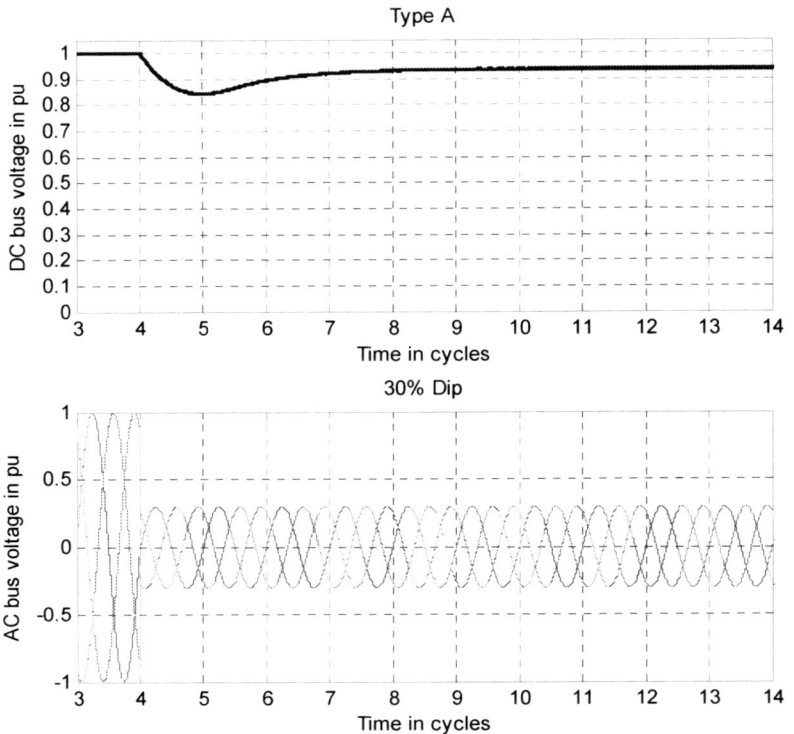

Figure 11.6. Voltage during a 30% three-phase unbalanced dip of type A: DC-side voltages (*top*) for a small capacitance and AC-side voltage (*bottom*)

It can be seen in Figure 11.7 that finally the minimum DC bus voltage reaches zero volts. In the same figure we can compare the response of the analysed PWM boost rectifier with the traditional diodes rectifier and with the same filter capacitor of 121 µF/kW.

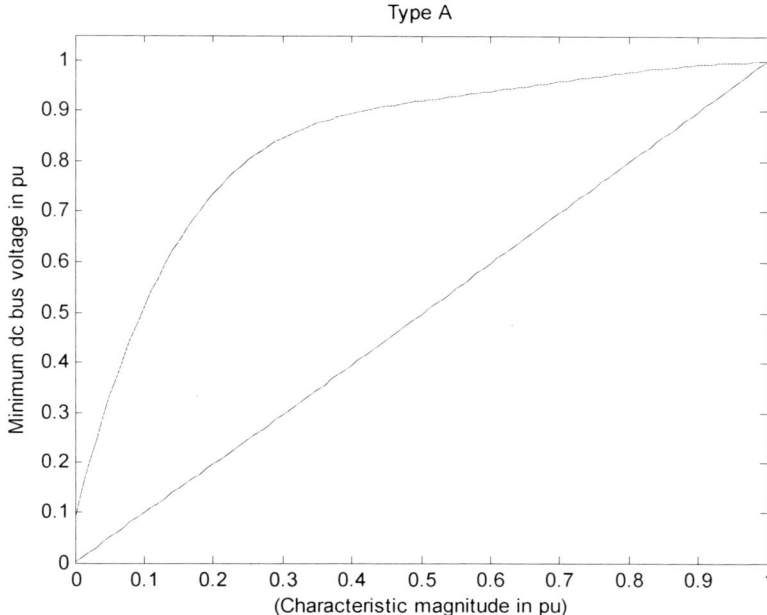

Figure 11.7. Minimum DC-bus voltage as a function of the characteristic magnitude of three-phase unbalanced dips of type A. *Top line*: pwm boost rectifier; *bottom line*: traditional diodes rectifier

It is obvious that a PWM boost rectifier presents the optimal results of a cost-effective solution because, as affirmed in [38], making a drive ride through three-phase balanced dips of 100-ms duration would require no feasible amount of capacitance or any other energy source. In addition, improved results can be obtained with advanced control techniques as in [39].

11.6.3 Behaviour Against Unbalanced Dips

In this section we analyse the effects of type-C and -D voltage-dips. The expressions for the three voltages become, for a type-C dip:

$$\underline{U}_a = U$$
$$\underline{U}_b = -\tfrac{1}{2}U - \tfrac{1}{2}jhU\sqrt{3} \quad , \tag{11.6}$$
$$\underline{U}_c = -\tfrac{1}{2}U + \tfrac{1}{2}jhU\sqrt{3}$$

and for a type-D dip:

$$\overline{U}_a = hU$$
$$\overline{U}_b = -\tfrac{1}{2}hU - \tfrac{1}{2}jU\sqrt{3}\ . \tag{11.7}$$
$$\overline{U}_c = -\tfrac{1}{2}hU + \tfrac{1}{2}jU\sqrt{3}$$

The behaviour of the DC-bus voltage and, thus, of the rectifier here is completely different from the behaviour for a balanced dip. The DC-bus voltage has been calculated for two capacitor sizes. A large capacitor, the initial rate of decay of the voltage is 10% per cycle; for a 592-V rectifier this corresponds to 590 µF/kW. A small capacitor, initial rate of decay of the voltage is 75% per cycle; for a 592-V rectifier this corresponds to 121 µF/kW.

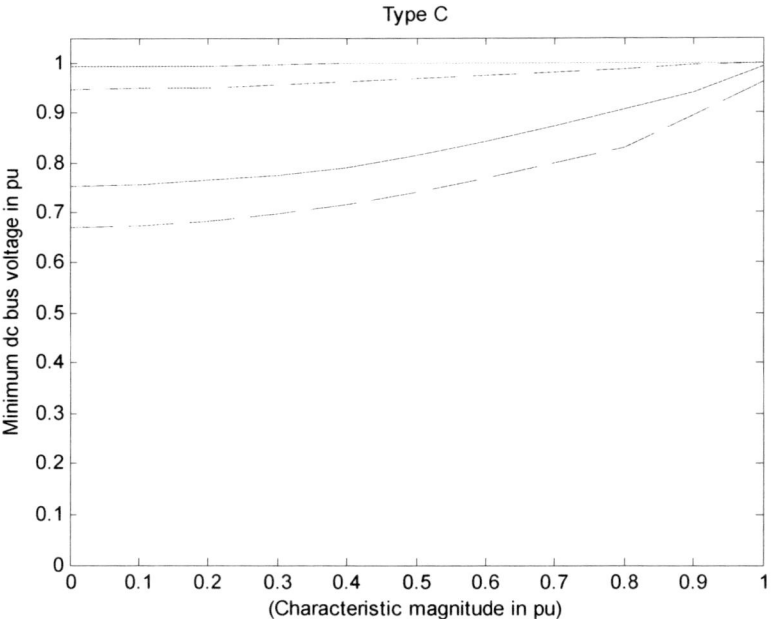

Figure 11.8. Minimum DC-bus voltage as a function of the characteristic magnitude of three-phase unbalanced dips of type C. *Top solid line*: PWM boost rectifier with large capacitance; *top dashed line*: PWM boost rectifier with small capacitance; *bottom solid line*: traditional diodes rectifier with large capacitance; *bottom dashed line*: traditional diodes rectifier with small capacitance.

Figure 11.8 shows the influence of the capacitor size on the minimum DC-bus voltage for a type-C dip. The DC-bus undervoltage protection normally uses this value as a trip criterion. We can observe that the PWM boost rectifier presents similar results for both capacitors, slightly better for the large capacitor. Below are presented the results for the traditional diode rectifier. For large capacitance, the drop in DC-bus voltage is very small. The smaller the capacitance, the larger the drop in DC-bus voltage.

The minimum DC-bus voltage for a dip of type D is shown in Figure 11.9. The PWM boost rectifier presents practically identical results for both capacitors. For the traditional diode rectifier and in comparison with type-C responses, here the minimum DC-bus voltage continues to drop with lower characteristic magnitude, but always above 70% for a small capacitor. For a rectifier with large capacitance, the DC-bus voltage does not drop below 80%, even for the deepest unbalanced dip.

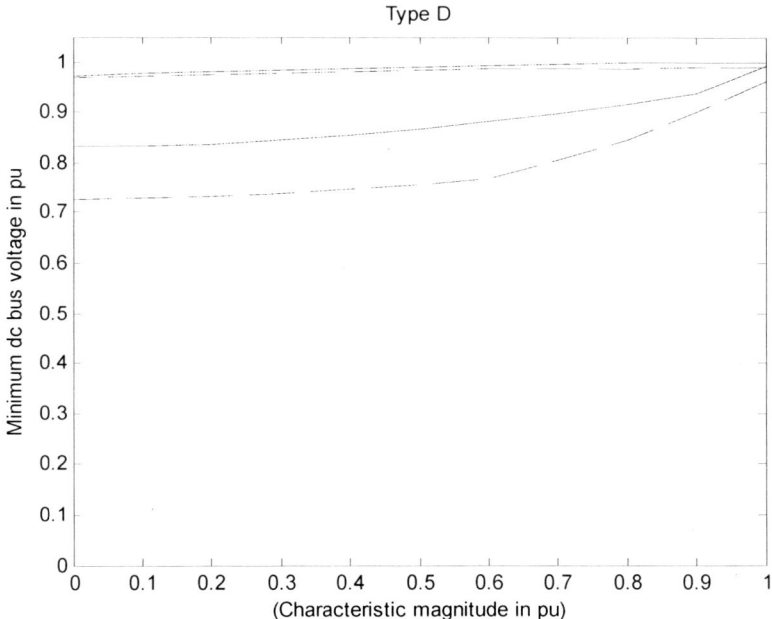

Figure 11.9. Minimum DC-bus voltage as a function of the characteristic magnitude of three-phase unbalanced dips of type D. *Top solid line*: PWM boost rectifier with large capacitance; *top dashed line*: PWM boost rectifier with small capacitance; *bottom solid line*: traditional diode rectifier with large capacitance; *bottom dashed line*: traditional diodes rectifier with small capacitance.

11.7 Conclusion

The massive penetration of electronically controlled devices and equipment in low-voltage distribution networks (the "digital society") could be responsible for the worsening of power-quality problems.

It has been seen [40] that the level of immunity for some power-quality phenomena would be insufficient to adequately protect terminating equipment from the disturbances defined in EN 50160. The 95% per week basis for assessing most parameters means that the actual power-quality could result in considerable disruption in equipment performance and yet still meet EN 50160.

The power supply is a component many users ignore when acquiring a microprocessor-based system and it is therefore one that some system vendors might choose to skimp on. After all, a dealer is far more likely to be able to increase the price of a computer by spending money on additional memory or a larger hard drive than by installing a better power supply.

Active PFC offers several benefits, including marked improvement of AC main power harmonics in medium- and high-power applications, stabilisation of the power-supply bulk bus voltage and automatic line voltage-dip compensation. These features, in turn, improve cost factors and the holdup time against the traditional diode rectifier [41].

While the PFC boost converter is basically a simple topology, the inherently wide range of operating input voltage requires careful evaluation in the design process. In particular, the wide input range relates to stress factors in a switch and in diodes. Then, if a voltage-dips deeper than 40% have to be compensated, the converter can be derated.

References

[1] Gulachenski EM. The low cost alternative to UPS. In: Proceedings of the Electro-95 International 1995; 97–107.
[2] Ward DJ. Power quality and the security of electricity supply. Proceedings of the IEEE 2001; 89 (12): 1830–1836.
[3] Pramod P, Edwin L. Power Quality Services: Technologies and Strategies for Energy Providers in the Deregulated Market. The Electricity Journal 1999; 12 (9): 79–84
[4] Robert E B, Scott P, Vivian WD. After the lights went out. The Electricity Journal 2004; 17(1): 11–15.
[5] Heydt GT. Grand challenges in electric power engineering: extreme system reliability. In: Proceedings of the IEEE Power Engineering Society Summer Meeting 2002; 3: 1695–1697.
[6] Allen GW, Segall D. Monitoring of computer installations for power line disturbances. In: Proceedings of the IEEE PESC 1974; 199–205.
[7] Goldstein M, Speranza PD. The quality of U.S. Commercial AC power. In: Proceedings of the INTELEC 1982; 28–33.
[8] Field Handbook of Power Quality Analysis. Dranetz-BMI; 1998.
[9] Cumbria N, Deregt M, Rao ND. Effects of power disturbances on sensitive loads. In: Proceedings of the Canadian Conference on Electrical and Computer Engineering 1999; 1181–1186.

[10] Koval DO. Computer performance degradation due to their susceptibility to power supply disturbances. In: Proceedings of the Industry Applications Society Annual Meeting 1989; 2: 1754–1760.
[11] Information Technology Industry Council ITIC Curve Application. Available from: http://www.itic.org/iss_pol/techdocs/curve.pdf
[12] Arrillaga J, Bollen MHJ, Watson NR. Power quality following deregulation. Proceedings of the IEEE 2000; 88 (2): 246–261.
[13] Moreno-Muñoz A, Redel M D, González M. Power quality in high-tech campus: a case study. Proceedings of the Institution of Mechanical Engineers, Part A: Journal of Power and Energy 2006; 220 (3): 257–269.
[14] Moreno-Muñoz A, Pallarés V, Galisteo P, De-la-Rosa JJG. Study of voltage sag in a highly automated plant. In: Proceedings of the IEEE MELECON 2006; 1060-1063
[15] IEEE Std 493. IEEE recommended practice for the design of reliable industrial and commercial power systems (1997).
[16] IEEE Standard 1346. IEEE Recommended Practice for Evaluating Electric Power System Compatibility With Electronic Process Equipment (1998)
[17] Dugan RC, McGranaghan M, Santoso S, Beaty HW. Electrical Power System Quality. McGraw-Hill 2002.
[18] McGranaghan M, Roettger B. Economic evaluation of power quality. In IEEE Power Engineering Review 2002; 22(2): 8–12.
[19] Standler RB. Protection of small computers from disturbances on the mains. In: Proceedings of the IEEE Industry Application Society Conference 1988; 2:1482–1487.
[20] Stebbins WL. Power line disturbances: a user's perspective on the selection and application of mitigation equipment and techniques. In: Proceedings of the IEEE Textile Industry Technical Conference 1989; 4: 1–7.
[21] Schenung S. Overcoming Cost Reduction Barriers for Advanced Power Quality Mitigation Systems. EPRICSG, 1999. TP-114370.
[22] Moreno-Muñoz A, Oterino D, González M, Olivencia FA, De-la-Rosa JJG. Study of sag compensation with DVR. In: Proceedings of the IEEE MELECON 2006; 990–996.
[23] Available from: http://www.cpccorp.com/tips.htm
[24] Three Phase Power Source Overloading Caused by Small Computers and Electronic Office Equipment. ITI Information Letter, Available from: http://www.itic.org/technical/3phase.htm
[25] Moreno-Muñoz A, De-la-Rosa JJG. Analysis of voltage dips in PWM AC-DC converters. In: Proceedings of the International Symposium on Power Electronics, Electrical Drives, Automation and Motion, SPEEDAM 2006; S41: 6–9.
[26] Silva CS. Power factor correction with the UC3854. Unitrode Product and Applications Handbook 1995–1996, Unitrode Corporation 1995; 10: 303–322.
[27] Kazerani M, Ziogas PD, Joos G. A novel active current wave shaping technique for solid-state input power factor conditioners. IEEE Transactions on Industrial Electronics 1991; 38 (1): 72–78.
[28] Prasad AR, Ziogas PD, Manias S. An active power correction technique for three-phase diode rectifiers. In: Proceedings of the IEEE Power Electronics specialists conference 1989; 58–66.
[29] Moreno-Muñoz A, Pallarés V, Luna J, Flores JM. Switching mode power supply e-learning toolbox. In: Proceedings of the EPE 2003 congress [CD-ROM]: 1–6.
[30] Van-Zyl A, Spée R, Faveluke A, Bhowmik S. Voltage sag ride-through for adjustable-apeed drives with active rectifiers. IEEE Transactions on Industry Application 1998; 34 (6): 1270-1277.

[31] Durán-Gómez JL, Enjeti PN, Ok-Woo B. Effect of voltage sags on adjustable-speed srives: A critical evaluation and an approach to improve performance. IEEE Transactions on Industry Application. 1999; 35 (6): 1440–1449
[32] Xu, J.; Al-Haddad K.; Sicard, P. and Rajagopalan, V. (2001). A novel combined approach to voltage sag ride-through and current waveform improvement for adjustable-speed drives. In: Proceedings of the Power Electronics Specialists Conference, 2001; 3: 1315–1320.
[33] Montero-Hernandez OC, Enjeti PN. Ride-through for critical loads. IEEE Industry Application Magazine 2002; 45–52.
[34] Bollen MHJ. Understanding power quality problems – voltages and interruptions, New York: IEEE Press; 1999.
[35] Todd PC. UC3854, Controlled Power Factor Correction Circuit Design. U-134, Unitrode Application Note 1994; 3: 269–288.
[36] Mohan N, Undeland TM, Robbins WP. Power Electronics: Converters, Applications, and Design. New York: Wiley 2003.
[37] Stockman K, D'hulster F, Verhaege K, Didden M, Belmans R. Ride-through of adjustable speed drives during voltage dips. Electric Power Systems Research 2003; 66: 49–58.
[38] Bollen MHJ, Zhang LD. Analysis of voltage tolerance of AC adjustable-speed drives for three-phase balanced and unbalanced sags. IEEE Transactions on Industry Application 2000; 36 (3): 904–910.
[39] Zhang W, Feng G, Liu YF, Wu B. A New Power Factor Correction (PFC) Control Method Suitable for Low Cost DSP. In: Proceedings of the INTELEC 2002; 407–414
[40] Annex B2. Available from: http//www.iee.org.uk/PAB/EMC/core.htm
[41] Moreno-Muñoz A, Redel M^{a}D. Calm in the campus: power disturbances threaten university life. IEE Power Engineer 2005; 19 (4): 34–37.

12

Power-quality Factor for Electrical Networks

Juan-Carlos Montaño, María-Dolores Borrás[*] and Juan-Carlos Bravo[*]

Laboratorio de Electrónica,
Institute for Natural Resources and Agricultural Research (IRNAS)
Spanish Research Council (CSIC)
Reina Mercedes Campus, POB 1052,
41080-Sevilla, Spain
Email: montano@irnase.csic.es

[*]Dpto. Ingeniería eléctrica
Escuela Universitaria Politécnica
Universidad de Sevilla
Virgen de África 7
41011-Sevilla, Spain
Email: carlos_bravo@us.es, borras@us.es

The ever-growing proliferation of power-switching devices for source conditioning and motion control in single-phase and three-phase modern industrial applications has increased the occurrence of unbalanced currents, unacceptable harmonic levels and poor power factor in three-phase distribution systems. The harmful and costly effects of harmonics have been discussed extensively [15, 40] and spurred stringent requirements by international institutions regarding the allowed levels of harmonics at the point of connection to the power supply [19, 24, 32]. Unbalanced loading of the three-phase supply has other no less detrimental effects such as the underutilisation of the power-supply equipment and overloading of neutral conductors with fundamental frequency in addition to harmonic currents. Also, as is well known, a phase displacement between corresponding voltages and currents indicate both a low utilization of the generation and distribution equipment and increased line losses for the same power-consumption level.

The above disturbances occurring in the power system can be considered as stationary in a certain sense. They take place during sufficiently long time intervals and do not change appreciably for an observatory window. Thus classical Fourier methods of signal analysis can be used in the frequency domain to extract the stationary components present on the observed signal.

In order to analyse the transient and time-varying nature of the disturbance signals in power systems, other signal-processing techniques have been employed for assessment, detection, localisation and classification purposes [14, 17, 29, 30]. Recently, the wavelet transform [27, 28, 34, 35], and the short-time Fourier transform [18] have been frequently utilised in the study of power quality, which are characterised by the "time&scale" and "time&frequency" localisation of the transient signals, respectively.

This chapter is concerned with investigating the joint treatment for signal analysis based on both the time domain and the frequency domain. Distinguishing between stationary and transient components present on an electrical signal, the former part can be extracted by means of the classical Fourier methods, while the latter can be subjected to joint time frequency analysis (JTFA).

The Fourier transform has been used to gain knowledge of spectral components existing in a waveform; irrespective of the moment when they happen. However, when the time localisation of the spectral components is needed, the *JTFA* can be used to obtain the optimal time-frequency representation of the signal.

A general integral assessment of the power-transfer quality of a three-phase network by means of a new indicator designated the power-quality factor (PQF) is suggested using the Fourier techniques. The *PQF* considers various quality aspects (*QA*) notably the current and voltage harmonic levels, the phase displacements between corresponding phase voltages and currents at the fundamental frequency, and the degree of unbalance in the different phase voltages and currents. The different *QAs* are distinctly measured so that, if necessary, the specific quality aspect that needs correction can be readily identified. The *PQF* offers the convenience of assessing the power quality by means of a single indicator rather than requiring comparisons of a multitude of factors with their respective recommended values.

In the same sense, a single indicator, designated the voltage quality factor (*VQF*), in the range between zero to one, is also studied in this chapter to integrally reflect the voltage quality of a general three-phase network. Prominent voltage-quality aspects considered in the *VQF* are the power-system-frequency variations, the voltage harmonic levels and the degree of unbalance in the different phases at the fundamental frequency. A network supplying balanced sinusoidal voltages with no frequency variations would yield a *VQF* of unity.

The applicability of the power-quality factor and the voltage-quality factor are illustrated for typical source-load configurations. The simplicity offered by these new factors allows the development of efficient rate structures and eventual enforcing policies that aim at the enhancement of the reliability and power quality of the mains.

The *JTFA*-based power-quality indices discussed in this chapter can be interpreted as a generalisation of the Fourier power-quality indices. Besides the time–frequency-based transient power-quality indices developed in this chapter, one can develop other transient power-quality indices by calculating appropriate moments of the time–frequency distribution. The *JTFA*-based transient power-quality indices discussed are expected to play an important role in resolving various assessment and measurement issues for transient phenomena in power systems.

12.1 Quality of the Electrical Signal

Power systems operate with a constant line voltage, supplying power to a wide variety of load equipment. Power levels range from a few watts to megawatts, and the voltages at which the energy is generated, transported, and distributed range from hundreds of volts to hundreds of kilovolts. Transmission and primary distribution of this power are made at high voltages, tens to hundreds of kilovolts, in order to provide efficient and economic transportation of the energy over long distances. Final utilisation is generally in the range of 120 V (or 220 V) (typical residential) to less than one thousand (industrial), and a few thousands for large loads.

At all these voltage and power levels, no matter how high, the equipment is dependent upon maintenance of a normal operating voltage because it has only limited capability of withstanding voltages exceeding the normal level. At lower than normal levels, the equipment performance is generally unsatisfactory, or there is a risk of equipment damage. These two disturbances, excessive voltage and insufficient voltage, are described with different names depending on their duration. There are also types of disturbances, as described in [4] involving waveform distortion and other deviations from the expected sine wave.

Emerging concerns over these issues has resulted in attention being focused on the quality of the power necessary for successful operation of diverse loads, on practical limits to the capability of delivering power of high quality to diverse customers, and on the economics of the producer–user partnership. The term "power quality" is now widely used, but objective criteria for measuring the power quality – a prerequisite for quantifying this quality – needs better definition. A high level of power quality is understood as a low level of disturbances; agreement on acceptable levels of disturbances is needed.

12.2 Quantitative Formulations of Power-quality Aspects

The term "power quality" is defined as "set of parameters defining the properties of power quality as delivered to the user in normal operating conditions in terms of continuity of supply and characteristics of voltage (frequency, magnitude, waveform, symmetry)" in IEC Standard [24]. However, in IEEE Standard [32], "power quality" is defined as "The concept of powering and grounding electronic equipment in a manner that is suitable to the operation of that equipment and compatible with the premise wiring system and other connected equipment". As shown in these definitions, "power quality" includes considerations of all aspects of power supply. In this chapter, we focus on the effect of the quality of power on sensitive loads. Under steady-state conditions, three power-system parameters – frequency, waveform distortion, and symmetry – can serve as frames of reference to classify the disturbances according to their impact on the quality of the available power. The phase displacement between voltage and current has been added as another important quality aspect.

12.2.1 System-frequency Variations

Frequency variations are rare on utility-connected systems, but engine-generator-based distribution systems can experience frequency variations due to load variations and equipment malfunctions.

Several procedures exist for measuring system-frequency variations in the case of single- and three-phase networks [1, 33]. A virtual instrument for the measurement of the instantaneous power-system -frequency was proposed [25], which is based on the frequency estimation of the voltage signal using three equidistant samples.

In general, voltage waveforms derived from the electric wires are sampled and converted. Then, the instantaneous-frequency calculation of the voltage signal is performed avoiding errors due to lack of synchronisation between the signal period and the sampling sequence. Thus corrected samples of the set of input signals are obtained for further digital processing based on the FFT.

The change of frequency Δf can be defined as a difference between two frequency values

$$\Delta f = f - f_N \quad \text{Hz} \qquad (12.1)$$

where f is the current frequency value and f_N is the nominal frequency value (50 or 60 Hz). The relative frequency variation $\Delta f / f_N$ is defined as,

$$RFV = \frac{\Delta f}{f_N} = \frac{f - f_N}{f_N}. \tag{12.2}$$

The first quality aspects QA_1 is identified with the relative change of frequency RFV

$$QA_1 = RFV = \frac{\Delta f}{f_N}. \tag{12.3}$$

12.2.2 Total Current and Voltage Harmonic Distortion

The total harmonic distortion of the voltage and current VTHD and ITHD, respectively, for single-phase (or polyphase balanced) networks have been conventionally defined in the literature as

$$\text{VTHD} = \frac{\sqrt{\sum_{h \neq 1} V_h^2}}{V_1}, \qquad \text{ITHD} = \frac{\sqrt{\sum_{h \neq 1} I_h^2}}{I_1}, \tag{12.4}$$

where V and I denote rms values and 1 and h denote the fundamental and the harmonic order, respectively.

To account for the fact that higher-order harmonic currents cause greater losses than lower-order harmonics of the same amplitude, harmonic-adjusted total voltage and current harmonic distortions are, respectively, defined as

$$\text{VTHD}_{\text{H-A}} = \frac{\sqrt{\sum_{k=1}(C_h V_h)^2}}{V_1}, \qquad \text{ITHD}_{\text{H-A}} = \frac{\sqrt{\sum_{k=1}(D_h I_h)^2}}{I_1}, \tag{12.5}$$

where C_h and D_h are appropriate weighting factors, greater than one, that monotonically increase with the harmonic order h. Various mathematical expressions, by no means exhaustive, for these weighting factors as a function of the harmonic order h are suggested in [9].

An extension of the above concepts to unbalanced polyphase networks has been suggested in [21]. To this end a single "equivalent" harmonic rms voltage V_{eH} and harmonic rms current I_{eH} for the three-phase system a,b,c are defined as

$$V_{eH}^2 = \sum_{h \neq 1} \frac{V_{ah}^2 + V_{bh}^2 + V_{ch}^2}{3}, \qquad I_{eH}^2 = \sum_{h \neq 1} \frac{I_{ah}^2 + I_{bh}^2 + I_{ch}^2}{3} \tag{12.6}$$

Similarly to definitions of Equation 12.2 individual harmonics can be separately weighted yielding adjusted harmonic equivalent values as follows

$$V_{eH-A}^2 = \sum_{h \neq 1} C_h^2 \frac{V_{ah}^2 + V_{bh}^2 + V_{ch}^2}{3}, \qquad I_{eH-A}^2 = \sum_{h \neq 1} K_h^2 \frac{I_{ah}^2 + I_{bh}^2 + I_{ch}^2}{3} \qquad (12.7)$$

The second quality aspects QA_2 and QA_3, are identified with the total harmonic distortions VTHD and ITHD for a three-phase unbalanced system and are given by

$$QA_2 = \text{VTHD} = \frac{V_{eH-A}}{V_{el}}, \qquad (12.8)$$

$$QA_3 = \text{ITHD} = \frac{I_{eH-A}}{I_{el}}, \qquad (12.9)$$

where V_{el} and I_{el} denote the fundamental equivalent phase voltage and current defined, similarly to Equation 12.4, from the fundamental components of the generally unbalanced three-phase voltages and currents as

$$V_{el}^2 = \frac{V_{a1}^2 + V_{b1}^2 + V_{c1}^2}{3}, \qquad I_{el}^2 = \frac{I_{a1}^2 + I_{b1}^2 + I_{c1}^2}{3}. \qquad (12.10)$$

Definition of Equations 12.4, 12.5 and 12.7 satisfy the particular requirement that the total losses in the equivalent balanced three-phase system be equal to those in the actual system considered. The underlying assumptions are that a) the line and equipment losses for a certain frequency are proportional to either the square of the voltage or the square of the current and b) the line and equipment parameters are symmetrical.

12.2.3 Degree of Unbalance

As is well known, unbalanced voltages and currents in polyphase networks affect the quality of power transfer in many aspects, such as increased line losses for the same power-transfer level, extra rotating losses in drives and overloading of neutral conductors in four-wire distribution systems.

Applying the theory of symmetrical components [13], an unbalanced three-phase sinusoidal voltage system $[V_a, V_b, V_c]$ can be decomposed into a positive-sequence three-phase balanced system V^+, a negative-sequence system V^-, and a zero-sequence system V^0 according to

$$\underline{V}^+ = \frac{1}{\sqrt{3}}(\underline{V}_a + a\underline{V}_b + a^2\underline{V}_c)$$

$$\underline{V}^- = \frac{1}{\sqrt{3}}(\underline{V}_a + a^2\underline{V}_b + a\underline{V}_c), \qquad (12.11)$$

$$\underline{V}^0 = \frac{1}{\sqrt{3}}(\underline{V}_a + \underline{V}_b + \underline{V}_c)$$

where $\underline{V}$ denotes the phasor of V and the factor $a = \exp(j2\pi/3)$. The symmetrical components of the currents $\underline{I}^+$, $\underline{I}^-$ and $\underline{I}^0$ at the fundamental frequency are derived similarly. From Equations 12.7–12.9, after some manipulations, the equivalent voltage can be expressed as [10]

$$V_{e1}^2 = \underline{V}^{+2} + \underline{V}^{-2} + \underline{V}^{0^2}. \qquad (12.12)$$

Identical considerations apply to the three-phase currents where V is exchanged by I.

Pertinent quality aspects QA_4 and QA_5 are the voltage and current unbalance factors VUNB and IUNB, respectively, defined as

$$QA_4 = \text{VUNB} = \frac{\sqrt{V_{e1}^2 - (V^+)^2}}{V_{e1}}, \qquad QA_5 = \text{IUNB} = \frac{\sqrt{I_{e1}^2 - (I^+)^2}}{I_{e1}}. \qquad (12.13)$$

The voltages and currents in the definition of Equation 12.10 denote absolute values of the corresponding phasors as defined in Equations 12.7 and 12.8. For balanced three-phase voltages and currents it can readily be shown from Equation 12.8 that $\underline{V}^-$, $\underline{V}^0$, $\underline{I}^-$ and $\underline{I}^{0+}$ are equal to zero, $V_{e1} = V^+$, $I_{e1} = I^+$, and hence $QA_4 = QA_5 = 0$.

12.2.4 Phase Displacements Between Corresponding Fundamental Voltage and Currents

In a power-transfer context, a phase displacement between the fundamental frequency voltage and current in a source or load indicates a less than full utilisation of the generation and distribution equipment as well as increased line losses for the same level of power transfer. This has been traditionally expressed through the concept of a power factor K defined in single-phase situations as the ratio between the average active power P and the apparent power S, the latter being the product of the rms values of the voltage and current.

In a nonsinusoidal situation, however, the concept of power factor has been proved to be misleading [11, 37]. Moreover, a universally accepted definition of the power factor in polyphase situations has been lacking and hotly debated for almost a century [10, 12]. This stems from the different definitions and interpretations of the apparent power in these situations [3, 7]. It should be noted in this context that only the instantaneous power, *i.e.* the product of the instantaneous values of the voltage and current, has the physical meaning of power. The conventional active power P derives its physical relevance from being the time average of the instantaneous power. In contrast, the apparent power S has no physical meaning since it does not consider the corresponding values of the voltage and current at each instant. Only in the special case when the instantaneous voltage bears a constant ratio to the instantaneous current does the apparent power equal to P and hence acquires the physical meaning of the latter [11].

All the inadequacies and ambiguities above can be circumvented when the physical entity directly responsible for a poor utilization of the power capacity at the mains fundamental frequency is considered. This entity is the phase difference between the fundamental frequency voltage and current at the terminals of the source or load. To reflect this aspect of power quality, a new factor, designated the orthogonal current factor (OCF), is suggested as follows

$$QA_6 = \text{OCF} = \frac{I_{1a}\sin\varphi_{1a} + I_{1b}\sin\varphi_{1b} + I_{1c}\sin\varphi_{1c}}{I_{1a} + I_{1b} + I_{1c}}, \quad (12.14)$$

where I_{1a}, I_{1b} and I_{1c} denote the rms values of the phase currents at the fundamental frequency and φ_{1a}, φ_{1b} and φ_{1c} denote the phase differences between the fundamental frequency components of the corresponding phase voltages and currents.

The rationale behind Equation 12.14 is that the orthogonal current component $I\sin\varphi$, of a phase current wrt the corresponding phase voltage, is directly related to the effort required to ideally reduce the phase φ to zero. For example, in a single-phase situation, the capacitance required to connect in parallel to an inductive load to bring φ to zero is given by [36]

$$C = \frac{I\sin\varphi}{\omega V}. \quad (12.15)$$

In particular instances lagging and leading orthogonal components in different phases may partially or totally cancel each other in Equation 12.14. This means however that the polyphase system is strongly unbalanced and the unbalance quality aspects in Equation 12.14 would prominently reflect it.

12.3 Voltage-quality Factor and Power-quality Factor

Power quality, like quality in other goods and services, is difficult to quantify. There is no single accepted definition of "quality power". There are standards for voltage and other technical criteria that may be measured, but the ultimate measure of power quality is determined by the performance and productivity of end-user equipment. If the electric power is inadequate for those needs, then the "quality" is lacking.

While the common term for describing the subject of this chapter is power quality, it is actually the quality of the voltage that is being addressed in most cases. Technically, in engineering terms, power is the rate of delivery or energy and is proportional to the product of the voltage and current. It would be difficult to define the quality of this quantity in any meaningful manner. The power supply system can only control the quality of the voltage; it has no control over the currents that particular loads might draw. Therefore, the standards in the power-quality area are devoted to maintaining the voltage within certain limits.

Alternating-current power systems are designed to operate at a sinusoidal voltage of a given frequency (typically 50 or 60 Hz) and magnitude. Any significant deviation in the magnitude, frequency, or purity of waveform is a potential power-quality problem. Of course, there is always a close relationship between voltage and current in any practical power system. Although the generators may provide a near-perfect sine-wave voltage, the current passing through the impedance of the system can cause a variety of disturbances to the voltage. For example, the current resulting from a short-circuit causes the voltage to sag, or disappear completely, as the case may be. Distorted currents from harmonic-producing loads also distort the voltage as they pass through the system impedance. Thus a distorted voltage is presented to other end-users.

Therefore, while it is the voltage with which we are ultimately concerned, in some situations we must address phenomena in the current to understand the bases of many power-quality problems.

12.3.1 Definition of the Power-quality Factor

A single measurable indicator, designated the power quality factor (PQF), is suggested to integrally reflect the different power-quality aspects formulated in the last section. This is expressed as

$$PQF = \sum_i w_i \left(1 - QA_i\right), \qquad (12.16)$$

where w_i are judiciously selected weighting factors that sum up to one and QA_i are the different quality aspects formulated in Section 12.3.

A balanced loaded network, with sinusoidal currents and voltages and zero phase displacements yield an ideal *PQF* of unity. Conversely, a low value of *PQF* would indicate a low degree of utilization of the power capacity of the source and/or a high level of harmonics and/or a high degree of unbalance between the phases with the contribution of each aspect well defined and measurable as illustrated later.

The weighting factors w_i together with the recommended value of PQF are optimally selected so as to reflect the economic and technical importance of a high power quality in each of its aspects. Any particular recommendation in practice should be supported by an appropriate economic study. The considerations in such a study are, on the one hand, the relative cost and availability of balancing circuits, reactive power compensation and harmonic passive or active filters for harmonic mitigation and, on the other hand, the economic gains resulting from a high-quality power transfer, such as a high degree of utilisation of the source capacity, prolonged life of system components and minimisation of system outages due to resonance phenomena caused by harmonics. These considerations are much the same as those required to specify acceptable limits to the voltage and current distortions [20] or the minimum tolerated power factor, deviations from which are usually penalised in the rate structure.

Notable features of Equation 12.16 are
a) It does not make use of controversial or misleading quantities such as apparent power and power factor in nonsinusoidal situations.
b) It can be expanded to include additional power-quality aspects (QAs) such as voltage sags and swells adequately defined in terms of their frequency and/or amplitudes.
c) The weighting factors are selected to reflect technical priorities and objectives with due consideration to economic and practical issues. These factors may (and should) be different in different situations and environments.

12.3.2 Definition of the Voltage-quality Factor

A single measurable indicator, designated the voltage quality factor (VQF), is suggested to integrally reflect the different voltage quality aspects QA_i (i= 1, 2, 4) formulated above. This is expressed as

$$VQF = \sum_{i=1,2,4} w_i \left(1 - QA_i\right), \tag{12.17}$$

where w_i are judiciously selected weighting factors of the voltage signal that sum up to one. Balanced loaded networks, with sinusoidal voltages yield an ideal *VQF* of unity. Conversely, a low value of *VQF* would indicate a high level of harmonics and/or a high degree of unbalance between the phases with the contribution of each aspect being well defined and measurable.

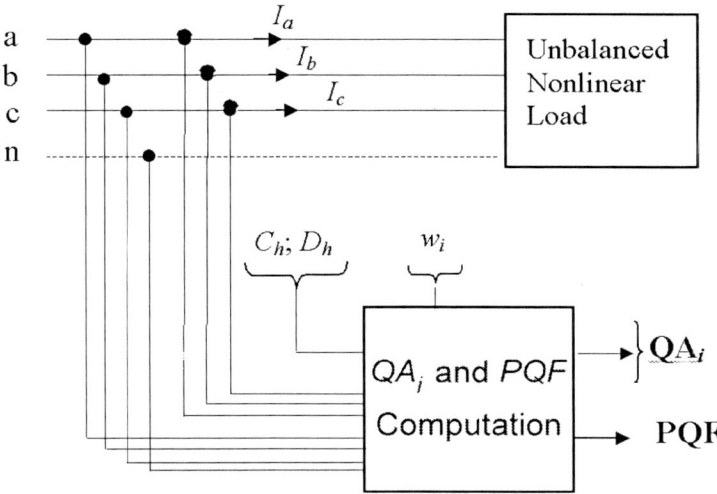

Figure 12.1. Conceptual block diagram for the measurement of the power quality factor and assessing of the different quality aspects in a three–phase supply

12.4 Measurement of *PQF* and *VQF*

A conceptual block diagram of a device that measures the *PQF* or *VQF* as defined in Equations 12.16 and 12.17 at a chosen location in a power system is shown in Figure 12.1. The device offers the additional feature of distinctly measuring the different quality aspects formulated in Section 12.3 so that, if necessary, the power-transfer aspect that needs correction is identified. The direct measurement of *PQF* or *VQF* offers the convenience of assessing the power-transfer quality by means of a single indicator rather than requiring separate measurements of various entities and individually comparing them with their respective recommended values.

When an unbalance situation is measured and identified, a relatively small additional effort is needed to assess if the unbalance is due to a voltage supply asymmetry or to an unbalanced load. In connection to harmonic pollution, the development of a practical method that equitably allocates the responsibility and cost of the harmonics between the utility and consumers is much needed. This is crucial to create the right incentives to mitigate the harmonics for the benefit of both utility and consumer. Many efforts have been invested in this direction ([8] and references therein).

A virtual instrument, the power-quality analyzer [28] and the voltage-quality analyser [27] can be constructed to operate online, with three voltage transducers, a data-acquisition card adapted to a personal computer (PC) and a program of control, installed in the PC.

These analysers offer the additional feature of distinctly measuring the different quality aspects formulated in the above section so that, if necessary, the power-transfer aspect that needs correction is identified. The direct measurement of *PQF* or *VQF* offers the convenience of assessing the power-transfer quality by means of a single indicator rather than requiring separate measurements of various entities and individually comparing them with their respective recommended values.

Voltage and current sensors were built with Hall-effect voltage and current transducers, type LV 25-P and LA25-NP, respectively. The data-acquisition card consists of a PCI-MIO 16E-4 inserted in the expansion bus of the PC. It can acquire eight differential inputs using CMOS analog input multiplexers with overvoltage protection. Analog inputs are converted with 12 bits of resolution; sampled at 1.25 MS/s. The device is connected to the mains to measure quality aspects QA_i, [27, 28].

A graphical programming language (LabVIEW 6.1) was used to create the user interface that gives interactive control of the software system. LabVIEW is integrated with the data-acquisition card for data development, analysis and presentation solutions.

The measurement results are stored in different files according to the type of information recorded. These include

- the instantaneous voltage and current values of input signals;
- the values of the different quality aspects QA_is;
- the weighting factors w_i;
- the resultant *PQF* or *VQF*.

Examples of recorded data of the *PQF* and *VQF* are shown in Figures 12.2 and 12.3, respectively.

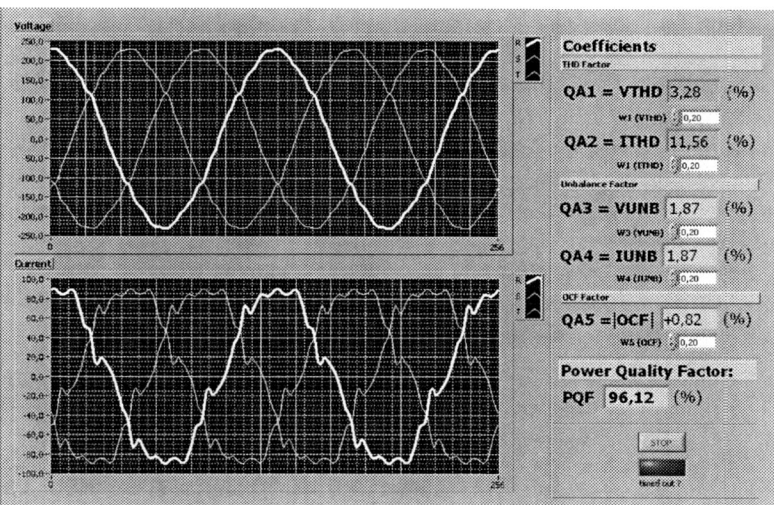

Figure 12.2. Presentation of actual results using the virtual realisation of the *PQF* meter

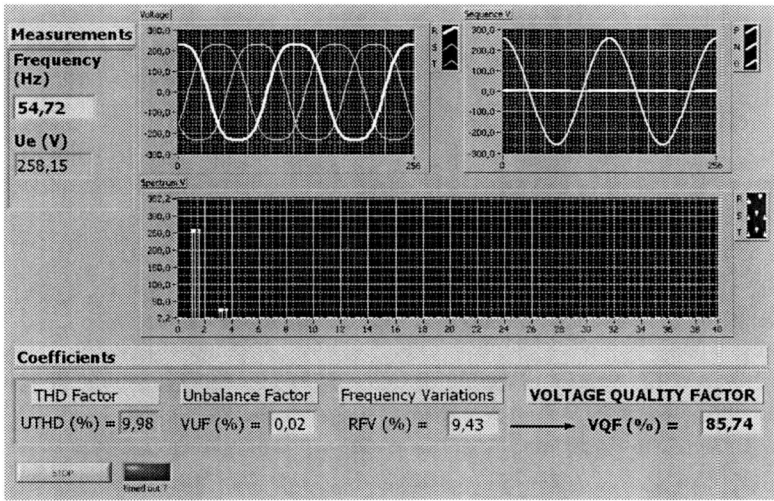

Figure 12.3. Presentation of actual results using the virtual realisation of the *VQF* meter

12.5 Illustrative Use of the Power-quality Factor

The use and relevance of the power-quality factor suggested in the last section are illustrated in the following through numerical examples of various three-phase source/load configurations. Four cases are analyzed for the *PQF* measurement:

a) an unbalanced resistive load supplied by a balanced three-phase sinusoidal voltage source;
b) a balanced resistive load supplied by an unbalanced three-phase sinusoidal voltage source;
c) a balanced resistive load supplied by an unbalanced three-phase nonsinusoidal voltage source;
d) a balanced inductive load is connected to the voltage source of case c);
e) an AC–DC phase-controlled three-phase power converter connected to a balanced sinusoidal voltage source.

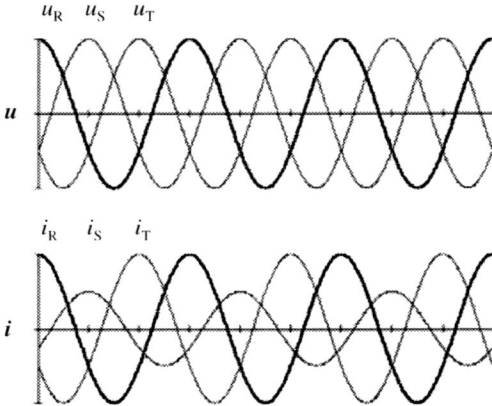

Figure 12.4. Voltage and current waveforms for case a) balanced source voltage and asymmetric resistive load (*PQF*= 0.955)

In case a) an unbalanced resistive load defined by R_a= 0.1 Ω; R_b= 0.2 Ω; R_c= 0.1 Ω, is connected to a balanced three-phase sinusoidal source defined by V_{phase1}= 220 V. The weighting factors C_h and D_h in Equation 12.5 are given by 1 and $h^{0.5}$, respectively. The calculated values of the quality aspects QA_1 to QA_6 as defined in Section 12.3 are 0, 0, 0, 0, 0.272, and 0, respectively. For equal weighting factors w_i= 0.2 for all *i* in Equation 12.16, this yields a power quality factor *PQF* of 0.955.

In case b) a balanced resistive load defined by R_a= R_b= R_c= 0.1 Ω, is connected to an unbalanced three-phase sinusoidal source defined by V_a= 220 V; V_b= 237.60 V and V_c= 220 V. The calculated values of the quality aspects QA_1 to QA_6 are 0, 0, 0, 0.0367, 0.0367, and 0, respectively. This yields a power-quality factor *PQF* of 0.988.

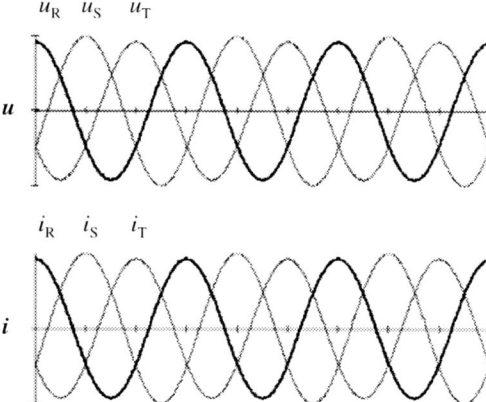

Figure 12.5. Voltage and current waveforms for case b) unbalanced source voltage and symmetric resistive load (*PQF*= 0.988)

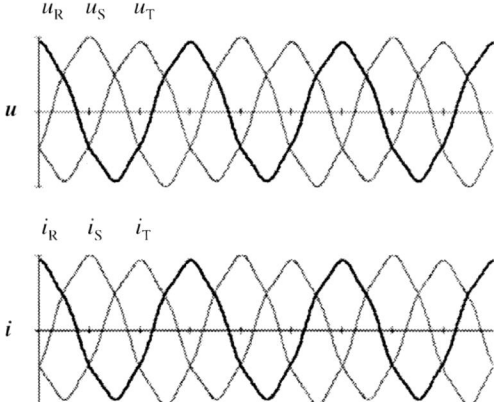

Figure 12.6. Voltage and current waveforms for case c) unbalanced nonsinusoidal source voltage and symmetric resistive load (*PQF*= 0.971)

In case c) the balanced resistive load defined above is connected to an unbalanced nonsinusoidal source. The nonsinusoidal voltage source was implemented by adding to the above-defined three-phase voltages a fifth-order harmonic having an amplitude of 5% of the fundamental. The calculated values of the quality aspects QA_1 to QA_6 for case c) are 0, 0.049, 0.049, 0.037, 0.037, and 0, respectively, which yields a *PQF* of 0.971.

The voltage and current waveforms for cases a), b) and c) are depicted in Figures 12.4, 12.5 and 12.6, respectively.

In case d) a balanced inductive load defined by $R_a = R_b = R_c = 1\ \Omega$ and $L_a = L_b = L_c = 1$ mH, is connected to the voltage source of case c). The current lags the voltage by 32°, which implies an orthogonal current factor of 0.532 and a degradation of the power quality factor *PQF* of 0.888. Corresponding waveforms are shown in Figure 12.7.

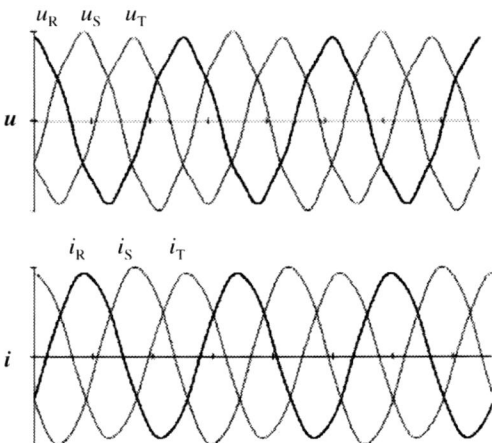

Figure 12.7. Voltage and current waveforms in case d) unbalanced nonsinusoidal source voltage and symmetric inductive load (*PQF*= 0.888)

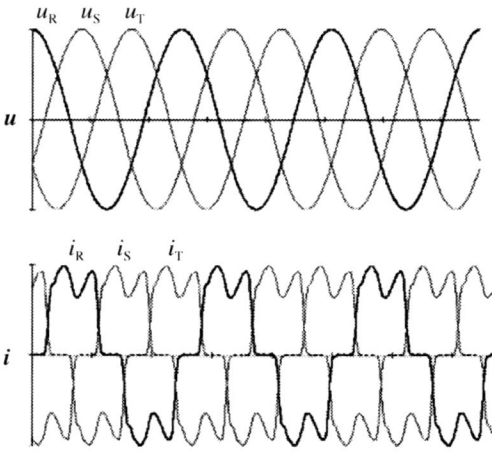

Figure 12.8. Voltage and current waveforms of a three-phase thyristor phase-controlled AC to DC converter (*PQF*= 0.930)

Finally, in case e) a three-phase thyristor phase-controlled AC to DC converter is connected to a 220 V balanced sinusoidal voltage source. The voltage and current waveforms for a thyristor-firing angle of 60° are shown in Figure 12.8. The power-quality factor is now 0.930 owed to the equivalent current distortion (*ITHD*= 0.304).

12.6 Quantitative Formulations of Power-quality Aspects Under Transient State Conditions

One of the most interesting applications is the assessment of power quality by re-defining the power-quality indices for transient events using signal-processing techniques. Due to the limitations of the traditional power-quality indices for aperiodic signals, new definitions and application of power-quality indices based on the short-time Fourier transform have been suggested [16]. However, as discussed in [26] the short-time Fourier transform requires a time-localisation window with time duration that prohibits a more general application of the power-quality indices to "nonstationary" signals [22]. In addition, the potential applicability of time–frequency analysis to transient power quality is mentioned in [26].

12.6.1 Fourier Analysis Versus Time–Frequency Analysis for Power Quality

Based on the Fourier coefficients, various types of frequency-domain quality aspects (*QAs*) and power-quality indices (*PQF* and *VQF*) were obtained in the above paragraphs. Once the Fourier coefficients are calculated, the *QAs* can be directly calculated.

However, the evaluation of the Fourier series requires a periodicity of the disturbance signal with respect to the fundamental frequency ($\omega_0/2\pi$). Hence, the treatment of transient disturbances, whose periodicity with respect to the fundamental frequency cannot be defined, via the Fourier-series-based power-quality indices is inappropriate from a signal analysis point of view. Therefore, we will utilise time–frequency analysis, which provides simultaneous time and frequency information for the analysis and assessment of transient disturbance signals.

Time–frequency analysis is motivated by the analysis and representation of nonstationary signals whose spectral characteristics change in time. Various types of time–frequency distributions, *e.g.*, the spectrogram, Wigner–Ville distribution and Choi–Williams distribution, Gabor spectogram, wavelets, *etc.*, have been studied in this book (Chapter 3) for improvement of the time–frequency resolution. The various types of time–frequency distributions have been generalised by the following equation known as "Cohen's class" [6]

$$C_s[i,k;\Phi] = \sum_{m=-L/2}^{L/2} \sum_n \Phi[n,m] R[i-n,m] e^{-j\frac{2\pi km}{L}} \quad (12.18)$$

where $\Phi[n,m]$ is the kernel of the time–frequency distribution and is different for each member of the class, $R[i, m]$ is the instantaneous correlation given by $R[i, m] = z[i + m] \, z^*[i - m]$, where $z[i]$ is the interpolated form of a given signal $s[i]$. The signal $s[i]$ is the analytic (complex) version of the signal to be analysed $x(t)$. Variables i, k denote a time–domain shift, and a frequency–domain shift, respectively.

The Cohen's class defined in Equation 12.18 is a bilinear transformation so that interference terms will appear if a signal is composed of multiple components. The time–frequency characteristic of the interference term for a given signal depends upon the selection of the kernel. Hence, the arguments regarding the selection of the proper kernel in time–frequency analysis are also applicable to the case of power–quality analysis. As the disturbance signal in power systems is characterised by the presence of multiple frequency components over a short time duration, interference is also problematic and a high-resolution time–frequency distribution is required [5, 23, 31, 38].

12.6.1.1 The Wavelet Transform

The wavelet transform (WT) has many applications due to its powerful features of time–frequency analysis of nonstationary signals. WT has the potential for efficiently assessing the spectrum information of aperiodic and time-varying power-system waveforms. It overcomes the shortcomings of the Fourier transform (FT) and avoids the above interference problems of the Cohen's class. This can be of great assistance in constructing new wavelet-based power-quality indices analogous to the currently used FT-based ones.

For the wavelet expansion, the square-integrable density function $f(t)$ can be analysed or processed by the decomposition [39]

$$f(t) = \sum_k a_{J_0,k} \varphi_{J_0,k}(t) + \sum_{j=J_0}^{\infty} \sum_k d_{j,k} \Psi_{j,k}(t), \quad t \in R, \quad (12.19)$$

where j, k, and J_0 are non-negative integers. This expansion is similar to that by Fourier showing a linear combination of wavelet coefficients, $(a_{J_0,k}, d_{j,k})$ and a set of basis functions $\varphi_{J_0,k}(t)$, called *scaling functions*, and *wavelet functions* $\Psi_{j,k}(t)$. Sets $a_{J_0,k}$ and $d_{j,k}$ are the *discrete wavelet transform* (DWT) [39] of $f(t)$ and can be calculated by

$$a_{J_0,k} = \langle f(t), \varphi_{J_0,k}(t) \rangle \qquad d_{j,k} = \langle f(t), \Psi_{j,k}(t) \rangle. \quad (12.20)$$

In expression $f(t)$, the first sum is a *coarse representation* of $f(t)$, where $f(t)$ has been replaced by a linear combination of 2^{J_0} translations of the scaling function $\varphi_{J_0,0}$. The remaining terms are the *detail representation*. For each j level, 2^j

translations of the wavelet $\psi_{j,0}$ are added to obtain a more-detailed approximation of $f(t)$ (Chapter 3).

The scaling function $\varphi_{j_0,k}$ and the wavelet bases $\psi_{j,k}$ are orthonormal bases and have the following properties:

$$\langle \varphi_{j_0,k}, \varphi_{j_0,k} \rangle = 1, \quad \langle \varphi_{j_0,k}, \psi_{j,k} \rangle = 0, \quad \text{where } j \geq j_0,$$
$$\langle \psi_{j,k}, \psi_{j,k} \rangle = 1, \quad \langle \psi_{j,k}, \psi_{i,k} \rangle = 0, \quad \text{where } j \neq j_0. \quad (12.21)$$

As the power of a periodic signal is preserved in both the time and frequency domains via Parseval's theorem, rms calculation of $f(t)$ can be described as follows, based on the above wavelet properties of Equation 12.21,

$$F_{rms} = \sqrt{\frac{1}{T} \int_0^T f(t)^2 dt} = \sqrt{\frac{1}{T} \sum_T a_{j_0,k}^2 + \frac{1}{T} \sum_{j \geq j_0} \sum_k d_{j,k}^2}. \quad (12.22)$$

12.6.2 Time–Frequency-based Transient Quality Aspects (*TQA*)

For the discussion of the transient power-quality assessment, the time–frequency-based transient quality aspect (*TQA*) is defined in this section. The DWT is implemented using a multiresolution signal analysis (MRA) algorithm [2], to decompose a given signal into its constituent wavelet subbands or levels (scales) with different time and frequency resolution. Each of the signal scales represents that part of the original signal occurring at that particular time and in that particular frequency band. In the common dyadic decomposition to be used, the scales are separated from adjacent scales by a frequency octave. These decomposed signals possess the powerful time–frequency localisation property, which is one of the major benefits provided by the wavelet transform. That is, the resulting decomposed signals can then be analysed in both the time and frequency domains.

The MRA is an adequate and reliable tool to detect sharp signal changes and clearly display high-frequency transients. The CWT is as reliable as this method but has the advantage of giving directly the magnitude of the 50-Hz signal. This method is suitable for quantifying power quality when detecting and measuring voltage sags, transients, overvoltages or flicker.

12.6.2.1 Instantaneous Transient Distortion Ratio

The instantaneous transient distortion ratio (ITD) is the transient version of the total harmonic distortion (THD) given in Equation 12.4 and it is defined as follows in terms of the time-scale distributions of the MRA components:

$$\text{ITD}(j;\Phi) = \sqrt{\frac{\sum_{k \in Kd} d_{j,k}^2}{\sum_{k \in Ka} A_k^2}}, \qquad A_k^2 = \frac{1}{N}\sum_{j=1}^{N} a_{j,k}^2, \qquad (12.23)$$

where we denote the time-scale distribution of a disturbance as $d_{j,k}$ and that of the fundamental component as $a_{j,k}$, Kd is the set of scale numbers corresponding to the transient disturbance and Ka is the set of scale numbers corresponding to the fundamental component.

The definition of the ITD($j;\Phi$) can be interpreted as a "time-varying" power-quality assessment determined by the time-frequency localised energy ratio of the transient-disturbance events to the fundamental frequency energy for a given kernel Φ. Note that in ITD($j;\Phi$), the energy of the disturbance is calculated not just from the harmonics but from all continuous scales. The magnitude of the coefficients depends on the decomposition frequency level. The lowest level that contains the finest signal features, *i.e.* the highest frequency content, has coefficients with the lowest energy content and *vice versa*. To remedy this, different weights to the coefficients of each level can be assigned, similarly to the definitions of Equation 12.5. The higher the level, the larger the weight.

Based on the types of time–frequency distribution given in Chapter 3, several formulations of the JTFA, for example definition 12.18, can be considered for the definition of the ITD($j;\Phi$). In general, it can be defined in terms of the time–frequency distributions of the disturbance and fundamental frequency components.

12.6.2.2 Normalised Instantaneous Transient Distortion Ratio

The transient disturbance energy $\sum_{k \in Kd} d_{j,k}^2$ can be normalised by the sum of the transient disturbance itself and fundamental energy $\sum_{k \in Ka} A_k^2$ as shown in the following definition of the normalized instantaneous transient distortion ratio (NITD).

$$\text{NITD}(j,\Phi) = \sqrt{\frac{\sum_{k \in Kd} d_{j,k}^2}{\sum_{k \in Ka} A_k^2 + \sum_{k \in Kd} d_{j,k}^2}}. \qquad (12.24)$$

Therefore, the NITD(j,Φ) increases with the transient disturbance energy, however, it cannot exceed a maximum value of 1.

A large-amplitude transient disturbance may result in a large value of ITD($j;\Phi$); however, for NITD(j,Φ), the maximum value is bounded by 1 so that the variations of the transient power-quality index is limited.

Transient disturbance parameters ITD and NITD can be used as individual transient quality expressions given in per cent values. This corresponds to data of the application example of the last section.

12.6.3 Procedure to Obtain the Transient Quality Aspect

Time–frequency-based transient quality indices, ITD($j;\Phi$) and NITD(j,Φ), can be introduced to quantify the time-varying power-quality aspect (*TQA*).

The transient power-quality indices provide useful information about the time-varying signature of the transient disturbance for assessment purposes. However, if the time-varying signature can be quantified as a single number, it would be more informative and convenient for an assessment and comparison of transient power quality. Therefore, a *"transient-interval average"* of the transient power-quality indices, <ITD($j;\Phi$)>, can be defined as an average of the time–frequency-based power-quality index function ITD($j;\Phi$) over a time interval T_0 as follows:

$$\langle \text{ITD} \rangle = \frac{1}{T_0} \sum_{j=t_0}^{j=t_0+T_0} \text{ITD}(j,\Phi), \qquad (12.25)$$

where t_0 is the start time of the transient disturbance. The selection of the time interval for the evaluation of the principal average <ITD> can be determined by the time index of the first peak value of the ITD($j;\Phi$), t_0, and the time index of the last peak value of the ITD($j;\Phi$), $t_0 + T_0$. Consequently, the principal average of the transient power-quality aspect is the local average over a T_0 s duration. A similar formulation for <NITD> can be considered using the power-quality index function NITD($j;\Phi$) over the time interval T_0.

<NITD> can be considered as a new transient quality aspect (TQA), QA_7, for computing a global power-quality factor. Thus, stationary and transient components present on an electrical signal could be included in *PQF* and *VQF* definitions. The first quality aspects are extracted by means of the Fourier methods and the transient quality aspect is subjected to the joint time–frequency treatment.

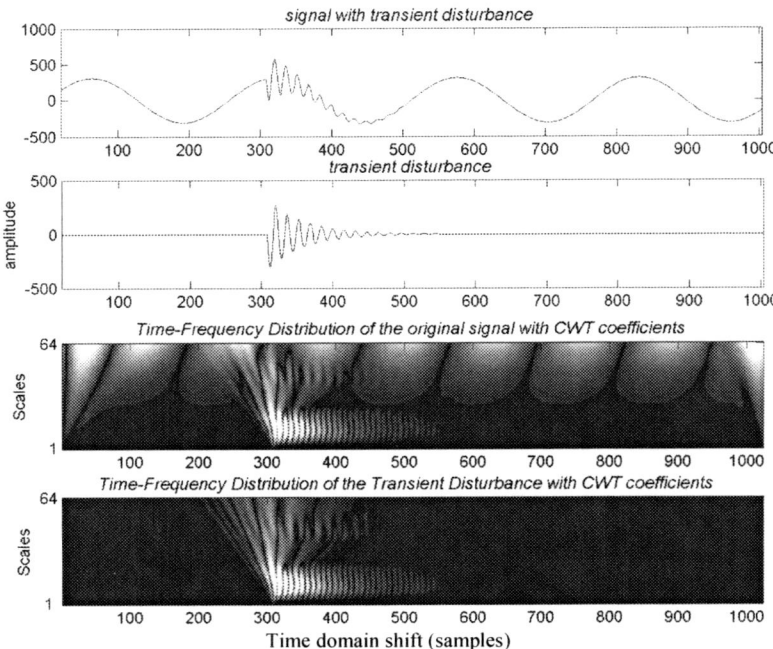

Figure 12.9. Time and time–frequency distribution of a capacitor-switching disturbance

12.6.4 Application Example of Transient Disturbance

In Figure 12.9, a fast capacitor-switching disturbance is provided with the time–frequency distribution, and the corresponding transient power-quality indices are shown in Figure 12.10. The waveform at the top of Figure 12.9 is the original waveform and the waveform in the middle is the extracted disturbance waveform obtained by Equation 12.21. The time–frequency distribution of the fast capacitor-switching disturbance is provided in the bottom of Figure 12.9. It is known that the disturbance generated by the capacitor-switching restrike on opening exhibits a transient oscillation with a natural frequency determined by the capacitance and inductance of the system. Therefore, the disturbance signal is more transient and oscillatory than normal capacitor energising, which will be considered next.

The time–frequency distribution in Figure 12.9 shows that the transient energy of the disturbance occupies approximately 500 Hz to 1000 Hz during 25–44 ms. The instantaneous transient disturbance ratio ITD in Figure 12.10c shows a peak value of 124.27% at 24.22 ms while the peak of the normalised instantaneous disturbance energy ratio NIDE(j,Φ) in Figure 12.10d shows a peak value of 77.91% at the same time. The principal averages, <IDE> and <NIDE>, are 24.96% and 21.46%, respectively, for $T_0 = 16.95$ ms.

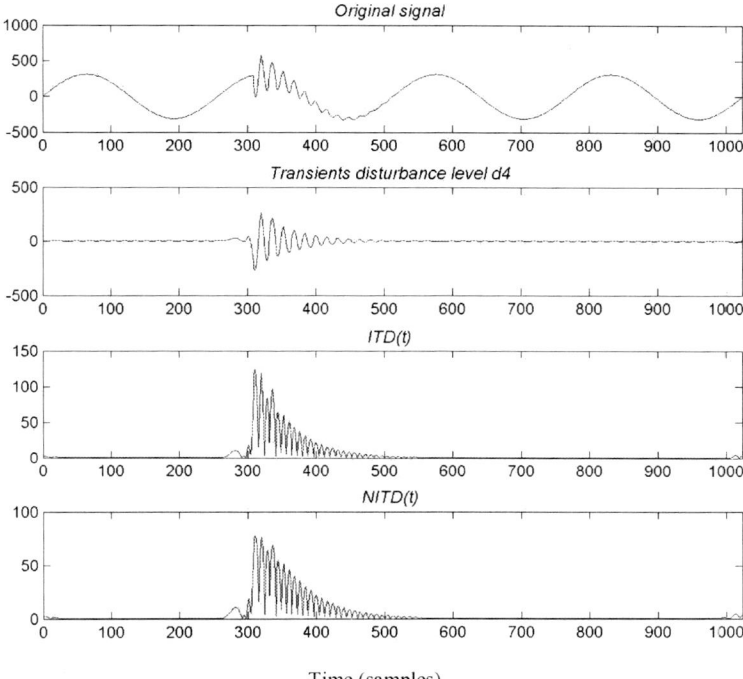

Figure 12.10. Time–frequency based transient power-quality indices of the capacitor switching: (**a**) Disturbance waveform. (**b**) Separated disturbance waveform. (**c**) Instantaneous disturbance energy ratio (ITD($j;\Phi$)). (**d**) Normalised instantaneous distortion energy ratio (NITD(j,Φ)).

References

[1] Adelson RM, (1997), Frequency Estimation from Few Measurements, Digital Signal Processing, 7, 47–54, London, Academic Press.
[2] Akansu A.N, Smith MJT, (1996), Subband and Wavelet Transforms Design and Application, Norwell, MA; Kluwer.
[3] Arrillaga J, Bradley DA, Bodger PS, (1985), Power System Harmonics, John Wiley & Sons, Chichester.
[4] Bollen M, (2000), Understanding Power Quality Problems Voltage Sags and Interruptions, New York, IEEE Press.
[5] Choi HI, Williams WJ, (1989), Improved time-frequency representation of multicomponent signals using exponential kernels, IEEE Trans. Acoust., Speech, Signal Process., 37, 6, 862–871.
[6] Cohen L, (1995), Time-Frequency Signal Analysis, Upper Saddle River, NJ: Prentice-Hall.

[7] Czarnecki LS, (1994), Comments on 'Apparent and reactive powers in three-phase systems: In search of a physical meaning and a better resolution', ETEP Eur. Trans. Elec. Power. Eng., vol. 4, 421–426.
[8] Davis EJ, Emanuel A.E, Pileggi DJ, (2000), Evaluation of single-point measurements method for harmonic pollution cost allocation, IEEE Trans. Power Delivery, 15, 1, 14–18.
[9] Mc Eachern A, Grady WM, Moncrief WA, Heydt GT, McGranaghan M, (1995) "Revenue and Harmonics: An evaluation of some proposed rate structures," *IEEE Trans. Power Delivery*, 10, 1, 474–482.
[10] Emanuel AE, (1993), On the definition of power factor and apparent power in unbalanced polyphase circuits with sinusoidal voltage and currents, IEEE Trans. Power Delivery, 8, 7, 841–852.
[11] Filipski PS, (1991), Polyphase apparent power and power factor under distorted waveform conditions, IEEE Trans. Power Delivery, 6, 7, 1161–1165.
[12] Filipski PS, (1993), Apparent Power – A misleading quantity in the nonsinusoidal power theory: Are all nonsinusoidal power theories doomed to fail?, ETEP Eur. Trans. Elec. Power. Eng., vol. 3, 21–26.
[13] Fortescue CL, (1918), Method of symmetrical coordinates applied to the solution of polyphase networks, Trans. AIEE, pt. II, vol. 37, 1027-1140.
[14] Gaouda AM, Salama MM, ,Sultan MR, Chikhani AY, (2000), Application of multiresolution signal decomposition for monitoring short duration variations in distribution systems, IEEE Trans.Power Del., 15, 2, 478–485.
[15] Henderson RD, Rose PJ, (1994), Harmonics: the effects on power quality and transformers, IEEE Trans. Ind. Applicat., 30, 3, 528-532.
[16] Herraiz S, Heydt GT, O'Neill-Carrillo E, (2000), Power quality indices for aperiodic voltages and currents, IEEE Trans. Power Del., 15, 2, 784–790.
[17] Heydt GT, Galli AW, (1997), Transient power quality problems analyzed using wavelet, IEEE Trans. Power Del., 12, 2, 908–915.
[18] Heydt GT, Fjeld PS, Liu CC, Pierce D, Tu L, Hensley G, (1999), Applications of the windowed FFT to electric power quality assessment, IEEE Trans. Power Del., 14, 4, 1411–1416.
[19] Heydt GT, (1991), Electric Power Quality, West Lafayette, IN, Stars in a Circle Publications.
[20] Heydt GT, (2000), Problematic Power Quality Indices, Panel Session – The need for future harmonic standards, IEEE/PES Winter Meeting, Jan., Singapore.
[21] IEEE Working Group on nonsinusoidal situations, (1996), Practical definitions for powers in systems with nonsinusoidal waveforms: A discussion, IEEE Trans. Power Delivery, 11, 1, 79–101.
[22] Jaramillo SH, Heydt GT, ONeill-Carrillo E, (2000), Closure to discussion of Power quality indices for aperiodic voltage and currents, IEEE Trans. Power Del., 15, 4, 1334–1334.
[23] Jeong J, Williams WJ, (1992), Kernel design for reduced interference distributions, IEEE Trans. Signal Process., 40, 2, 402–412.
[24] Limits for Harmonic Current Emissions, (1995), IEC 61000, Part 3, Section 4.
[25] López A, Montaño JC, Castilla M, Gutiérrez J Borrás D, Bravo JC, (2003), A virtual instrument for the instantaneous line-frequency measurement under nonstationary situations, 8th Portuguese-Spanish Congress in Electrical Engineering (8CLEEE), Vilamoura, Portugal.
[26] Mariscotti A, (2000), Discussion on "power quality indices for aperiodic voltage and currents, IEEE Trans. Power Del., 15, 4, 1333–1334.
[27] Montaño JC, Bravo JC, Borrás D, Castilla M, López A,. Gutiérrez J, (2005), Voltage quality analyzer, Scientific Database IEEE Xplore, Compatibility in Power

Electronics, IEEE, June 1, 25–29, Digital Object Identifier 10.1109/CPE.2005.1547541.

[28] Montaño JC, López A, Gutierrez J, Castilla M, Borrás D, Bravo JC, (2004), Power quality factor and line-disturbances measurements in three-phase systems, IEEE Power Electronics Specialists Conference (PESC'04), 20–25 June, Aachen, Germany.

[29] Parameswariah C, Cox M, (2000), Frequency characteristics of wavelets, IEEE Trans. Power Del., 17, 3, 800–804.

[30] Poisson O, Rioual P, Meunier M, (2000), Detection and measurement of power quality disturbances using wavelet transform, IEEE Trans. Power Del., 15, 3, 1039–1044.

[31] Qian S, Chen D, Discrete Gabor transform, (1993), IEEE Trans. Signal Processing, 41, 7, 2429–2439.

[32] Recommended Practices and Requirements for Harmonic Control in Power Systems, (1992), IEEE 519 Standard.

[33] Salcic Z, Li Z, Annakkage UD, Pahalawaththa N, (1998), A Comparison of Frequency Measurement Methods for Underfrequency Load Shedding, Electric Power System Research, 45, 3, 209–219.

[34] Santoso S, Powers E J, Grady WM, Hofmann P, (1996), Power quality assessment via wavelet transform analysis, IEEE Trans. Power Del., 11, 2, 924–930.

[35] Santoso S, Powers EJ, Grady WM, Lamoree J, Bhatt SC, (2000), Characterization of distribution power quality events with Fourier and wavelet transforms, IEEE Trans. Power Del., 15, 1, 247–253.

[36] Sharon, D., (1973), Reactive power definitions and power factor improvement in nonlinear systems, Proc. Inst. Elec. Eng., 120, 8, 704–706.

[37] Sharon D, (1996), Power factor definitions and power transfer quality in nonsinusoidal situations, IEEE Trans. Instrum. Meas., 45, 3, 728–733.

[38] Shin YJ, Parsons A.C, Powers EJ, Grady WM, (1999), Time-frequency analysis of power system fault signals for power quality, Proc. IEEE Power Eng. Soc. Summer Meet., Edmonton, AB, Canada, 402–407.

[39] String G, Nguyen T, (1997), Wavelets and Filter Banks. Cambridge, MA: Wellesley-Cambridge Press.

[40] Wagner VE, Balda JC, Barnes TM, Emannuel AE, Ferraro RJ, Griffith DC, Hartmann DP, Horton WF, Jewell WT, McEachern A., Phileggi DJ, Reid WE, (1993), Effects of harmonics on equipment, IEEE Trans. Power Delivery, 8, 2, 672–680.

13

IEC 61850 and Power-quality Monitoring and Recording

Alexander Apostolov

OMICRON Electronics,
2950 Bentley Ave., Unit 4,
Los Angeles,
California, USA
Email: alex.apostolov@omicronusa.com

13.1 Introduction

IEC 61850 is an approved international standard for communications in substations that is creating opportunities for a revolution in electric power-systems monitoring, control and protection. It represents the next step in the integration of multifunctional intelligent electronic devices (IEDs) based on the development and implementation of standardised abstract object models and advanced distributed applications.

IEDs are the standard in new or upgraded integrated substation automation systems. Power-quality monitoring IEDs are sophisticated multifunctional devices designed to detect and record the effects of different abnormal system conditions. Since short-circuit faults, power interruptions and system disturbances are rear in the system, the IEDs take advantage of their data acquisition and processing capabilities and include multiple non power-quality monitoring functions like measurements, waveform, disturbance and event recording and some builtin analysis tools. This makes them an important device that can be used for many different tasks at the process level of a substation automation system.

Control, protection and disturbance-recording devices may complement the power-quality monitoring IEDs by providing some specific functionality that may not be available. This allows the optimisation of the integrated substation automation system, while at the same time meeting the strict requirements for reliability and security.

The selection of the communications protocol used at the substation level is one of the critical factors to consider in the design of the power-quality-monitoring

system. The protocol should provide all required services that will allow the optimal implementation of different substation functions. This requires:
- proper definition of the functional and performance requirements;
- good understanding of the substation communications protocol.

As can be seen from the following sections, IEC 61850 is designed to meet all the requirements of power-quality-monitoring and recording applications, as well as the challenges, benefits and opportunities for future developments.

13.2 What is IEC 61850?

According to the names of the different parts of IEC 61850 it is a standard for communication networks and systems in substations. It was developed with the goal of meeting the requirements of all different functions and applications in the substation, such as:
- Protection
- Control
- Automation
- Measurements
- Power-quality monitoring
- Recording

At the same time it should support different tasks related to the above-listed substation functions, such as:
- Engineering
- Operations
- Commissioning
- Testing
- Maintenance
- Event analysis
- Security

IEC 61850 was developed over a period of about 10 years and was the result of the combined efforts of numerous industry experts from around the world. Initially there were two separate activities:
- The development of GOMSFE (Generic Object Models for Substation and Feeder Equipment) as part of UCA (the Utilities Communications Architecture).
- The IEC 61850 project for development of a standard substation communications protocol under IEC Technical Committee 57.

In 1997 a conclusion was reached that due to the similarities of both activities it will be beneficial to the industry to have a single standard for substation communications and the members of the UCA working group were integrated into the IEC TC 57 working groups.

So, the standard was completed with the efforts of three working groups:
- Working group 10 focused on the definition of the functional architecture and general requirements.
- Working group 11 addressed the communications within and between unit and station levels that are now know as the station bus.
- Working group 12 developed the communications within and between process and unit levels known as process bus.

Since the publication of the standard and its widespread application in hundreds of substations the UCA International Users Group is working on resolving the different technical issues. The solutions and new developments, such as the modelling of some power-quality-event detection functions addressed by WG 10 will be included in amendments and later in version 2 of the standard.

IEC 61850 was developed on the basis of some key requirements:
- It should be technology independent.
- It should be flexible.
- It should be expandable.

By meeting the above requirements the standard allows us to meet the changing needs of the electric power industry and take advantage of the developments in computers, communications and sensors technology.

The IEC 61850 standard consists of fourteen different documents that cover a wide range of issues and make it clear that it is much more than a communications protocol definition. It defines not only how to communicate over the substation local area network, but also what to communicate. It provides an abstract model of the substation equipment and functions that can be used as the foundation of the development of different tools.

The standard also addresses the substation integration and automation engineering process and specifies the conformance testing for devices that support it.

It needs to be well understood that the IEC 61850 standard does not specify individual implementations, communication architectures or products. It also does not attempt to describe any details of the functionality of the different devices, such as algorithms, but focuses only on the specification of the externally visible functionality of primary or secondary equipment, functions or implementations in substation protection, control and automation systems.

The IEC 61850 standard is a collection of 14 documents grouped in 10 different parts:

Part 1 - Introduction and Overview

This document, as is clear from its name, is an introduction and overview of the concepts and documents in the IEC 61850 standard. It is a summary that includes text and figures from other parts of the standard.

Part 2 - Glossary

This part represents a collection of specific terminology and definitions from other standards or terms defined in different parts of IEC 61850 that are used in the context of substation automation systems within the various parts of the standard.

Part 3 - General Requirements

Like any other technical solution, IEC 61850-based devices and systems need to be developed to meet specific requirements. Environmental, reliability, maintainability, system availability, security and other requirements are defined in this part.

Part 4 - System and Project Management

Management of large substation power-quality monitoring, protection and automation systems is a complex multiphase process that presents many challenges to all participants. This part of the standard addresses some engineering issues such as parameter classification, engineering tools and documentation. Quality-assurance responsibilities, type tests, system tests, factory and sight acceptance tests, as well as system lifecycle issues are discussed.

Part 5 - Communication Requirements for Functions and Device Models

This part of the standard includes the principle communications requirements related to the functional and device models that are later detailed in the several sub-parts of Part 7 of the standard. It defines the specific logical interfaces and identifies those out of the scope of IEC 61850. It covers the requirements for interoperability, lists substation automation system functions, their specifications and performance requirements. The LN (logical nodes) and PICOM (piece of information for communications) concepts are presented, including LN categories, list of logical nodes, their use and interaction.

Specification of message types with performance requirements, list of PICOMs and classification of PICOMs to message types and performance calculations for typical substation configurations are defined.

Part 6 - Substation Automation System Configuration Language

The substation configuration language is one of the most important differentiating components of IEC 61850. It defines a file format based on XML (including the schema) that can be used for system-parameters exchange related to:
- substation primary system schematic description (single line diagram)
- communication connection description;
- IED capabilities and configuration description;
- allocation of logical instances to primary system;
- allocation of logical nodes to physical devices.

The different types of files defined in Part 6 are used at different stages of the engineering process.

Part 7 Basic Communication Structure for Substation and Feeder Equipment

Part 7 actually contains 4 subparts that define the details of the abstract model used in IEC 61850 to meet the requirements of all functions and applications in the substation protection and automation domain.

Part 7-1 Principles and Models

This is the introduction to the remaining parts 7-x and describes the concepts of communications modelling in IEC 61850.

Part 7-2 Abstract Communication Service Interface

This is one of the key documents of the IEC 61850 standard. It includes the specification of abstract communication models and services required by substation automation and protection systems.

Part 7-3 Common Data Classes

This part defines the common data classes (CDC) that are necessary to implement the concepts of the hierarchical object model.

Part 7-4 Compatible Logical Node Classes and Data Classes

Part 7-4 includes 92 logical node classes grouped based on their association with some of the basic substation functions, such as protection, control, measurements, etc. The logical node classes are represented as collections of data objects of specific data classes.

Part 8-1 Specific Communication Service Mapping (SCSM) – Mappings to MMS (ISO 9506-1 and ISO 9506-2) and to ISO/IEC 8802-3

Since in the previous parts of IEC 61850 the abstract models of the substation automation system were addressed, Part 8-1 specifies the mapping of these models to the selected MMS and ISO/IEC 8802-3 protocol. The transmission of sampled values is, however, excluded from this mapping.

Part 9 Process Bus Mapping

This part is split into two subparts that define two different implementations of the IEC 61850 Process Bus.

Part 9-1 Sampled values over serial unidirectional multi-drop point to point link.

Mapping of the core elements from the model for transmission of sampled measured values in a point-to-point link was historically based on the state-of-the-art of the technology at the time of the development of the standard. No known use of this Process Bus mapping exists today.

Part 9-2 Specific Communication Service Mapping (SCSM) – Sampled values over ISO/IEC 8802-3

This part covers the specific mapping of the complete model for transmission of sampled measured values and the model for generic object-oriented system events (GOOSE).

Part 10 - Conformance Testing
Part 10 defines the procedures for conformance (but not functional) testing of IEC 61850 compliant devices. It covers:
- quality assurance and testing;
- required documentation;
- device related conformance testing;
- certification of test facilities, requirement and validation of test equipment.

13.3 Logical Interfaces and Distributed Applications

The selection of IEC 61850 as the communications protocol used for substation communications is one of the critical factors that affect the design of the substation protection and control system. It allows the development of a distributed function resulting in improved performance and reduced costs. A function can be divided into subfunctions and functional elements.

A "distributed function" is one that requires exchange of data between two or more functional elements located in different physical devices. The exchange of data is not only between functional elements, but also between different levels of the substation functional hierarchy. It should be kept in mind that functions at different levels of the functional hierarchy can be located in the same physical device, and at the same time different physical devices can be exchanging data at the same functional level.

IEC 61850 defines interfaces that may use dedicated or shared physical connections – the communications links between the physical devices. The allocation of functions between different physical devices defines the requirements for the physical interfaces, and in some cases may be implemented into more than one physical LAN.

Functions in the substation are performed by the protection, control, power-quality-monitoring and recording systems. A function can be divided into subfunctions and functional elements. The functional elements are the smallest parts of a function that can exchange data.

The allocation of functions between different physical devices defines the requirements for the physical interfaces, and in some cases may be implemented into more than one physical LAN.

IEC 61850 defines functions of a substation automation system (SAS) related to the protection, control, monitoring and recording of the equipment in the substation.

The functions in the substation can be distributed between IEDs on the same, or on different levels of the substation functional hierarchy. IEC 61850 defines three such levels:
- station;
- bay/unit;
- process.

These levels and the logical interfaces are shown by the logical interpretation of Figure 13.1.

IEC 61850 focuses on a subset of the interfaces shown in Figure 2 and listed below:

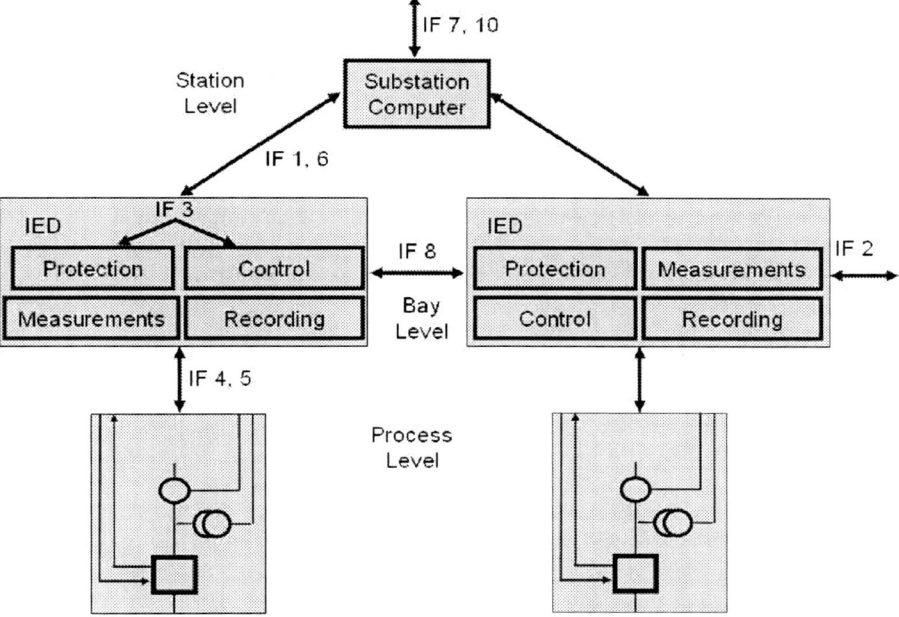

Figure 13.1. Logical interfaces in Substation Automation Systems

The logical interfaces shown above are defined as:
IF1: protection-data exchange between bay and station level
IF2: protection-data exchange between bay level and remote protection
IF3: data exchange within bay level
IF4: CT and VT instantaneous data exchange (especially samples) between process and bay level
IF5: control-data exchange between process and bay level
IF6: control-data exchange between bay and station level

IF7: data exchange between substation (level) and a remote engineer's workplace
IF8: direct data exchange between the bays especially for fast functions like interlocking
IF9: data exchange within station level
IF10: control-data exchange between substation (devices) and a remote control center

Interfaces 2 and 10 have been identified as outside of the scope of IEC 61850 at the time of the development of the standard. However, the availability of the ethernet interface in multiplexers over SONET rings or other communication links results in applications using high-speed peer-to-peer communications between relays in different substations, for example in a directional comparison transmission line protection.

In order to better understand the distributed functions as they are defined in an IEC 61850-based system, we need first to clarify the meaning of some of the terms used in the logical interfaces listed above.

13.4 Functional Hierarchy

A complex substation automation system has an hierarchical structure that may have a different number of levels from the communications and logical point of view. The three typical levels in the functional hierarchy are defined as follows:

Process-level Functions: These are all functions interfacing to the process (the primary equipment in the substation), such as analog signals, binary status signals or binary control signals. In the conventional substation this is hard-wired current and voltage circuits, as well as the hard-wired auxiliary contacts of switching devices and trip or close coils.

These functions communicate via the logical interfaces 4 and 5 to the bay level. In the IEC 61850-based substation these functions are performed by a sensor IED with a digital interface to the bay level defined in the standard. This interface can be used for distributed applications based on analog-sampled values, but are not a subject of this chapter.

Bay-level Functions are those using mainly the data of one bay and acting mainly on the primary equipment of one bay. A substation consists of closely connected subparts with some common functionality. Examples are the breakers and switches between the substation bus and a transmission or distribution line, a transformer with its related switchgear between two or three buses with different voltage levels, etc.

Typically in conventional systems such a part of the substation is protected or monitored by a single device. This subpart is defined in IEC 61850 as a "bay", being protected and controlled by "bay protection" and/or "bay controller" devices. The functionality of these devices represents an additional logical control level below the overall station level that is called the "bay level". Power-quality-monitoring functions can also be defined at this level of the substation functional hierarchy.

The bay-level functions can be implemented in physical devices that also perform process-level functions or substation-level function, i.e. there is a difference in the logical and physical functional hierarchy.

These functions communicate via the logical interface 3 within the bay level and via the logical interfaces 4 and 5 to the process level, and 1 and 6 with the substation level.

Station-level Functions are related to the overall operation of equipment in the substation. They are divided into two groups:

Process-related station-level functions are functions using the data of more than one bay or of the complete substation and acting on the primary equipment of more than one bay or of the complete substation. These are functions that in conventional systems use hard-wired connections between relay outputs of one device and opto inputs of another device. An example of such an interface will be the triggering of the recording of a voltage dip by multiple power-quality-monitoring devices. In IEC 61850-based systems these functions communicate mainly via the logical interface 8.

Interface-related station-level functions are functions representing the interface of the SAS to the substation HMI (human machine interface), to SCADA or to a remote engineering station. These functions communicate via the logical interfaces 1 and 6 with the bay level and via the logical interface 7 and the remote control interface to the outside world.

13.5 The IEC 61850 Model

The foundation of the IEC 61850 is the concept of virtualisation, i.e. providing a virtual representation of the behaviour of real primary or secondary substation devices.

As mentioned earlier, the virtualisation covers only the relevant and communications visible components of the model. Figure 13.2 shows the use of this process to model a power-quality-monitoring function, such as harmonics measurements of a power-quality-monitoring device as an IEC 61850 logical node.

The modelling approach in the standard uses the principles of functional decomposition and UML notation. It is used to understand the logical relationships between components of a distributed function and is presented in terms of the model hierarchy that describes the functions, subfunctions and functional interfaces.

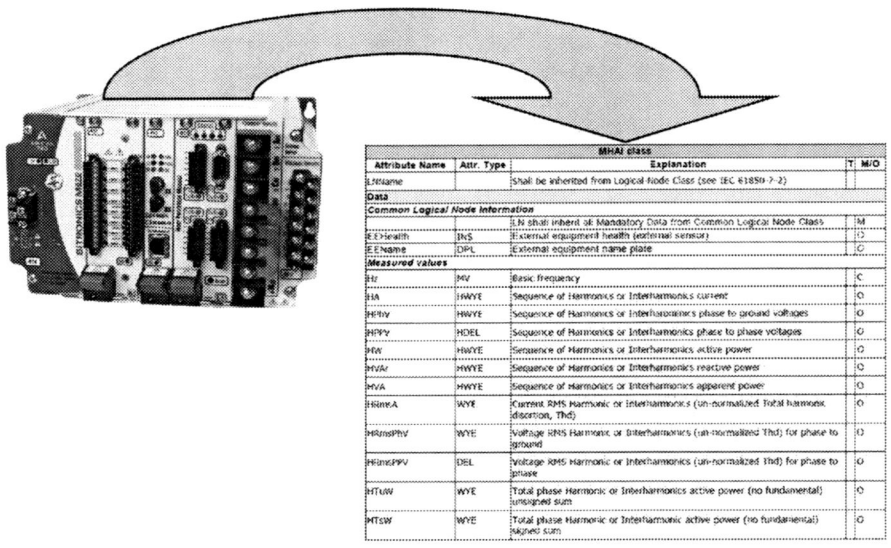

Figure 13.2. Virtualisation of a power-quality-monitoring IED

The data flow is used to understand the communication interfaces that must support the exchange of information between the distributed functional components for different applications. The information modelling on the other hand is used to define the abstract syntax and semantics of the information exchanged. It is presented in terms of the data object hierarchy that includes data object classes, types and attributes.

The modelling of complex multifunctional power-quality-monitoring IEDs from different vendors that are also part of distributed functions requires the definition of a modelling concept, as well as basic elements that can function by themselves or communicate with each other. These communications can be between the elements within the same physical device or in the case of distributed functions between multiple devices over the substation local area network.

The communications visible behaviour of any IED in a substation automation system (including power-quality-monitoring devices) in the model is represented by a Server. The server interfaces with different clients, for example by responding to their requests. It also communicates with its peers – by sending them information of the change of state of a functional element or sampled analog values.

The basic functional elements defined in IEC 61850 are the logical nodes. They form the foundation of the modelling concept of the standard. A logical node is defined in the standard as "the smallest part of a function that exchanges data". It is an object that is defined by its data and methods. When instantiated, it becomes a logical node object. Multiple instances of different logical nodes are components of different power-quality monitoring, protection, control or other functions in a substation automation system. Different data objects in a logical node provide means to specify typed information, for example, the state of a voltage sag detection element, including attributes providing quality information and timestamp.

Multiple logical nodes can be grouped in a logical device based on some similarity or if they can be enabled as a group.

Each of the above-described information models is defined as a class that comprises attributes and services. The conceptual class diagram of the abstract model is shown in Figure 13.3.

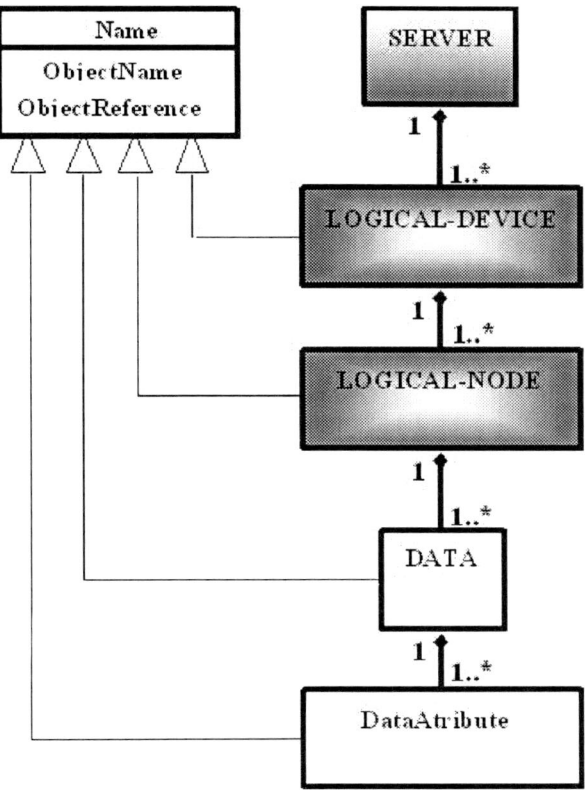

Figure 13.3. IEC 61850 abstract model class diagram

A logical device contains several logical nodes of three different types. It has a single logical node zero, a single logical node physical device, plus one or more other functional logical nodes.

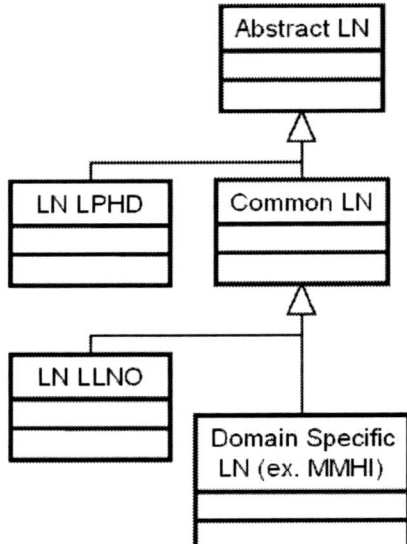

Figure 13.4. Logical nodes inheritance

As can be seen from Figure 13.4, all logical nodes inherit from the abstract logical node. Logical node zero and all domain-specific logical nodes (including power-quality measurements and events logical nodes) inherit from the common logical node.

One of the most important concepts that needs to be understood at the very beginning of the IED modelling process is that the model includes only objects that are visible to the communications. The IED may contain a lot of data internal to the device, such as data exchanged between elements of a specific power-quality event-detection logic – for example power interruption. If this logic is represented to the outside world as a black box with certain inputs and outputs, these internal signals are not visible and as a result they are not included in the model.

In order for the logical nodes to interoperate over the substation LAN, it is necessary to standardise the data objects that are included in each of them. IEC 61850 considers three levels of data and services for the modelling of different substation automation functions.

The first level is the abstract communication service interface (ACSI). It specifies the models and services for access to the elements of the specific object model, such as reading and writing object values or controlling primary substation equipment.

The second level defines common data classes (CDC) and common data attribute types. A CDC specifies a structure that includes one or more data attributes.

The third level defines compatible logical node classes and data classes that are specialisations of the common data classes based on their application.

Logical nodes typically include not only data, but also data sets, different control blocks, logs and others as defined by the standard. The DATA represents domain-specific information that is available in the devices integrated in a substation automation system. It can be simple or complex and can be grouped in data sets as required by the application.

Any DATA should comply with the structure defined in the standard and should include DataName, DataRef, Presence and multiple DataAttributes. The DataName is the instance name of the data object, while the DataRef is the object reference that defines the path name of the DATA object instance, while the Presence is a Boolean-type attribute that states if the data object is Mandatory or Optional.

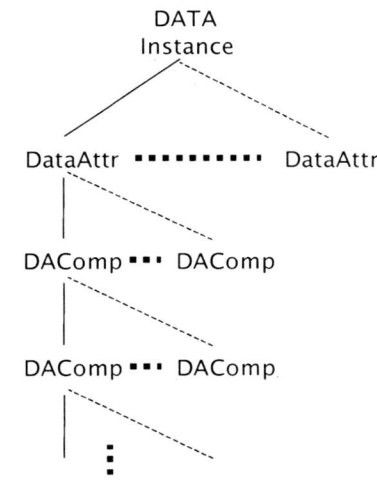

Figure 13.5. Data-object hierarchy

Each instance of a DATA class object must contain at least one DataAttribute. Instead of a DataAttribute it is possible to have a SimpleCDC or Composite CDC (both are specialisations of the DATA class). DataAttribute's can be simple or nested. If they are nested, at each nesting level other than the first, the DataAttributeName is called DAComponentName (see Figure 13.5). The DataAttributes are of a certain data type that can be primitive (BasicType) or composite (DAType).

The different DataAttributes can be grouped based on their specific use. For example some indicate the status of the logical node, while others are used for configuration or measurements. The property of DataAttribute that shows its use is a functional constraint (FC). The standard defines many different functional constraints. Some more commonly used ones are:

CO – control
SP – set point
CF – configuration
DC – description
SG – setting group
MX – measurements

The SP functional constraint is used for settings that are global to the IED, while SG applies to settings that may have different values in the different setting groups. The SP and SG data attributes can be read and set.

The MX functional constraint is used to indicate that the data attribute represents measuring information. The value of this data object can be read, substituted, logged or reported. The values of these DataAttributes are normally based on processed data from the IED.

After describing some of the typical data objects used to model measured values, we are finally at a point when we can give an example of a data path (the DAComponentRef) for a single-phase measurement of the current in phase B represented as a floating point:

MMXU1.A.phsB.cVal.mag.f,

where:

> **MMXU1** is an instance of the Compatible LN class MMXU defined in Part 7-4
> **A** is an instantiation of the Composite DATA class WYE (defined in 7-3) used to represent the three phase currents and the neutral current
> **phsB** is the value of the current in phase B as a Simple Common DATA class of type CMV (defined in 7-3)
> **cVal** is the complex value of the current in phase B (of the Common DataAttribute type Vector)
> **mag** indicates that this object represents the magnitude of the complex value (of type AnalogValue - defined in 7-3)
> **f** is a DataAttributeComponent that is of the basic type FLOATING POINT (defined in 7-2)

All measurements in multifunctional IEDs are modelled in a similar way and grouped into special logical nodes.

13.6 Distribution and Modelling of Functions in Power-quality Monitoring Devices

A very important differentiating factor of IEC 61850 compared to other communication protocols is that everything in this model has a name. This allows the definition of standard device models that support self-description and use of metadata for development of different engineering tools. At the same time, even though the standard defines the model hierarchy discussed earlier, it does not specify how it should be implemented in any specific device.

The models of multifunctional power-quality-monitoring IEDs can be based on one of these two basic modelling approaches:
- single logical device based model
- multiple logical devices based model

The modelling of a complex multifunctional power-quality-monitoring IED is possible only when there is not only good understanding of the IEC 61850 modelling principles, but also of the problem domain discussed in the previous chapters of the book. At the same time we should keep in mind that the models apply only to the communications visible aspects of the IED.

The functional hierarchy of the IED in some cases may be more complex than the modelling hierarchy of IEC 61850. This requires the use of some modelling aspects of the standard that allow the grouping of logical nodes different from in a logical device in order to provide some additional layers in the model.

Since the logical node name includes not only the logical node class name, but also an instance suffix and if necessary a prefix, the latter can be used to group logical nodes. Depending on the complexity of the model, the prefix can be split into two parts, to indicate a different level of functional grouping.

In the case of power-quality-monitoring devices with a more complex functional hierarchy it might be necessary to group together several logical nodes in a functional group such as voltage variation. It may contain several subfunctions, such as:
- voltage sag;
- voltage swell;
- voltage interruption.

Each of the subfunctions then may contain several instances of functional elements represented by logical node objects.

The fact that a logical node belongs to a function or sub-function group of logical nodes then can be represented by a functional group name as a prefix in the logical node name. If the device has a very complex functional hierarchy such as the one described above, it is possible to use external functional group name (EFGN) or an internal functional group name (IFGN) as shown in Figure 13.6

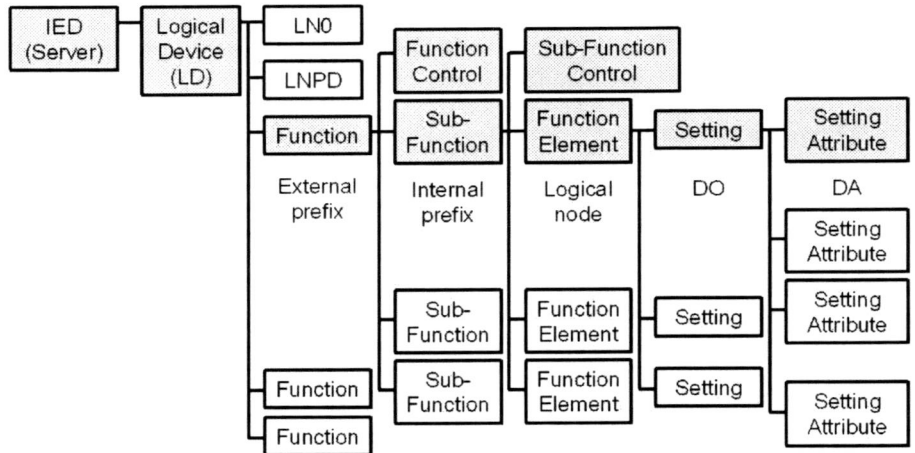

Figure 13.6. Device-functional hierarchy

As described earlier, Part 5 of IEC 61850 defines the logical node concept and the communications requirements for different functions and device models. Part 7-2 specifies the first level of modelling – ACSI. Part 7-3 covers the CDC, while Part 7-4 defines the compatible logical node and data classes used to build the model of the power-quality-monitoring IED.

The functions in a relatively simple IED (such as a low-end device that only measures the harmonic content of the voltage or detects voltage sags) are fairly easy to understand and group together in order to build the object model. That is not the case for the more complex devices that contain all different power-quality-monitoring and recording functions. They all have different components that need to be taken into consideration in the model. Complex to represent are also advanced power-quality-monitoring systems that typically exist in large substations or plants and implement distributed functions based on high-speed peer-to-peer communications between multiple IEDs.

The functional hierarchy of a modern power-quality-monitoring device to a great extent is dependent on the application and the design of the device. Since it is not expected that a very simple low-end device may have an ethernet interface, we are going to concentrate on IEDs that support IEC 61850 and will typically have a more complex functional hierarchy. For such high-end power-quality-monitoring devices the model also has to consider the availability of multiple-analog inputs – for example in the case of dual breakers, breaker-and-a-half or ring bus configurations.

A more complex example is a power-quality-monitoring IED installed on a transformer between the substation transmission and distribution buses that will interface the device with two or more voltage levels, as shown in Figures 13.7 and 13.8.

As mentioned earlier, the modelling of complex power-quality-monitoring devices can be done in different ways. One option is to model them as servers with a single logical device and multiple logical nodes. In this case, certain functional elements have to be grouped together using the available object hierarchy and the naming conventions for the data objects. A simplified block diagram of this approach is given in Figure 13.7.

The use of prefixes to group logical nodes can be illustrated in this example. All power-quality-related measurements for the high voltage side of the substation can be grouped together as shown with a dashed line in the figure and indicated by the letters hv that will become a prefix in each of the logical node names. In this way the user or any client application will know that logical nodes **hvMMXU1**, **hvMMTR1**, **hvMMHI1** and **hvMSTA1** all belong to the same functional group **hv** within logical device **LD1**.

This model can not really be applied for very complex power-quality-monitoring IEDs. The use of multiple logical devices offers better flexibility and provides the foundations for a future implementation agreement for the use of logical devices that may lead to the development of analysis tools based on the self-description support of IEC 61850.

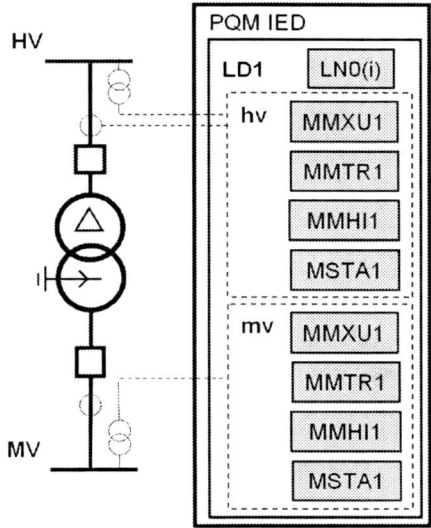

Figure 13.7. Single logical device object model

Figure 13.8 shows an object model of the same device, but using two separate logical devices – one representing the power-quality-monitoring functions for the high voltage level of the substation and the other the low side – medium-voltage level. The names of the two logical devices may reflect their association with each voltage level.

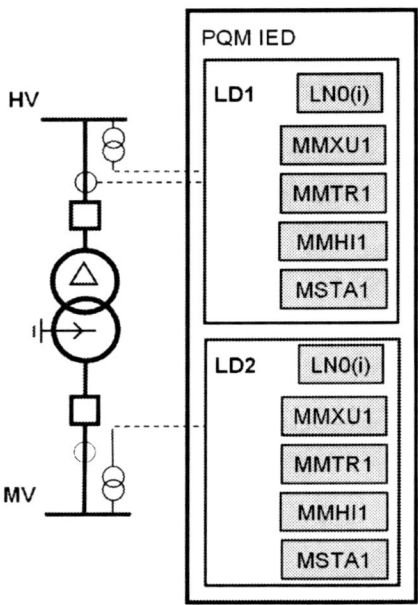

Figure 13.8. Multiple logical devices object model

The logical devices in power-quality-monitoring IEDs can also be based on functions. For example:
- Measurements – for grouping of logical nodes modelling power-quality measuring functions.
- Events – groups logical nodes used for the detection of power-quality events.
- Recording – this is a group of logical nodes related to the recording of transients, sags, swells and others.

13.6.1 Logical nodes for Measurements

The logical nodes for measurements are included in the standard in logical nodes group M. These are the most common logical node in all IEDs typically available in a substation automation system. The logical node that can be found in any multifunctional IED is mainly used to provide measurements to a substation HMI, another substation IED, any remote system operator or other corporate client. The name of this logical node is **MMXU** and it inherits all mandatory data from the common logical node class defined by the standard.

Table 1 shows the attribute names and types of the different optional measured values included in **MMXU**. As can be seen from the table most of the data attributes in this logical node are of WYE type, i.e. they model the three-phase and neutral values of measured voltages (*PhV*), currents (*A*), impedance (*Z*), etc. The phase-to-phase voltages are modelled using PPV, which is of DEL type.

Table 13.1. Measured values attributes in MMXU

Name	Type	Description
PPV	DEL	Phase-to-phase voltages
PhV	WYE	Phase-to-ground voltages
A	WYE	Phase currents
W	WYE	Phase active power (P)
VAr	WYE	Phase reactive power (Q)
VA	WYE	Phase apparent power (S)
TotW	MV	Total active power (total P)
TotVAr	MV	Total reactive power (total Q)
TotVA	MV	Total apparent power (total S)
TotPF	MV	Average power factor (total PF)
Hz	MV	Frequency
PF	WYE	Phase power factor
Z	WYE	Phase impedance

The total values of active, reactive and apparent power, as well as the total power factor are of type MV.

Power-quality-monitoring devices measure different system parameters that are used to determine unbalanced system conditions. Such measurements are modelled using logical node **MSQI**. Most of the measured values included in the logical node are of type MV. Some are WYE and DEL. The sequence components of the currents and voltages are modelled as attribute type SEQ.

Most of the commonly used measurements in the electric power- system are three phase. However, in some cases we may need to represent measurements that are not phase related. In this case we need to use logical node **MMXN** (non phase related measurements). Most of the values in that logical node are of the MV type.

In order to analyse the behaviour of the system, it is sometimes necessary to calculate the average, minimum and maximum values of a system parameter over a predefined period of time. These values of system parameters are modelled as MV type and are available in the metering statistics logical node **MSTA**.

Advanced metering and power-quality-monitoring devices, as well as specialised energy metering devices calculate the energy that is then used for billing or other purposes. Different energy values are available in the metering logical node **MMTR**.

All metered values are represented as attribute type BCR (binary counter reading).

Power-quality-monitoring devices calculate hundreds of different system parameters, such as harmonics or interharmonics. Their modelling is based on a logical node **MHAI** dedicated to these measurements. The attribute types of the different measured values included in **MHAI** are HWYE, HDEL, WYE, DEL or MV. This logical node can be instantiated for either harmonics or interharmonics depending on the value of the basic settings.

In the cases when the harmonics or interharmonics are calculated in a single-phase system with no phase relations, the **MHAN** logical node should be used in the device model. In this case the attribute types for the measured values are of type MV or HMV.

13.6.2 Logical nodes for Power-quality Events

The analysis of power-quality events in IEC 61850-based substation automation systems can be done based on the available functionality of the used IEDs. Power-quality-events related functions are represented by a new group of logical nodes Q. The following is a list of such logical nodes:
- **QVVR** for RMS voltage variations
- **QFVR** for frequency variations
- **QVUB** for voltage unbalance variations
- **QIUB** for current unbalance variations
- **QVTR** for voltage transients
- **QITR** for current transients

Since voltage variations are the most common power-quality events with significant impact on sensitive loads, we will consider as an example in more detail the logical node QVVR. Each instance of this logical node class represents a functional element that operates when its input voltage is outside of a predetermined value range, as is the case with voltage dips, voltage swells or voltage interruptions.

An instance of this logical node can be used to detect temporary overvoltage, while another instance will be used for the detection of undervoltage conditions. Instances of QVVR are also used to detect voltage interruptions, i.e. the complete loss of voltage. Separate thresholds for under and overvoltage, as well as for voltage interruption should be available as configuration data objects in this LN. This LN should determine the duration and level of the voltage variation.

The detection of voltage-variation events is typically based on the rms voltage measurements on a subcycle or full-cycle algorithm. In the case of bus VTs, voltage interruption should be detectable using other methods, such as breaker auxiliary contacts status.

Since it is possible (for example during single-phase-to-ground faults) to have simultaneous under and overvoltage condition in different phases, the LN should monitor voltage variations on a per-phase basis.

13.6.3 Logical nodes for Recording

The logical nodes for recording are included in the group R – protection related logical nodes. They are of several different classes and can be used to model recording functions in different types of devices that have recording capabilities.

RDRE is the logical node representing the acquisition functions for voltage and current waveforms from the power process (CTs, VTs), and for position indications of binary inputs. Calculated values such as frequency, power and calculated binary signals may also be recorded by this function if applicable. RDRE is used also to define the trigger mode, prefault, postfault, etc. attributes of a disturbance-recording function.

The logical node class **RADR** is used to represent a single analog channel, while **RBDR** is used for the binary channels. Thus the disturbance recording function is modelled as a logical device with as many instances of RADR and RBDR logical nodes as there are analog and binary channels available. Some examples are shown later in this chapter.

13.7 Power-quality-event Analysis in IEC 61850-based Systems

As discussed earlier in this chapter, one of the key differentiators between IEC 61850 and other communication protocols is that it not only defines how the data is transmitted over the media, but also an abstract model of what is communicated. The functional elements represent individual components that can exist independently of each other and not necessarily in the same device, thus allowing the development of distributed functions and applications. This approach can be used for power-quality-event analysis in IEC 61850-based substation automation systems and is described below.

The currents and voltages from logical nodes **TCTR** (representing a current sensor) and **TVTR** (representing a voltage sensor) accordingly are delivered as sampled values over a digital data bus. This data bus can be internal to an IED or can be the substation LAN. It provides the interface between the instrument transformer logical nodes and the different logical nodes that are used to model the functional elements of the IED.

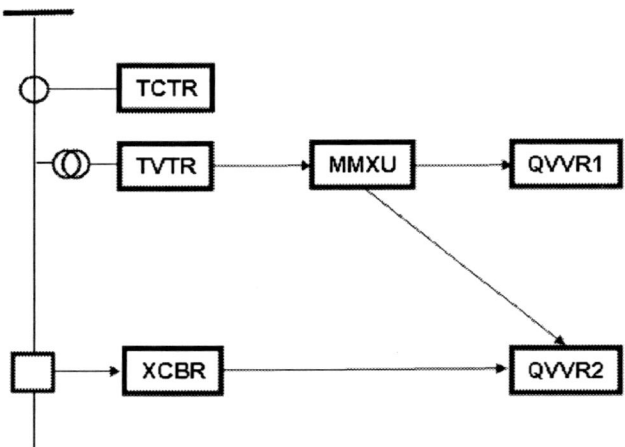

Figure 13.9. Logical nodes for power-quality analysis

The status of the breakers in the substation is modelled using the **XCBR** logical node. It will provide information on the three phases or single-phase status of the switching device, as well as the normally open or closed auxiliary contacts.

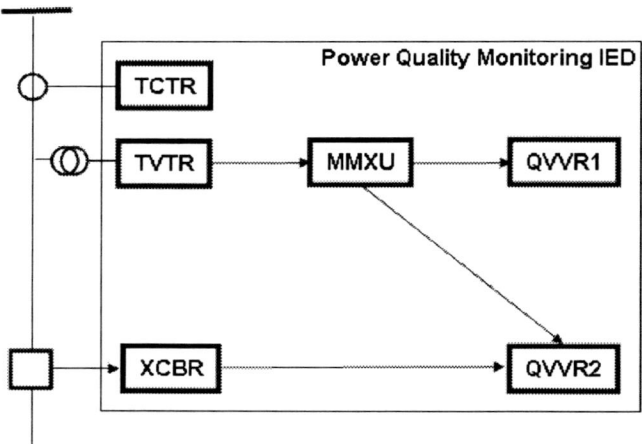

Figure 13.10. Logical nodes in power-quality-monitoring IED

We already know that MMXU is the logical node that represents the measurement of the three-phase voltages that are then used by the first instance of QVVR1 to detect voltage sag. A second instance QVVR2 is used to detect voltage interruption. This is why it requires also breaker status in the case of location of the voltage transformers on the bus side of the interrupted feeder breaker.

IEC 61850 does not define the location of the logical nodes, but only the fact that they exchange information with each other. The following are some examples of different distribution of the power-quality-event analysis logical nodes in a substation automation system.

Figure 13.10 shows the case of a power-quality-monitoring IED that includes within a single device all the logical nodes from Figure 13.9. If the substation automation system includes a process bus, then the power-quality-monitoring function becomes distributed, i.e. the same logical nodes are located in several different physical devices as shown in Figure 8.

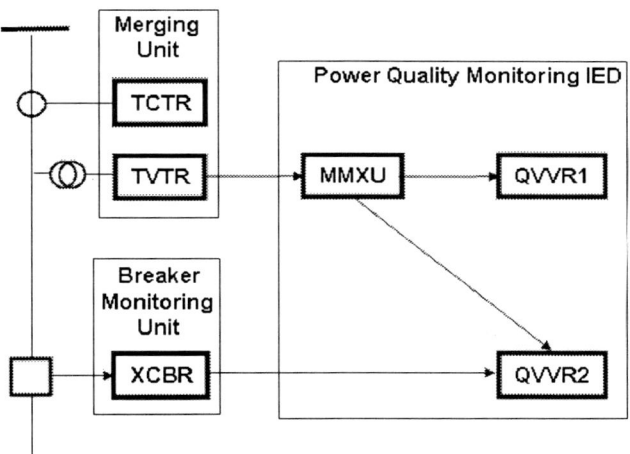

Figure 13.11. Logical nodes in distributed power-quality-monitoring system

As can be seen from the figure the TCTR, TVTR are implemented in a merging unit and XCBR is in a breaker monitoring unit. In some cases they can all be located in an interface unit. This name is used instead of merging units due to the fact that the devices have binary inputs in addition to the analog inputs typically available in merging units.

In both cases in Figures 13.10 and 13.11 we are talking about real-time detection of voltage-variation power-quality events.

Figure 13.12 shows the logical nodes of a distributed offline power-quality-event-analysis system. In this case the recording device (recording function represented by logical nodes RADR, RBDR and RDRE) creates one or more COMTRADE files that are automatically extracted and then analysed.

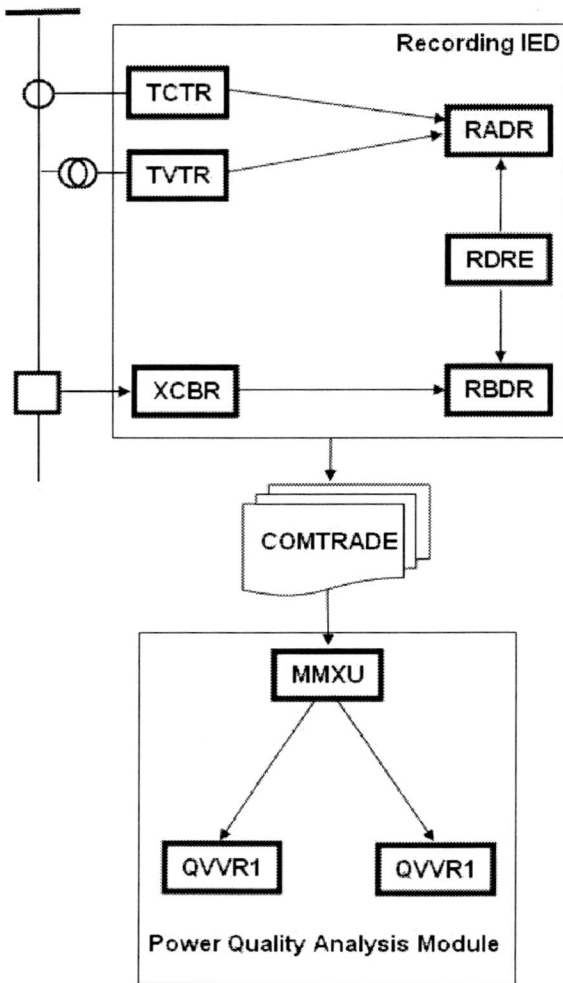

Figure 13.12. Logical nodes in distributed power-quality-monitoring system

Another approach that can be used for power-quality-event analysis in IEC 61850-based substation automation systems is based on report control blocks that control the procedures required for reporting values of event data from one or more logical nodes to one client. Instances of report control are configured in the IED at configuration time.

IEC 61850 defines two classes of report control: buffered report control block (**BRCB**) and unbuffered report control block (**URCB**).

Buffered report control blocks are used for sequence-of-event purposes. They define internal events (caused by trigger options data-change, quality-change, and data-update) that issue immediate sending of reports or buffer the events for transmission. This prevents data from being lost in the case of loss of connection.

Unbuffered report control blocks are quite similar to the **BCRB**. However they do not buffer the data, so event information may be lost in the case of communication problems. Obviously the unbuffered report control block does not support sequence-of-events reporting in case of loss of communications.

13.8 Recording of Power-quality Events

Since power-quality events include changes in the monitored system parameters that can range from fractions of a cycle to several minutes or even days (voltage interruptions following a snow storm or hurricane), it is impossible to record the variety of events using the most commonly available waveform capture. Many of the power-quality events are also based on the changes in the rms value of the voltages, so the waveform capture is not appropriate for the recording of such events. The same is true for the recording of different abnormal power-system conditions. That is why multifunctional protection and monitoring IEDs provide recording features that can be used to meet the primary and backup recording requirements of various utility departments.

The need for monitoring and recording at the transmission level has been recognised for a long time. The experience with centralised disturbance recording systems has shown how valuable this information is since it allows for a better understanding of the steady-state and dynamic behaviour of the system. More and more utilities and industries are realising that the same is true at the distribution level. The availability of multifunctional IEDs with advanced communications capabilities and a standard communication protocol leads to a new concept for distributed monitoring and recording not only in the substation, but throughout a complete electric power system.

The recording modes of multifunctional IEDs are determined by the requirements for recording of different system events. Some examples are wide-area-system disturbances that result in power swings or frequency variations, transients during a short-circuit on a high-voltage transmission line or the voltage sag at the distribution level, and load changes caused by time of day or meteorological condition variations. As can be seen from these examples, the recording requirements can vary significantly and cover a wide range from more than a hundred samples per cycle, to more than a minute between samples.

The analysis of these different types of events in some cases requires sampling of the waveform, while in other events they need a periodic log of the RMS value of the monitored parameter.

That is why state-of-the-art multifunctional IEDs with recording capabilities have multiple recording types that allow the coverage of any possible type of fault or power-quality event. In order to allow the user to "zoom-in," all recording types should run in parallel, as required by the application, power-system condition and triggering criteria specified by the user. This is possible, since the same triggers can be used for the different types of recording and also because all records have accurate timestamps based on the time-synchronisation feature in the IEDs.

13.8.1 Waveform Recording

Waveform recording in many cases is known as disturbance recording. It captures the individual samples of the currents and voltages measured by the IED with a high sampling rate that may be as low as 4 samples/cycle for some low-end protection IEDs to hundreds of samples per cycle for high-end monitoring and recording IEDs. Because of this, specialised recording devices provide the primary waveform recording function, while the protection relays are used for backup recording.

The user typically has options to define the triggering criteria, the pre-trigger or post-trigger intervals and if extended recording should be available in cases of evolving faults or other changing system conditions.

The trigger for waveform recording can be defined as a threshold on any measurement, operation of a protection or monitoring function as well as the output of a user-defined programmable scheme logic. External triggering of the recording should also be possible through the opto inputs of the device or based on communication messages from other IEDs or the substation computer.

13.8.2 High- and Low-Speed Disturbance Recording

High-speed or low-speed disturbance recording is intended for capturing events such as voltage sags or voltage swells during short-circuit faults on the transmission or distribution system. The disturbance recording IED stores the values of a user-defined set of parameters for every log interval. The setting range is dependent on the available memory in the IED, for example from 1 to 3600 cycles and can be changed with a step of 1 cycle. If the sampling rate is more than one cycle per sample, the user should be able to select the recording of minimum, maximum and average value through the specified sampling interval.

An option to trigger high-speed disturbance recording when a waveform capture is triggered is achieved by using the same trigger with different recording modes. The combination of waveform capture and high- or low-speed disturbance recording triggered by the same power- system event allows the recording of long events, while at the same time the details of the transitions from one state to another are recorded in the waveform capture.

Backup recording for some disturbances can be provided by the historical data logging features of the substation automation system that can store data with a sampling rate of 1 sample/s.

13.8.3 Periodic Measurement Logging

Planning studies and short-and long-term load forecasting require the recording of system parameters over long periods of time. The recording device should be able to store the values of a user-defined set of parameters for every log interval. This interval defines the sampling rate of a trend recording and the user should be able to change it as required by the application.

All records – waveforms, disturbances or trends – should be in a standard file format, such as COMTRADE. This allows the use of off-the-shelf programs for viewing and analysis of the records. Since this is a comma-separated text format, the files can easily be imported into other applications for further processing. Because of problems with the implementation of COMTRADE by many vendors, it is quite difficult today to read such files with many of the available viewers. A new working group in the IEEE PSRC is considering an update of COMTRADE in order to avoid such problems in the future and simplify the analysis of the data.

13.9 Recording Systems for Power-quality-event Analysis

If the IEDs in the substation automation system do not include specialised power-quality-monitoring devices, the analysis of power-quality events can be executed at the substation level by analysis tools based on waveforms and disturbance records of the events of interest. This requires the implementation of a distributed recording system dependent on the recording capabilities of the IEDs in the substation.

The recording of power-system events related to power-quality can be executed at different levels of the substation automation system functional hierarchy. This is especially valid in the case of IEC 61850 systems with a process bus. The fact that sampled current and voltage values are multicasted over the substation process bus makes them available to any equipment or bay-level IED, as well as to the substation computer or other devices at the upper level of the functional hierarchy.

Because of the high sampling rate (usually 128 or 256 samples/cycle) and the availability of multiple recording modes, it is obvious that power-quality-monitoring or specialised disturbance-recording devices will be used as the primary recording devices.

Multifunctional protection devices can be used as the backup recording devices. Their sampling rate is much lower – typically 16 to 64 samples per cycle for the waveform capture and without any disturbance-recording capabilities. However, some devices allow waveform capture of more than 10 s that will be sufficient for capturing typical system or power-quality events.

Since power-quality events are identified in most cases based on the rms values of the voltages or the harmonic content of the waveform, further processing of such waveform files will be required. On the other hand, analysis of power-system disturbances may be based on phasor information that can be directly recorded by the specialised recording devices or processed at the substation computer using software analysis tools.

The sampling rates of the relays might not be appropriate for some harmonic calculations, but they are still adequate considering that they are only performing backup monitoring functions. Even if the sampling rate is sufficient to meet the requirements of the power-quality event-analysis applications, we should still keep in mind that the analog signals are filtered before the analog-to-digital conversion and typically harmonics above the fifth will not be available even if the sampling rate of the relay is 128 samples/cycle.

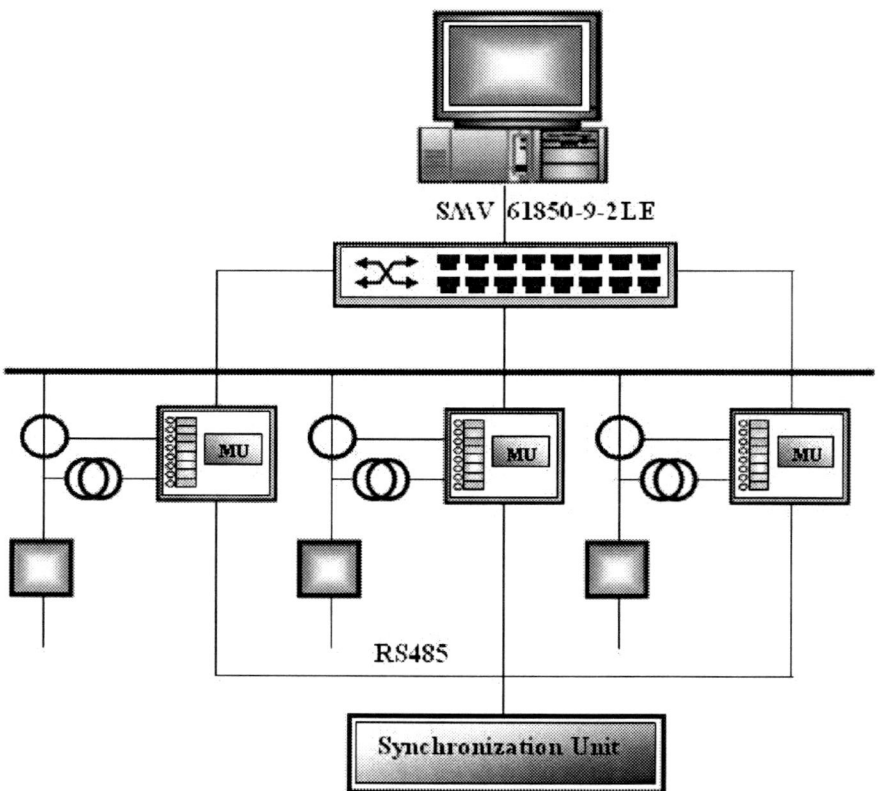

Figure 13.13. Centralised waveform-recording system based on sampled analog values

In substation automation systems, especially those based on IEC 61850, the substation level can perform the functions of a centralised waveform and disturbance recorder. This is due to the fact that it can subscribe to the sampled analog values messages coming from the merging units in the zone of recording. For power-quality purposes the sampling rate in the implementation agreement is defined as 256 samples/cycle at the nominal frequency.

Figure 13.13 shows a centralised waveform recording function based on the sampled analog values from multiple merging units at the process level of the substation automation system. The synchronisation unit allows submicrosecond synchronisation of the real-time clocks of the merging units.

The merging units sample 256 times per cycle the three-phase current and voltage inputs, and generate the ethernet messages that are sent using 100 Mb/s to the recording device. Eight sets of current and voltage samples are grouped in each ethernet frame. As a result, each interface unit sends 32 messages per cycle to the central recording unit.

Each merging unit is connected to an ethernet switch that in this case is dedicated to the process bus. The recording device receives from the switch all ethernet messages from the interface units included in the system. Considering the size of the ethernet frames a single 100-Mb/s port of the recording device can handle the traffic from up to seven interface units.

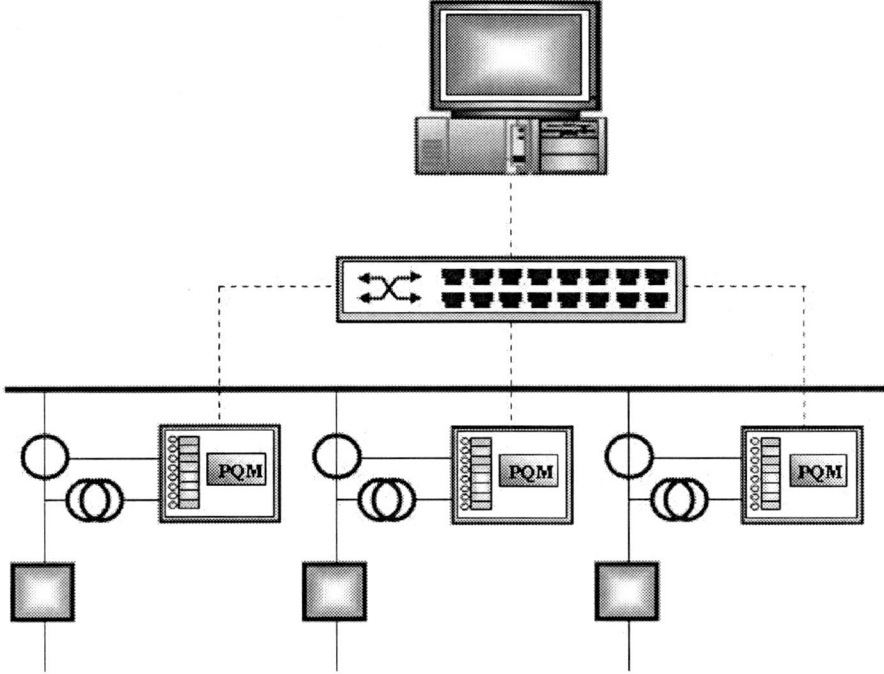

Figure 13.14. Centralised waveform-recording system based on distributed PQM devices

If the central recording unit needs to record currents and voltages from more than 7 interface units, a second ethernet port may be used to expand the distributed waveform recording system to a total of up to 14 interface units.

Another alternative solution for more than seven interface units is to use a computer with 1 Gb/s ethernet port connected to a 1 Gb/s Ethernet switch with 100 Mb/sec ports connected to the interface units. The architecture in this case will be exactly the same as the one shown in Figure 13.13.

The central recording function records power-quality or other events based on built-in or external triggers. It takes the samples from the individual merging units that it received and creates a combined waveform record based on user defined selection.

Another distributed recording system architecture is based on multiple recording IEDs and is shown in Figure 13.14. The difference between the sampled measured values-based system is that the recording is performed in the central unit based on a trigger criteria and a single record is created as a COMTRADE file, while in this case each IED records a set of current and voltage waveforms and creates its own COMTRADE file. The triggering of the recording needs to be synchronised and cross triggering using GOOSE messages over the substation bus is the preferred method of achieving it.

The central unit that performs the analysis functions needs first to perform automatic disturbance or waveform records extraction, followed by the integration of the sampled values from the individual records in a single disturbance or waveform record based on a user-defined selection.

The disturbance records are then available for analysis to determine if there was a power-quality event and what its characteristics are.

13.10 Performance Requirements

The different distributed functions impose different performance requirements that have to be considered in the design process of substation protection, control, monitoring and recording systems. IEC 61850 defines performance requirements for the typical substation functions. The transfer time definition is based on Figure 13.15:

There are two independent groups of performance classes:
- for control and protection
- for metering and power-quality applications

Since the performance classes are defined according to the required functionality, they are independent of the size of the substation.

The requirements for control and protection are higher, because of the effect of the fault-clearing time on the stability of the system or on sensitive loads.

IEC 61850 defines three performance classes for such applications:

P1 – applies typically to the distribution level of the substation or in cases where lower performance requirements can be accepted.

P2 – applies typically to the transmission level or, if not otherwise specified, by the user.

P3 – applies typically to transmission-level applications with high requirements, such as bus protection.

The transmission time is

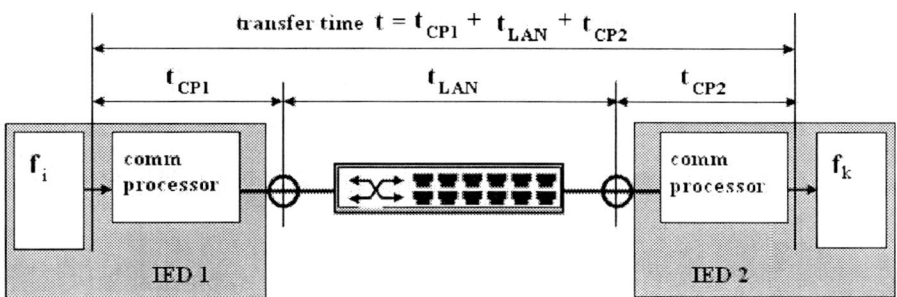

Figure 13.15. Transfer time definition

where:

T_{CP1} - time from the moment the sending IED 1 in Figure 13.15 puts the data content on top of its transmission stack until the message is sent on the network;

T_{LAN} - the time over the network (in reality the delay of the ethernet switch);

T_{CP2} - the time from the moment the receiving IED 2 in Figure 13.15 gets the message from the network until the moment it extracts the data from its transmission stack.

The overall performance requirements also depend on the message type. Type 1 is defined in the standard as fast messages. Since trip (type 1A) is the most important fast message in the substation, it has more demanding requirements compared to all other fast messages. The same performance may be requested for interlocking, intertrips and logic discrimination between protection functions.

For performance class P1, the total transmission time shall be of the order of half a cycle. Therefore, 10 ms is defined.

For performance class P2/3, the total transmission time shall be below the order of a quarter of a cycle. Therefore, 3 ms is defined.

All other fast messages are defined as type 1B. They are also important for the interaction of the automation system with the process but have less demanding requirements compared to the trip.

For performance class P1, the total transmission time shall be in this case less than or equal to 100 ms, while for performance class P2/3 the total transmission time shall be in the order of one cycle.- 16.6 ms (60 Hz systems) or 20 ms (50 Hz systems). These messages are typically used by interfaces IF3, IF5, and IF8 in Figure 13.1.

Type 2 defines medium speed messages for which the time at which the message originated (time tag included in the message) is important but where the transmission time is less critical. Normal "state" information or measurements belong to this type of message. These messages are typical for interfaces IF3, IF8, and IF9 and the total transmission time shall be less than 100 ms.

Low-speed messages are defined as Type 3 and include complex messages that are also usually time tagged. This type should be used for low-speed automation functions, changes of settings or other configuration parameters or transmission of event or fault records. The total transmission time for these messages shall be less than 500 ms.

Type 3 messages are typical for nearly all interfaces of Figure 2 IF1, IF3, IF4, IF5, IF6, IF5, IF7, IF8, and IF9.

Type 4 defines raw data messages such as the output data from digitising transducers and digital instrument transformers independent from the transducer technology (magnetic, optical, etc.).

The data will consist of continuous streams of synchronised data from each IED, interleaved with data from other IED. These messages are typical for interfaces IF4 and in some applications for IF8.

The transmission time for P1 is specified as 10 ms, while for P2 and P3 it should be 3 ms.

Type 5 of messages is used for the transfer of large files of data for recording, information purposes, settings, etc. Since the substation network is used for the transmission of all the other types of data listed above, the transferred file data must be split into blocks of limited length, to allow for other communication network activities. Typically, the bit lengths of the file type messages are equal to or greater than 512 bits.

Since transfer times are not critical there are no specific limits. Typically, the time requirements are equal to or greater than 1 s.

If the file transfer is requested by a remote client (located outside of the substation) it will require some form of access control, i.e. type-7 communications will be used.

Time-synchronisation messages are defined as Type 6 and are used to synchronize the internal clocks of the IED in the substation automation system. The requirements for accuracy of time synchronisation are very different for the different applications. As a result, different communications will be used as well.

Depending on the purpose (time tagging of events or sampling accuracy of raw data) different classes of time-synchronising accuracy are required.

Type 7 specifies Command messages with access control and is used to transfer control commands issued from a local or remote HMI. These are typically functions that require a higher degree of security and shall include some form of access control. These command messages go usually from the substation-level functions to the bay-or equipment/process level IEDs and cover IF1, IF6 and IF7.

13.11 High-speed Peer-to-Peer Communications Applications

The peer-to-peer communications in an integrated substation protection and control system are based on what is defined as a GSE. This is a generic substation event (GSE) and it is based upon the asynchronous reporting of an IED's functional elements status to other peer devices enrolled to receive it during the configuration stages of the substation integration process. It is used to replace the hard-wired control signal exchange between IEDs for interlocking and protection purposes. In order to meet the requirements of such critical applications it is time critical and must be highly reliable.

The information exchange is based on a publisher/subscriber mechanism. The publisher writes the values in a local buffer at the sending IED, while the receiver reads the values from a local buffer at the receiving IED. A generic substation event-control class in the publisher is used to control the procedure.

The associated IEDs receiving the message use the contained information to determine the required response to the indicated state change. The decision of the appropriate action to GSE messages and the behaviour should a message time out due to a communication failure is determined by local intelligence in the IED receiving the GSE message.

Considering the importance of the functions performed using GSE messages, IEC 61850 defines very strict performance requirements described earlier in this chapter. The idea is that the implementation of high-speed peer-to-peer communications should be equal to or better than what is achievable by existing technology. Thus, the total peer-to-peer time should be less than 4 ms, as discussed in the previous section.

Another key requirement for the GSE messages is very high reliability. Since the messages are not confirmed, but multicasted, and considering the importance of these messages, there has to be a mechanism to ensure that the subscribing IEDs will receive the message and operate as expected. To achieve a high level of reliability, messages will be repeated as long as the state persists. To maximise dependability and security, a message will have a time-to-live, which will be known as the "hold time". After the hold time expires, the message (status) will expire, unless the same status message is repeated or a new message is received prior to the expiration of the hold time.

The repeat time for the initial GSE message will be short and subsequent messages have an increase in repeat and hold times until a maximum is reached. The GSE message contains information that will allow the receiving IED to know that a message has been missed, a status has changed and the time since the last status change.

In order to achieve high-speed performance and at the same time reduce the network traffic during severe fault conditions, the GSE message has been designed based on the idea of having a single message that conveys a set of application-required information regarding an individual IED or a logical device that it contains. It represents a state machine that reports the status of the functional elements in the IED to its peers. The number n of members of the GSE dataset may include hundreds of entries.

GSE messages in IEC 61850 are defined as:
- generic substation state event (GSSE);
- generic object oriented substation event (GOOSE).

The GSSE messages were included in the standard to provide backward compatibility with the UCA (utility communications architecture) GOOSE. What is important to note is that the dataset in this message contains only state information.

The IEC GOOSE is quite different, because it allows improved flexibility based on the use of a dataset that can contain not only state information, but different data types supported by the standard. This allows the use of GOOSE messages not only to indicate tripping of a breaker, but also for recording of system events as described later in this chapter.

A typical application of GOOSE messages is in systems for distributed power-quality-event recording to initiate simultaneously the recording by multiple IEDs in the substation automation system in order to capture the effect of the event throughout the substation. Each recording device in this case will have to subscribe to GOOSE messages from multiple IEDs, so it can be triggered by these devices when necessary.

An example of the crosstriggered recording is the case when a protection IED acts as the Publisher and the power-quality-monitoring IEDs (PQM) are the Subscribers. The protection IED will detect the fault or other abnormal system condition and send a GSSE message to the PQM IEDs (see Figure 3) that needs to record it.

In the subscribing IEDs it will trigger waveform recording with a high sampling rate (for example 128 or 256 samples/cycle) to record the fault for future analysis. At the same time this may trigger high-speed disturbance recording (for example, 1 or more cycles/sample of the rms voltage profile) for analysis of the voltage sag caused by the fault.

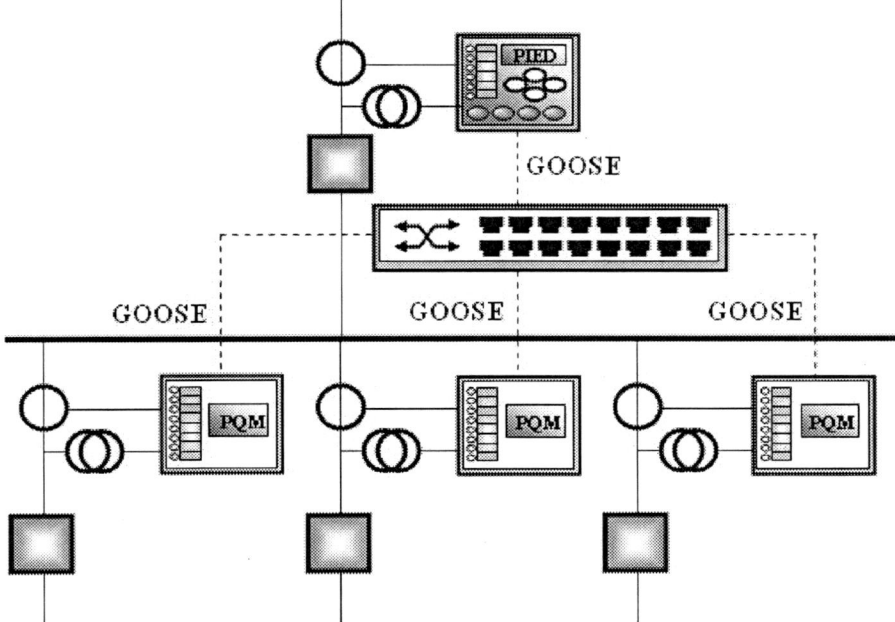

Figure 13.16. Cross triggering of power-quality-event recording

The benefit of the high-speed peer-to-peer communications-based distributed power-quality event recording is that it allows the integration of multiple power-quality-monitoring devices in a system that can be expanded as needed, while at the same time provides different types of records that look and feel like the records from a large centralised recording device.

References

[1] IEC 61850-1 Communication Networks and Systems in Substations, Part 1: Introduction and Overview
[2] IEC 61850-5 Communication Networks and Systems in Substations, Part 5: Communication Requirements for Functions and Device Models
[3] IEC 61850-7-2: Communication Networks and Systems in Substations, Part 7-2: Basic communication structure for substations and feeder equipment – Abstract communication service interface (ACSI)
[4] IEC 61850-7-3: Communication Networks and Systems in Substations, Part 7-3: Basic communication structure for substations and feeder equipment – Common data classes
[5] IEC 61850-7-4: Communication Networks and Systems in Substations, Part 7-4: Basic communication structure for substations and feeder equipment – Compatible logical node classes and data classes
[6] Integration of Power-quality-monitoring and Disturbance Recording Devices in UCA Based Systems, B. Muschlitz, A. Apostolov, Western Power Delivery Automation Conference, Spokane, WA, April 1–4, 2002
[7] Power-quality-monitoring in Modern Substation Automation Systems, A. Apostolov, IASTED EuroPES 2002, Crete, Greece, June 25-28, 2002
[8] Power-quality-monitoring in UCA Based Substation Automation Systems, A. Apostolov, 10th International Conference on Harmonics and Quality of Power, Rio de Janeiro, Brazil October 6–9, 2002
[9] Distributed Systems for Power-quality-monitoring in Substations, A. Apostolov, Power-quality Technology 2002, Rosemont, IL, October 29–31, 2002
[10] Distributed Monitoring and Recording in UCA/IEC 61850-based Substation Automation Systems, A. Apostolov, Biennial Southern African Conference on Power-system Protection, Johannesburg, South Africa, November 6–7, 2002
[11] Object Models for Power-quality-monitoring in UCA 2.0 and IEC 61850, A. Apostolov, DistribuTech 2003, Las Vegas, NV, February 4–6, 2003
[12] Detection and Recording of Power-quality Events in Distribution Systems, A. Apostolov, Georgia Tech Fault and Disturbance Analysis Conference, Atlanta, GA, May 5–6, 2003
[13] Fault and Disturbance Recording in Substation Automation Systems, A. Apostolov, Georgia Tech Fault and Disturbance Analysis Conference, Atlanta, GA, May 5-6, 2003
[14] Web-Enabled Power-quality-monitoring In Small Distribution Substations, A. Apostolov, 17th CIRED Conference & Exhibition, Barcelona, Spain, May 12–15, 2003
[15] Object Modelling of Measuring Functions in IEC 61850-based IEDs, A. Apostolov, B. Muschlitz, 2003 IEEE PES Transmission & Distribution Conference & Exposition, Dallas, TX, September 7–10, 2003
[16] Use of IEC 61850 Object Models for Power- system Quality/Security Data Exchange, A. Apostolov, C. Brunner, K. Clinard, IEEE/CIGRE Symposium on Quality and Security of Electric Power Delivery Systems, Montreal, Canada, October 7–10, 2003
[17] Fault, Disturbance and Power-quality Analysis in Substation Automation Systems, A. Apostolov, CIGRE Colloquium 2003 and Study Committee B5 Meeting, Sydney, Australia, October 19–24, 2003
[18] Requirements for Automatic Event Analysis in Substation Automation Systems, A. Apostolov, IEEE PES 2004 Meeting & Conference, Denver, CO, June 6–11, 2004
[19] Distributed Protection, Control and Recording in IEC 61850-based Substation Automation Systems, A. Apostolov, 2004 Southern African Power- system Protection Conference, November 3–5, 2004, Johannesburg, South Africa

[20] Detection and Recording of Voltage Variation Power-quality Events, A. Apostolov, Power-quality Conference & Exhibition, November 16-18, 2004, Chicago, IL
[21] IEC 61850 and Disturbance Recording, A. Apostolov, Georgia Tech Disturbance Recording and Analysis Conference, Atlanta, GA, April 25–26, 2005
[22] On the Detection and Recording of Voltage Dips and Interruptions, A. Apostolov, CIRED, Turin, Italy, June 6–9, 2005
[23] Recording of Power-quality Events in IEC 61850-based Systems, A. Apostolov, C. Ziegler, E. DeMicco, Power-quality 2005 Conference, Baltimore, MD, October 25–27, 2005
[24] A Distributed Recording System Based on IEC 61850 Process Bus, A. Apostolov, F. Auperrin, R. Passet, M. Guenego, F. Gilles, Power- system Conference 2006, Clemson, SC, March 14–17, 2006
[25] Power-quality Events Analysis in Substation Automation Systems, A. Apostolov, C. Ziegler, E. DeMicco, Western Power Delivery Automation Conference, Spokane, WA, April 11–13, 2006
[26] Combining Digital Fault Records from Multiple Devices (Virtual DFR), A. Makki, M. Rothweiler, A. Apostolov, J. Pond, Georgia Tech Disturbance Analysis Conference, Atlanta, GA, May 1–2, 2006
[27] Analysis of Voltage Variation Power-quality Events, A.P. Apostolov, Georgia Tech Disturbance Analysis Conference, Atlanta, GA, May 1–2, 2006

Index

ABC classification, 343
Abstract communication service interface (ACSI), 390
Abstract model, 381, 383, 389, 399
AC-DC converter, 342
Active Power-line Conditioners, 231
Adapted protective device, 311
Adaptive
    representation, 52, 56, 67
    spectrogram, 62, 67
    transform, 52, 56
Adjustable-speed drive (ASD), 103, 298, 325, 342, 354
Admittance
    mutual, 131
    self, 131
Akagi p–q theory, 262
Analog-to-digital converter, 30
Analysis function, 44, 47, 55
ANSI/IEEE C37.41, 308
ANSI/IEEE Standard 1001, 294
Arcing, 307
Automatic seccionalizer, 318
Automation, 379, 381, 382, 383, 385, 386, 388, 389, 390, 391, 396, 398, 399, 401, 402, 404, 405, 406, 407, 410, 411, 412
Availability, 326, 327

Basis function 41, 57

Bay, 385, 386, 387, 405, 411
Bay-level Functions, 386
Biorthogonal function, 47
Blackout, 295
Boost, 336, 342, 346, 348, 349, 350, 351, 352
    follower preregulator, 342
Breakers, 386, 394, 400
Breaking capacity 296
Bypass, 113, 115, 116, 129
    relay, 115

Capacitance, 347, 348, 349, 350, 351
Capacitor, 336, 342, 347, 348, 350, 351
    bank, 107, 112, 113, 124
Capacitor-switching transient, 103, 104, 105, 107, 121
CBEMA, 90, 297, 308, 328
CENELEC, 4
Characteristic impedance • 107, 112
Chemical charges, 318
Choi-Williams distribution, 65
Circuit breakers, 303, 304
Class, 389, 391, 392, 393, 398, 399, 411
Cohen class, 65
Common data
    attribute, 390

classes (CDC), 383, 390
Commutation frequency, 269
Compensation
    parallel, 141, 142, 144, 145, 160, 161, 221
    series, 141, 143, 144, 145, 160, 161, 221
Compensator
    central, 140, 141
    group, 140
    local, 140
Computer, 325, 327, 328, 329, 331, 332, 339, 340, 341, 352
COMTRADE, 401, 405, 408
Conic distribution, 66
Continuity of supply, 7
Continuous wavelet transform, 57
Correlation function, 49, 64
Crossterm interference, 65, 68
Current
    DC – offset, 134, 153
    disturbances, 130, 132, 133, 140
    harmonics, 141
    instantaneous,
    source, 130, 132, 133, 141
    THD, 331
    transformers, 25
    unbalance, 398
    unbalance factor, 359
Custom power, 183, 188

Damped overvoltage, 125
Damping effect, 105, 124
    ratio, 109
    resistance, 116, 117, 118, 119, 120, 122, 124, 125
DATA, 391, 392
DataAttribute, 391, 392
DataName, 391
DataRef, 391
Data-mining, 36
Daubechies function, 87
DC-AC converter, 342
DC-DC converter, 342
DC-link
    capacitor, 103, 121

inductance, 103, 121
reactors, 114
voltage, 103, 104, 109, 110, 111, 114, 117, 122, 123, 124, 125, 126, 127, 129
Delta-connected, 124
Deregulation, 293
DFT, 80
Digital clock blinking, 320
Diode, 342, 351, 352, 354
Dip, 8, 327, 331, 334, 335, 342, 343, 345, 346, 347, 348, 349, 350, 351
Dips, 329, 332, 333, 335, 336, 341, 342, 343, 345, 349, 350, 351, 352, 353, 354
Discrete wavelet transform (DWT), 46, 57, 86, 370
Dirac basis, 50
Directive 89/338/CEE, 16
Distributed
    energy sources, 327
    generation, 293, 296, 302
Distribution
    static compensator (DSTATCOM), 234, 336
    system, 133, 135, 137, 157, 161, 188
Disturbance, 2, 3, 15, 327, 328
    recording, 399, 403, 404, 413
DS1103 PPC controller board, 269
Dual function, 46, 54
Duration
    of dip, 74
    of interruption, 74, 90
Duty cycle, 117, 119
Dynamic Voltage Restorer (DVR), 164, 212, 213, 214, 215, 216, 217, 218, 219, 220, 221, 222, 227, 336, 353
Dynamic dip corrector, 335

Earth leakage current, 338
Earth-bound noise, 326
Effective
    capacitance, 107

Index 419

damping resistance value, 117
line inductance, 107
Electric power-quality, 231
Electric power grid, 1
Electromagnetic
compatibility, 2, 329, 338
disturbances, 2, 4
interference, 2
Electronic, 325, 329, 330, 341, 353
damping, 115, 122, 129
EMC Directive 89/336, 3
EMI filters, 338
EN 50160, 3, 5, 9, 11, 12, 13, 15, 18, 73, 96, 330, 352
EN
60950, 339, 340
61000-3-2, 342
61000, 3, 4, 5, 6, 16, 331, 342
61000-4-15, 6
61000-6-1, 4
61000-6-2, 4
61000-6-3, 3
61000-6-4, 4
Energy
concept, 299
deficit, 316
sources, 161, 162, 163, 164, 209
Equivalent harmonic rms
current, 357
voltage, 357
Ethernet, 407

Facility, 329, 331, 332, 333, 335, 337, 340, 341
Faraday effect, 28
Fault, 328, 337, 340, 341, 344, 345
Ferroresonant overvoltages, 322
FFT, 17, 48
Filter, 15, 53, 58, 325
Filters, 336, 339
Filtration, 150, 151, 153, 174, 208, 210
Flexible AC Transmission System (FACTS), 183, 206, 224

Flexible manufacturing system (FSM), 331, 332
Flicker, 5, 7, 8, 12, 13, 138, 139, 321
coefficient, 322
Flickermeter, 6
Fluctuations, 325
Flywheel, 336
Fourier
analysis, 80
basis, 42, 50
decomposition, 241
transform, 43, 47
Frequency
deviations, 8
resolution, 52, 59, 63
variation, 45, 63, 70, 356
Fuel cells, 336
Fundamental equivalent phase
current, 358
voltage, 358
Fuse coordination, 305
Fuse-saving operation, 319
Fuses, 295, 303, 307

Gabor
elementary function, 55
expansion, 52
spectrogram, 66
Gaussian function, 54, 56, 66
GOMSFE, 380
GOOSE, Generic Object Oriented Substation Event, 411, 412
Harmonic, 2, 6, 7, 8, 11, 12, 125, 135, 144, 147, 148, 149, 150, 151, 153, 154, 163, 173, 183, 186, 188, 192, 193, 195, 201, 203, 208, 209, 210, 211, 212, 320, 330, 331, 336, 337, 338, 340
distortion, 8, 15
tramp, 282

Harmonic-adjusted total
current harmonic distortion, 44, 63, 357
voltage harmonic distortion, 357

Heisenberg inequality, 43, 50
Holdup time, 347, 352
Human machine interface, 387
Hybrid methods, 89

IEC
    50160, 5
    60282, 308
    61000-3-3, 5
    61000-3-4, 5
    61000-3-5, 5
    61000-3-6, 5
    61000-3-7, 5
    61000-4-1, 5
    61000-4-4, 6
    61000-4-5, 6
    61000-4-6, 6
    61000-4-7, 6, 33
    61000-4-11, 5, 91
    61000-4-15, 6, 35
    61000-4-2, 6
    61000-4-3, 6
    61000-4-30, 32, 75
    61010-1, 22
    61850, 379, 380, 388
    Technical Committee 57, 380
IEEE 1159, 9, 10, 11, 73
    Standard 493, 333
    Standard 519, 321, 342
    Standard 1346, 299, 335, 353
    Standard 1547, 294
IGBT, 115, 116, 117, 119, 120, 121, 122, 124, 125, 129, 337, 338
Impedance
    internal, 130
    mutual, 131, 132, 133
    self, 313
Information and communication technology (ICT), 1
Inspections, 329, 332
Instantaneous transient distortion ratio, 372
Instrumentation, 15
Intelligent
    electronic device (IED), 379, 382, 386, 388, 390, 392, 393, 394, 396, 399, 400, 401, 402, 404, 405, 408, 409, 410, 411, 412, 413
    fuse, 317
Interconnection impedance, 306
Interface-related station-level functions, 387
Interharmonic, 6, 11, 138, 139, 150, 153, 398
Interruption, 10, 43, 63, 328
Islanding, 294, 305, 313
ITIC, 328, 331, 333, 353, 329
    curve, 90, 98, 297

Kalman filtering, 80, 81

LAN, 390, 399
Large switching loads, 325
LC
    filter, 149, 150, 151, 152
    resonant circuit, 103
Leakage current, 338
Lightning, 325, 327
Limiter, 318
Line
    conditioner, 335
    Inductance, 110
Line-to-line voltage, 104, 108, 110, 116, 120, 124, 125
Line-voltage regulators, 13
Load pickup, 303
Logical
    device, 393, 395, 396, 399, 412
    interfaces, 382, 385, 386, 387
    nodes, 382, 389, 390, 392, 393, 395, 396, 398, 399, 401, 402
Long interruptions, 14
Long-duration voltage variations, 8
Low – pass filter (LPF), 208

Medium-voltage fuses, 318
MEPERT, 18
Merging unit, 401, 407
MG set, 335
Mitigation

methods, 104, 120
   techniques, 112
Monitoring, 15
Mother wavelet, 57, 63
Motor, 133, 134, 152
   overheating, 136
   start, 303
Motor–generator set, 13
Multiresolution analysis, 59
Muth and Zimmermann, 318

Normalized instantaneous transient distortion ratio, 372
Nonlinear
   load, 325
   methods, 37
Notching, 11
Nuisance tripping, 104, 105, 114, 115, 116, 117, 120, 127, 129

Orthogonal expansion, 46
Orthogonal current factor, 360
Oscillatory transient, 60
Out-of-phase connection, 320
Overcompensation, 141
Overcurrent protection, 294, 296, 304, 307
Overvoltage, 11, 73, 74, 90, 97
   trip, 103, 123

Parseval's theorem, 371
Peak
   current, 107, 114
   time, 109
Perturbations, 325, 329
Phase displacement, 359
Phase-angle jump, 300
Physical interfaces, 384
PICOM, 382
Point of common coupling (PCC), 301
Postprocess, 15
Potential transformers, 25
Power
   active, 141, 142, 148, 153, 161, 163, 164, 190, 192, 200, 201, 203, 209, 213, 214, 215
   compensation, 147
   conditioner, 14
   factor Correction (PFC), 354
   generators, 295
   quality, 1, 2, 4, 6, 7, 8, 15, 16, 325
   quality definition, 356
   quality factor, 362
   quality and reliability (PQR) 1
   reactive, 145, 146, 148, 149, 150, 161, 185, 186, 190, 194, 198, 199, 201, 209, 213
   reserve, 164
   supplies, 328, 330, 331, 333, 338, 341, 342
   supply, 133
   System Blockset, 272
Power-distribution system, 105, 123
Power-quality 325, 326, 327, 332, 336, 341, 343, 352, 379
   measurement, 6
   surveys, 19
Power-quality-monitoring, 379, 380, 384, 387, 388, 393, 394, 395, 396, 398, 400, 401, 402, 405, 413
Power-system fault, 327
PQ-event-caused losses, 326
Prearcing, 307
Preinsertion
   inductors, 112
   resistors, 104, 112
Process, 379, 381, 382, 385, 386, 387, 390, 399, 401, 405, 407, 408, 410, 411
Process-level Functions, 386
Process-related station-level functions, 387
Programmable AC power source, 121, 124
Protection-data, 385
Protective device (PD), 294, 308
Pulse-width modulation (PWM), 252, 338
   ASD, 105, 115, 116, 117, 129

Quality of
  consumption, 7
  supply, 7

Radial distribution system, 301, 304
RCCB, 339
Records, 15
Recloser-fuse coordination, 319
Rectifier, 337, 342, 346, 347, 348, 349, 350, 351, 352
Reliability, 1, 326, 327, 329, 336, 337, 338, 352
Resistor
  approach, 115
  damping approach, 119
Resonant frequency, 107, 109
Ride-through, 103, 129
  capability, 298, 341, 342
Rms, 398, 399, 403, 405, 413
  variation, 73, 96
  voltage, 74, 75, 94

Sag, 8, 10, 14, 45, 67, 327, 353, 354
Saturable transformers, 318
SCADA, 387
Scaling function, 58, 371
Security, 412
Semi-rigid connection, 314
Sensitive equipment, 297
Sensitivity to voltage events, 90
Short
  duration variations, 10
  interruption, 5, 74, 91
Short-term power interruption, 103
Simulink® block diagram, 273
Smart fuse, 318
Soft-charge resistor, 105, 115, 116, 124
Space vector, 208
Specific Energy Concept, 298, 310
Spectrum analysers, 16
Static
  compensators, 13
  current Breaker (SCB), 158, 159
  current Limiter (SCL), 157, 158, 159
  series compensator, 336
  transfer Switch (STS), 158, 159, 337
  VAR systems, 13
Statistical classification, 36
Station, 385, 386, 387
Station-level Functions, 387
Steady-state event, 44
STFT, 52, 53, 80
STFT spectrogram, 62
Subharmonics, 138, 153
Substation, 379, 380, 381, 382, 383, 384, 385, 386, 387, 388, 389, 390, 391, 394, 395, 396, 398, 399, 400, 401, 402, 403, 404, 405, 406, 407, 408, 409, 410, 411, 412
  automation system (SAS), 385
Susceptance, 184
Susceptibility, 104
Swell, 8, 10, 14, 45, 67, 326, 327, 331
Switching devices, 240
Switching transient frequency, 113
Symmetrical components, 359
Syntactic classification, 37
Synthesis function, 54

THD, 330, 331, 337, 342
Time–current curves, 300, 303
Time–voltage curves, 300, 303
Time resolution, 51, 59, 67
Time-frequency atoms, 50
Traction, 145
Time-varying power-quality, 373
Total harmonic distortion (THD), 330, 331, 357
Transformers, 138, 151, 157, 159, 189, 206, 224, 225, 301, 329, 331, 335, 336, 338, 343, 344, 345
  ferro – resonant, 165
  Scott's, 145, 146

Transient, 8, 10, 13, 327, 328, 335, 339, 340, 396, 403
    behavior, 286
    events, 44, 67
    overvoltages, 103, 112
    quality indices, 373
    voltage-surge suppression (TVSS), 340
Trigger, 399, 402, 404, 408, 413

UCA, 380
Ultracapacitors, 336
UML, 387
Unavailability, 326
Unbalanced
    current, 358
    voltage, 358
Uncertainty principle, 49
Undervoltage, 10
    trip, 103, 114
Unified power-quality conditioner (UPQC), 336
Uninterruptible Power Supply (UPS), 13, 14, 15, 164, 165, 160, 167, 200, 202, 203, 213, 329, 336, 337
    delta – conversion, 166, 221
UNIPEDE, 97
UPS, 329, 335, 336, 337, 338, 339, 340, 341, 352
Urms(1/2) method, 75
Utility line inductance, 103

Varistor, 339
Verdet's constant, 28
Virtual instrument, 364
Voltage
    components, 131
    dip, 5, 12, 73
    dip, effects on computers, 90
    dip magnitude, 78
    disturbance, 130, 132, 136, 150, 155, 160, 163, 164, 183, 216, 227
    drop, 133, 136, 137, 146, 157
    event, 73
    event, effects on equipment, 90
    fluctuations, 8, 11, 139
    imbalance, 11
    instantaneous, 162
    interruption, 393
    level, 138, 144
    Norton – Thevenin, 132
    quality, 130, 155, 176, 183, 200, 205
    quality, 7
    quality factor, 363
    ripple, 270
    sag, 103, 294, 297, 302, 389, 400, 403, 413
    source inverter, 238
    supply, 133, 136, 144, 160, 174, 183, 207, 212, 218, 221, 222, 223
    suppressor, 339
    swell, 76, 103, 393
    tolerance, 90
    tolerance curve, 73, 90
    unbalance, 8, 12, 15, 398
    unbalance factor, 359
    unbalanced, 134, 144
Voltage-dip, 326, 332, 333, 335, 336, 341, 342, 343, 344, 345, 352
Voltage-dip contour chart, 335
Variations, 13, 398, 399

Wavelet
    analysis, 80, 86
    bases, 371
    expansion, 57
    function, 58, 86
    transform, 43, 57, 370
Wigner-Ville distribution, 64
Wind turbine, 321
Window function, 53, 65
Windowed Fourier transform, 52

XML, 382

Zero-voltage closing control, 105

Printed in the United States
76163LV00001B/116